The Grain Harvesters

Graeme R. Quick
and
Wesley F. Buchele

American Society of Agricultural Engineers

FIRST EDITION

LCCN: 78-65524
ISBN: 0-916150-13-5

Printed in the United States of America by the American Society of Agricultural Engineers, 2950 Niles Road, P.O. Box 410, St. Joseph, Michigan 49085.

Acknowledgments and Photo Credits

More than ten years of research and travel on five continents were forged into the preparation of *The Grain Harvesters*. Many individuals were most helpful and considerate during the endeavor. The authors particularly acknowledge:

Sam Rosenberg, ASAE Editor—Special Projects, for his initiative, consistent effort and encouragement throughout the preparation of the draft. It was Sam who, several years ago, foresaw the potential for such a publication following the success of *The Agricultural Tractor: 1855-1950*, a book which he revised and edited. Jimmy Butt, ASAE Executive Vice-President, became enthusiastic about the book and formal approval to proceed was not long in forthcoming.

Geoffrey F. Cooper of Massey-Ferguson Limited, provided constructive and scholarly editorial criticisms of the draft and fresh points of view. Dick V. Schiller, Allen E. Neal, Bob Ashton and Jack Spiden also provided useful inputs from Massey-Ferguson sources. They supplied many of the photographs in chapters 13 and 14.

Donald A. Murray, Kenneth J. DePaepe and Greg Lennes of International Harvester Company. They supplied many of the photographs in chapter 15.

George Griffin, Willard McCracken, Bill Bennett, Wayne E. Slavens, D. C. (Bud) Bichel and Ed Hengen of Deere & Company. Most of the photographs in chapter 16 came from Deere and Company corporate files.

Donald Thielke, Lyle M. Shaver, Irene Anderson, Ed Carlson and Jerry C. Boone of Allis-Chalmers, who provided many of the photos in chapter 17.

Wolfgang Scholich, Murray Mills, Dave Stewart and Gary Morten of White Farm Equipment.

Larry H. Skromme, Bill Rowland-Hill and Mark Branstetter of Sperry New Holland.

Others within the business: George Frushour, Louis Scheidenhelm, Archie Neal, Bill Sheardown, Harold DeMent and Tom Cruisenberry.

A singular source of firsthand information on the more recent combine developments are these men, among those who have retired after years of faithful service to the industry:

Elliott Adams, Horace D. Hume, Robin Miller, Stuart D. Pool, Archie Watts, Charles J. Scranton, Lee Oberholtz, Gene Allen, the late Tony Bakken and Wayne Worthington.

From the universities and other institutions: Clarence W. Bockhop, Roger Ditzel (assisted by Donald Zarley), Joseph Campbell, Howard McColly, Bill Splinter, Malcolm Wright, John Goss and Roy Bainer, Glen Downing, Amir U. Khan and John A. McMennamy, Louis Liljedahl and Wayne C. Rasmussen of USDA.

Among the museum specialists in North America: the late F. Hal Higgins, Peter Cousins, Alexis Prauss, and Carl Hugh Jones.

Pertinent photo and illustration credits are noted throughout the book. In certain cases the full title of the source has been abbreviated, for example: material reproduced from the F. Hal Higgins Library, Davis, California—by courtesy F. Hal Higgins Library of Agricultural Technology, University of California Library, Davis, California.

Henry Ford Museum—Courtesy Greenfield Village and Henry Ford Museum, Dearborn, Michigan.

Nebraska SHSM—Courtesy The Nebraska State Historical Society Museum, Lincoln, Nebraska.

Johnson, 1976—Paul C. Johnson's 'Farm Inventions in the Making of America', published by the Wallace-Homestead Book Company, Des Moines, Iowa.

John Deere Service Publications—Deere and Company, Moline, Illinois.

While the senior author was residing in Norway, Egil Oyjord, Joar Heir, Kristian Aas and H. V. Micklestad were most helpful. Others who assisted with European source material were Mrs. Lesley West of London's Science Museum; John Creasey, Reading; G. E. Fussell, Roger Arnold and Alexander Fenton in the U.K.; Hans Cuppers of Trier, and Ludwig Caspers of Claas, Germany; Grith Lerche and Axel Steensberg of the International Secretariat for Research on the History of Agricultural Implements, Denmark.

In Australia, John Kendall and Martin Hallett of the Science Museum, Victoria; Alan Thompson and the late S. S. McKay. Colleagues of the Commonwealth Scientific and Industrial Research Organization, Division of Mechanical Engineering, Melbourne: Peter Fricke, Neil Hamilton and Elizabeth Gray and the Chief of the Division, Barry Rawlings, for permission to use certain Divisional facilities.

Finally to our wives, Marlene and Mary respectively, for their patience as we have labored long hours at home on this work. To Mary for her valued editorial work on the final draft, but particularly to Marlene who bravely tackled the typing and undertook much correction on the several rough drafts that passed through her hands. To them, we dedicate this work.

Every care has been taken during the preparation of *The Grain Harvesters* to ensure accuracy.

The authors would welcome comments and would appreciate new source material that might be worked into future editions.

Graeme R. Quick, Melbourne, Victoria, Australia.

Wesley F. Buchele, Ames, Iowa, USA

May, 1978

Preface

Seeds are many things. Seeds are food. Seeds are storehouses of energy. Seeds are embryonic plants. Seeds are trade. But above all, seeds are life (Boswell 1960). If mankind were suddenly deprived of all sources of grain and seeds for even a year, all human and most animal life would vanish from the face of the earth.

Each of these vital grains must be harvested to be of use to man.

The Grain Harvesters is the story of the tools, machines and systems used throughout history to harvest grain.

The generic term "grain" includes "seeds", "corn" (a term used in the Old World for locally-grown grain), "kernels", or "fruit", that is, the reproductive part of seed-producing plants or spermatophytes. It will be applied without distinction to the fruits of grasses, cereals (small- or coarse-grained) and legumes.

The mainspring of modern power farming can be traced back to the grain harvest. Much of the world's farm equipment industry—production, distribution and service—had its origin in grain harvesting machinery. Indeed, certain landmarks in engineering progress began with the grain harvest. Here is to be found the first crank handle; the first centrifugal impeller; the first wheeled farm machine; the first fully automatic processing system; interchangeable parts; and the first

farm machinery service and credit organizations.

On these pages you can read about the beginnings of agriculture, when sickles were used as currency, and of the first "custom cutters" who followed the grain harvest northwards in ancient Egypt. And the energy-saving combine that used the residue of the crop for fuel, even as it was harvesting, and of the 50-ft cut California Leviathan. In Chapter 23 the full story of the axial revolution is traced as far back as 1785.

In 1776, at the time of the American Revolution, it took 13 farmers to feed one city dweller. Two centuries later, the average US farmer produced enough food for his family and 46 others.

In 1976, without any fanfare, the world's population slipped past four billion. Possibly one fourth of these people were fed in part by US agriculture. In many other parts of the world, it is as much as farmers can do to gather enough to feed themselves and their families. The harvest is often the major bottleneck.

The Grain Harvesters is about those determined individuals who have persevered to eliminate that bottleneck, who have reduced the drudgery of farming and who have given greater dignity to the occupation of farming. Their contributions have a more fundamental influence on our future than armaments or most political ideologies.

About The Authors

Graeme R. Quick (left) and Wesley F. Buchele, are shown here at the International Grain and Forage Harvesting Conference at Ames, Iowa, September 1977.

Both active participants in ASAE activities over the years, it was largely their effort that assured the success of this conference, the first of its kind. The Conference brought together representatives from all aspects of the industry and scientists from 27 countries.

Dr. Quick is Principal Research Scientist, Agricultural Engineering Group with the Division of Mechanical Engineering, Commonwealth Scientific and Industrial Research Organization, Highett, Victoria, Australia. Dr. Buchele is Professor, Department of Agricultural Engineering, Iowa State University, Ames, Iowa.

The Cyrus H. McCormick Medal. Eight times awarded by ASAE for meritorious service to contributors and pioneers in the grain harvesting business: To the late J. Brownlee Davidson (1933), the late Harry C. Merritt (1941), Charles J. Scranton (1952), Wayne H. Worthington (1954), the late Thomas Carroll (1958), Howard F. McColly (1968), Edgar L. Barger (1969) and Carlton L. Zink (1970).

THE GRAIN HARVESTERS
TABLE OF CONTENTS

"God also said, I give you all plants that bear seed everywhere on earth and every tree bearing fruit which yields seed: they shall be yours for food."—Genesis 1:29. (the New English Bible Second Edition © 1970 by permission of Oxford and Cambridge University Presses)

1

Harvest Tools of Antiquity

Direct Harvesting. Chippewa Indians harvesting wild "rice" (zizania aquatica) (Edwin Tunis. "Indians", 1959—Courtesy Thomas Y. Crowell Company Inc., used by permission).

Over 80 percent of mankind's diet is provided by the seeds of less than a dozen plant species. And to be of use to man all of these seeds must be harvested, either directly by hand or indirectly by machine. Without seeds or grain all human and most animal life would perish within a year. Thus the harvest becomes "the most important event on earth every year" (Canby 1975).

Direct harvesting involves plucking, combing, gleaning, gathering or recovering the grain from the plant or from the ground. It is a laborious, time-consuming process. *Indirect* harvesting involves detachment of the entire ear, head, stalk or seed-bearing organs of the plant. Subsequent operations, also usually mechanized, are needed to recover the grain from the husks, pods or seed capsules.

Reaping with Hand Tools

The first step in grain harvesting is commonly called "reaping". Reaping is a combined cutting and gathering process. In earliest times the cutting was done with a straight grain knife made of flint (Petrie 1917). Flint knife edges used on grain acquire a unique gloss or sheen, characteristic of the abrasive action of silicaceous cereal straws on the blade. Flint blades used on wood or bone do not acquire this gloss (Steensberg 1943). Tools found with gloss have provided important clues to the culture of ancient tribes. In the 1930's, for example, ancient knives of the Natufian culture were found in the Middle East in caves near Jericho. They were reaping knives made of bone, delicately carved with deer- and rams-head designs. The cutting edges, flint microliths, were secured in the bone with mineral pitch. They displayed the tell-tale grain gloss. The cave dwellers of the Natufian culture were evidently grain-gatherers.

A wood handled reaping knife of similar construction and wear pattern was also found in a basket of preserved grain in a Fayum prehistoric pit granary dated 3500 BC (Braidwood 1975).

J. R. Harlan, professor of Agronomy at the University of Oklahoma, recently tested this tool in a stand of Einkorn in Turkey. He harvested 6 1/4 pounds of ears containing two pounds of high protein grain in an hour from the wild ancestral cereal variety (Leonard 1973).

Angular Sickles

The earliest angular sickle designs, with blades angled forward from the line of the handle, appeared in Egypt and in Babylonia about the same time. The Egyptian sickles were often composites, with flint blades set in handles of wood or bone. The Ubaid craftsmen of Southern Mesopotamia, with less access to wood and flint, became adept at producing axes and sickles made of very hard baked clay. They were able to produce a finely serrated edge on the sickle sharp enough to cut dry straw. Clay sickle finds have been dated from as early as 3700 BC to as late as the Age of Metals (Braidwood 1975).

The Legacy of Ancient Egypt

Egypt's benign climate and the preoccupation of the wealthy with life after death have combined to produce the most complete record extant of any ancient civilization. Objects of wood and metal and delicate

Upper Left: One-piece grain knife of flint with serrated edge. Gloss effect, characteristic of cereal straw cutting is illustrated. Cereal straw may contain as much as 50 percent abrasive silicon dioxide in the ash (Steensberg 1943); Lower Left: Reaping knife of the Natufian culture with hafted bone handle and flint microlith cutting edge secured with mineral pitch (Hodges 1970); Upper Right: Wood handled reaping knife found in prehistoric pit granary in the Fayum of North Africa, ca. 3500BC (Forbes 1973); Lower Right: Egyptian short-handled angular sickle with flint blades set with mastic in two-piece wooden handle (Petrie 1917).

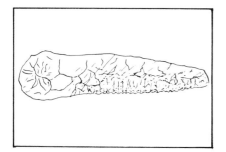
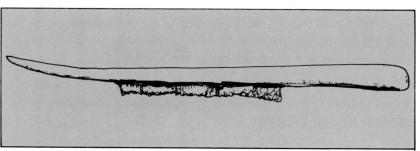
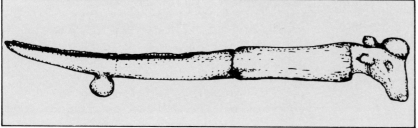
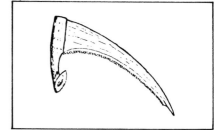

works of art have survived the millenia in the dry air. The ruling classes wanted the afterlife to be a repetition of their earthly life. They supervised the construction of gigantic and elaborate funerary monuments, with galleries and chambers colorfully decorated with scenes from everyday life along the sacred River Nile. Every care was taken in these representations for the imagined enjoyment of their souls during visits to the offering chambers.

Egypt was then, as it is today, an agricultural country. Most of the people who toiled in the fields of the wealthy were peasants. Numerous harvest scenes may be seen in tomb paintings and on carved reliefs. Reaping, flax pulling, head stripping, threshing by animal treading, winnowing, sifting, grinding, baking—all are depicted. Books of the Dead, written on papyrus, were often profusely illustrated. Some portray harvest scenes with the landowners wearing their distinctive smartly-pleated robes, themselves harvesting grain. More often the landed class was shown overseeing the harvest. Such was Menna, a wealthy field scribe for the Pharaoh around 1420 BC. Menna's tomb at Thebes contained a huge wall painting which showed him engaged in many aspects of his work and thus incidentally recorded the life of the peasants.

There was not always enough local help for the harvest and on larger estates gangs of mobile laborers were engaged. These crews began work in the region of the Southern Nile and gradually moved Northward with the harvest towards the Delta, the forerunners of today's custom cutters! Harvest help was exempt from military service.

The reapers of dynastic times used short-handled angular sickles, the shape sometimes suggestive of an animal jawbone and having an individual hieroglyphic sign. The reaper stood upright, grasped a bunch of grain in his left hand (reapers were invariably shown as right-handed) cut it just below the heads and laid the heads on the ground. The stubble was left standing, to be cut in a later operation. Wheat was rarely cut near ground level. The reapers were followed by women who picked up the bundles and carried them in baskets to one end of the field. Where the crop was put up in sheaves, the ears were shown protruding from both ends. Finally, the baskets or tied sheaves were taken to the threshing floor where the contents were dumped in a heap. Sometimes the heads were taken instead to a storehouse to be cured for a month or so prior to threshing.

Grain and Flax Threshing in Ancient Egypt

Threshing was effected by animal treading. The oxen or donkeys were driven around over the grain spread on the threshing floor, a reserved circular area of sun-baked earth packed rock hard by centuries of use. Laborers with wooden forks pitched and turned the straw into the path of the hooves.

Top: The Lady Anhei and her husband at work in the "Elysian Fields". Anhei appears to be worshipping the sacred blue heron (Re). The River Nile, also sacred, is shown in stylized form at the bottom of the register. Sketched from the original in the British Museum, London (Quick 1978); Bottom: Straight bladed, angular and balanced designs compared. Sickle types are classified according to the blade root curvature and the line of the handle.

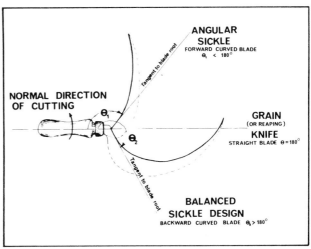

Flax grew thick and tall and was usually harvested while still at the flowering stage (Montet 1958). It was carefully pulled up by the roots, so that the valuable fibers were not broken, shaken free of dirt, and bound in the field. When the heads were dry enough, the sheaves were untied and the plants pulled by hand through a spiked combing tool to detach the heads. The flax heads were then threshed by treading (Rawlinson 1880).

Winnowing in Ancient Egypt

Before the crop was winnowed, a forking operation removed the coarse straw. Then winnowing scoops, of a type unique to Egypt, were used in pairs to ladle the grain from the threshing floor. The grain was scooped up and raised overhead, the scoops were then parted, allowing the grain to trickle into the breeze. As the grain fell onto the pile the chaff was blown aside and when the grain was clean the scribes appeared with their bushel measures. Once recorded, the grain was sacked or carried in containers to the granaries.

*Top to Bottom: Field Operations.
Wall painting from the Tomb of
Nakht at Thebes, 1420BC (Photo by
Egyptian expedition, The Metropolitan
Museum of Art, New York); Reaping
and animal treading scenes from a
Papyrus Book of the Dead. Sketched
from original in the British Museum,
London (Quick 1978); Conveying
bundles of ears to the threshing floor
in panniers lashed to the back of a
donkey. Tomb of Panhesy, Thebes
(Davies 1936); Reapers work to the
music of a vertical flute. Relief from
the tomb of Mereruka, Saqqara,
2300BC. The shape of the sickle
suggests an animal jawbone. Bone
handles were sometimes used. The
angular sickle had its own
hieroglyphic sign* ⟋ *(Oriental
Institute, University of Chicago).*

Left: Reaping and pulling crops of wheat and flax. From wall painting, Tomb of Sen-nuden, Thebes. ca 1200BC.

Winnowing operations. Tomb of Menna. Winnowing was a dusty job and the girls are illustrated wearing linen kerchiefs to protect their hair (Oriental Institute, Chicago).

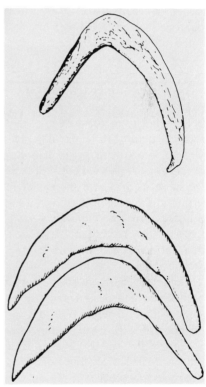

Above: Angular or crescentic sickles of baked clay from Babylonia, ca 3700-2600BC. The edges were finely serrated for cutting dry straw.

Donkeys treading grain and men thrashing donkeys (Baumann 1965).

The Age of Metals—Metal Reaping Tools

The earliest metal sickles were direct descendents of the clay versions. Cast in copper alloy or bronze, these angular sickles from Babylonia had spiked tangs for hafting. Bronze blades required working to maintain a sharp edge. Cast blades were sharpened by beating the lower side with a hammer or stone, then the edge was whetted with soft sandstone. Traces of hammer-sharpening can be seen on museum specimens.

Bronze sickles were used as currency in Anatolia (Turkey) around 2000 BC. This explains why hoards of sickles in large quantities have been unearthed. In certain areas of Europe, the sickle was also associated with the gods of soil fertility.

The coming of the Iron Age about 1200 BC brought little change in the shape of the sickle. The crescentic sickle blades were typically set into the crook of a wooden handle, as demonstrated by Steensberg (1943). In countries that imported their metals, flint-working continued into the Iron Age.

Development of the Balanced Sickle

The distinguishing feature of the balanced sickle design is that the blade curves backward at the blade heel, then curves forward in a sweep. The balanced sickle is used with a circular wrist motion, distinct from the drawcut or straight pulling movement of the reaping knife. The balanced sickle was a notable advance. The more natural arm movement reduced fatigue and permitted the reaper to cut grain more than four times as fast as could be cut with a grain knife. The varied shapes of this class of tool found around the world derive from different methods of hafting and attempts to distribute the weight of the blade more evenly between heel and tip. There are also regional preferences within countries in the design and use of balanced sickles. Stephens in 1851 referred to that certain curvature that will give the arm muscles the least possible cause for exertion as the "curve of least exertion" (Stephens 1851).

In the ancient East, the balanced sickle is attested to as early as any other type of sickle, other than the reaping knife (Childe 1951). The use of wrought iron and the opening of trade with the Far East were factors in the later spread of the balanced sickle design. Chinese discoveries have been dated back as far as the balanced sickle in Europe, to at least 1100 BC.

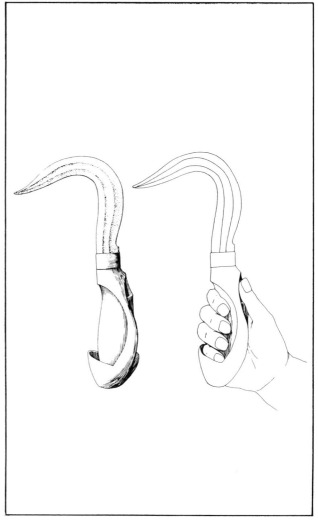

Top: Balanced sickle with elaborate wooden handle, shaped to fit the hand, Late Bronze Age, Morigen, Switzerland. Swiss lake dwellers of 1000BC may have used this bronze bladed sickle for cutting reeds (Hodges 1970); Left: modern sickle with cranked handle shaft and (right) stork's bill sickle from Mosul, Iraq.

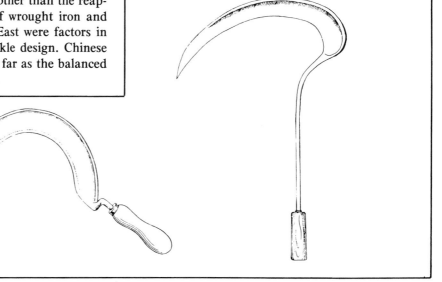

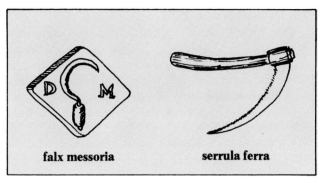

falx messoria   serrula ferra

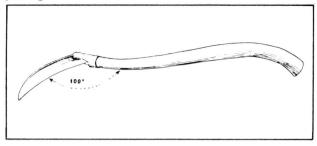

Left: Roman reaping tools (White 1968); Below: Chinese pulling sickle.

Wrought iron was readily forged and shaped and, being harder than bronze, could hold a better edge. The Roman legions carried both iron-working and the balanced sickle design to the outposts of their empire.

The Romans deliberately cut their wheat before it was ripe. They did this partly to minimize shatter loss, and partly from a notion that early cutting would increase crop yield. Marcus Varro (116-27 BC) listed three modes of cutting with the grain sickle (*falx messoria*):

- reaping close to the ground and leaving the "corn" on the ground in handfuls for subsequent gathering, heading and threshing.
- severing ears only and leaving the straw standing (this may have required a different tool, *serrula ferrea*)
- cutting the stalks midway between ground level and ear.

Sharpening and Use of the Balanced Sickle

A smooth bladed sickle works well in damp crops. Serrated blades were more typical of drier cereal regions such as those found along the Mediterranean and Sub-Sahara. In hand-formed iron sickles the serrations were formed by hammer and chisel, the chisel being applied obliquely to the edge. When the blade was filed or whetted with a stone, the characteristic serrations emerged to give the edge the extra sawing action needed for cutting dry stems.

The Short-Handled Scythe

The emergence of the scythe in the late Bronze Age appears to have been linked with a need to harvest winter forage for cattle at a time when the climate in Europe had entered a cooling cycle. Early scythes had a sickle-like crescentic blade hafted to a short handle.

In the Iron Age, scythes with a more slender, straighter blade appeared. The blade was hammered and whetted in the same way as a bronze blade. The early straight-handled scythe was probably used with both hands, since it would have been unbalanced and awkward to wield with one hand. The short scythe of the Middle Ages, however, is depicted in use with one hand but has an angular, balanced handle and is often portrayed in conjunction with a mowing crook in the left hand. The purpose of the crook was to divide the crop and move the severed grain away from the standing crop.

Roman soldiers foraging in a foreign land. Trajan's Column, 111BC.

Illustration from a French manuscript, 14th century (British Museum, London).

Top: Hainault scythe, mat hook, sharpening hammer and anvil from Flanders. The steel anvil was driven into the ground up to the tang and then used to support the blade as the underside of the scythe edge was hammered to an edge, or straightened if damaged. Sharpening with a whetstone followed (Quick 1978); Bottom: Hainault scythes were in use in England during the 1800's (Stephens 1851).

The long-handled scythe, recorded in Northern Europe in the 9th century at first had a plain straight handle. Hommel reported on the use of straight-handled scythes in China in 1937.

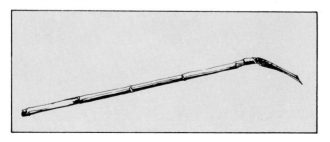

The Hainault Scythe

This Flemish grain cutting tool gained widespread use in the low countries of Europe during the Middle Ages. It had a curved handle about three feet long for balanced one-handed use. It was held in the right hand, with the forefinger in a leather loop. The grain was gathered by a hook (mat hak) held in the left hand. The Hainault Scythe was said to be more efficient for grain cutting than the sickle or even the straight-handled long scythe, then beginning to find some use in grain crops.

Long-Handled Scythes

Long-handled grass scythes were known in Roman times, but during the Dark Ages animal husbandry again became the dominant form of European agriculture. Political and social uncertainties abounded. Animals were a movable asset, but crops? Who would sow grain if there was the imminent risk of being unable to reap the grain in the event of aggression or invasion? With animals needing winter feed, there was a demand for a better mowing tool than the sickle or short scythe. Around the 9th century, the more productive long-handled scythe in use had a plain straight handle, or thole. The curved handle and now familiar short projecting hand-posts were refinements added around the 12th century (Singer, et al. 1956). But it was still a heavy tool, requiring strong arms and a strong back. In 1834 a patent was issued in the US for raising ridges along the blade by means of a press die to make the blade much stiffer and lighter.

The Cradle Scythe

Probably the earliest example of a cradle scythe is that seen in an illuminated Psaltery of the 13th century. It showed a simple loop attached to the handle just above the blade, similar to that illustrated by Brueghel in his classic engraving "Spring" (1565). The purpose of the cradle attachment was to enable the reaper to deposit the bunch or gavel to the side of the swath in one action, clearing the way for the next stroke.

The cradle scythe was imported into the American colonies sometime before 1646 (Van Wagenen 1930). Adoption was immediate and this indispensible tool was one of the earliest objects to be manufactured in the New World. Eventually "American Cradles" evolved distinctive shapes. A 19th century writer described the use of the cradle thus:

"On harvesting wheat with the cradle:
The stroke should be an even one, not too quick or by jerks, and from the entrance of the scythe in the grain to the end of cutting, the scythe should be kept even with the ground. At the close of cutting, the left hand may be drawn to the left hip, which prepares the scythe for the delivery, and a slight cant places it where it ought to lie."

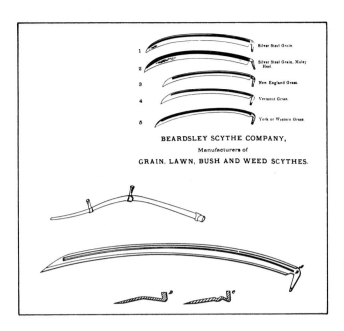

BEARDSLEY SCYTHE COMPANY,

Manufacturers of

GRAIN, LAWN, BUSH AND WEED SCYTHES.

Left: "Scythes for every kind of grass and grain". US advertisement for Beardsley Scythe Co., 1876; Below: Pieter Brueghel's engraving "Spring", (1565) shows the cradle scythe in use.

The cradle scythe was not used extensively in Britain until the 19th century. Although the adoption of the cradle would seem logical, because it increased productivity by a factor of eight or better, the transition from the sickle to the scythe and cradle encountered several obstacles in Britain between the Middle Ages and the 19th century. First when the judicial problem arose where cereal stubble was common property, the low cutting scythe was forbidden. Then there was tradition and, finally, shattered grain losses were perhaps twice as high. Collins (1969) cites sickle losses in grain at 5 to 10 percent and for the scythe and bagging hook 10 to 20 percent. Small crop patches that were inaccessible or on hillsides were better managed with the sickle, whereas the scythe required a flat stone-free surface.

Winter-sown cereals were ridged to help drain the soil. Under these conditions the scythe had to be whetted more often, reducing productivity. Finally, a person had to be robust to handle the scythe hour after hour. Women and children, who were cheaper to employ, could wield a sickle, but not a scythe.

Between 1790 and 1870, some English farmers adopted a heavier, smooth-edged sickle or 'hook', used in a slashing motion aimed at the base of the straw. The cut straw was gathered in with the aid of a crook and rolled under the foot to form a sheaf-sized bundle. This method was known variously as "bagging", "flagging", "hewing" or "hacking". It was up to four times as fast as the sickle, but less tidy and caused higher losses than the sickle (Collins 1969).

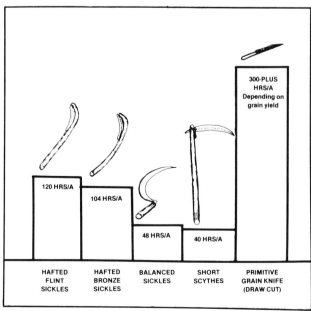

Left: An American cradle of the Drummond type (USDA); Above: Detail of an American cradle. In the Atlantic region up to 20 percent of grain was hand harvested before World War II (Nebraska SHSM); Below: A comparison of the different productivity rates among hand reaping methods (Quick 1978).

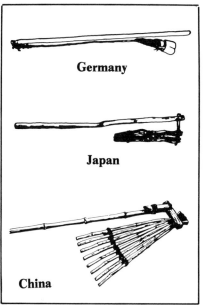

Hand Threshing Tools

There is historical evidence that the earliest grain harvesters had an understanding of plant structure and adapted threshing techniques to the individual species. About 720 BC Isaiah wrote:

"Then, fennel is not threshed with sledges,
 Cummin never needs a cartwheel;
Men thresh fennel (peas) with a stick,
 and cummin (herbs) with a flail;
Bread corn is not ground to pieces
 no one threshes it forever
but, once the cartwheel passes over it
 we spread it out, instead of crushing
'tis the Eternal who this lore supplies,
 so great a guide, so wonderfully wise."
  —Isaiah 28:27, 28.

Possibly the earliest illustration showing a hand beating tool in use for threshing was found in an agricultural scene on the wall of the tomb of Petosiris in Egypt, dated 325 AD. Men were shown administering blows to the sheaves with clublike threshing sticks.

The Jointed Flail

The earliest reference to the jointed flail occurred in St. Jerome's (d 420 AD) commentary on Isaiah and it has been conjectured that the jointed flail was invented in Gaul (Singer, et al. 1956). The flail consisted of two parts: a handle, typically about five feet long and the beater, about three feet long, flexibly connected and made of compact wood often thicker at the free end. The practice of threshing with flails was well established in England before the Middle Ages (Fussell 1959). Flails are as varied in design as the places they have been used.

Agricultural scene from the tomb of the high priest Petosiris, Egypt (Hermopolis West), 325 AD. The men appear to be administering blows to the sheaves with club-like threshing sticks.

Above Left: Threshing with the jointed flail. Fifteenth century calendar of the "Playfair Book of Hours". Original in the Victoria and Albert Museum, London; Above: Jointed flails. The handle was known as a "hand-staff" or "helve", the beater as the "whipple" or "supple".

Oriental Threshing

Rice and soybeans have been staples in China and Japan for millenia. Both of these "sacred grains" of the Orient thresh readily. Among hand threshing techniques that were adapted in China the whipping of bundles against a ribbed wooden surface was favored. The use of the Japanese rice threshing comb "Semba-koki" is reminiscent of Pliny the Elder's comment: "panicum et milium singillatim pectine manueli legunt Galliae . . ." (In the Gallic provinces they gather both varieties of millet ear by ear with a comb held in the hand.)

The Japanese preferred the more intensive action of the flail for threshing wheat and other grain.

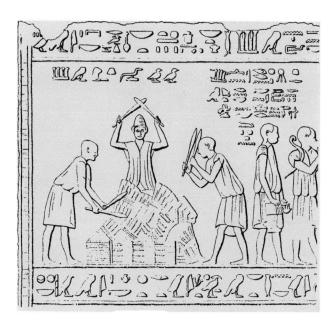

Above: Scenes of early hand threshing in Asia; Right: Scenes of peasant women stripping grain from the stalk, while others use flails; Below: Iranian farmer sharpening his sickle with a file (Lerche 1968).

Time-Honored Hand Tools Still Used Today

Possibly half the people on Earth today are living on grain harvested by hand. The simple tools of the peasant or small-scale farmer are valued possessions and show remarkable adaptation to local socio-economic needs. Hopfen (1960) has considered how hand tools may be adapted to the particular region and discusses the practicability of time-honored harvesting tools.

The sickle is still used around the world to reap cereals, particularly paddy rice which has soft but tough straw and is easily shattered. While the use of the sickle may be a slow method, it allows careful reaping and prevents shattering the precious grain from the ears. The scythette is still in use in the Benelux countries of modern Europe and bears a striking resemblance to the Hainault Scythe. Modern scythes and sickles come in two forms—with ground or with hammered blades. In industrialized cultures where these tools are used for odd jobs or recreation, the blades of higher carbon steel tend to be somewhat thicker and heavier, and the edge is produced on a grindstone. The hammered blade, on the other hand, tends to be lighter and the cold-worked edge is less brittle and tougher. Besides, the hammer and the anvil used to produce the edge are cheaper than a grinder. Both types of blade are field-sharpened with a whetstone.

Summary: Traditional Tools Have Their Place

Social and economic circumstances suggest that performances or effectiveness of one tool or harvesting system over another cannot be judged by productivity along. In societies everywhere the grain producers have seen the need to use tools and mechanisms for "the most important event on earth every year"—the harvest.

According to an ancient Egyptian legend, all men were cannibals until the great god Osiris instructed them how to make agricultural tools . . .

"Galliarum latifundiis valli praegrandes, dentibus in margine insertis, duabus rotis per segetum impellunter, iumento in contrarium iuncto; ita dereptae in vallum cadunt spicae."—Pliny (Natural History, 18:296.)

2

To The Unknown Gaul

Soldier from Gaul defends his home and country from Roman invaders

Their enemies spoke well of the Gauls. To die like a Gaul was to fight and die with dignity and courage. Julius Caesar had great respect for these worthy foes. At sea, his galleys were no match for their sturdily built ships. The Gauls were so skillful as artisans that the early dawn of machine harvesting can be traced to them. Pliny, Roman historian and farmer, wrote the quotation at the top of the page around 70AD. It says: "On the vast estates (latifundiis) in the provinces of Gaul, large frames (valli) fitted with teeth at the edge and carried on two wheels are driven through the corn by a pack animal pushing from behind; the ears thus torn off fall into the frame".

Pliny must be given credit for much that is known of Roman agriculture. But not all that Pliny wrote has been accurate. He wrote of one-legged Indians who used the other leg as a parasol; he wrote of Atlantis, and of the Phoenix bird; writing about these legends with the same degree of conviction as he used in his daily narrations on farming.

Later historians who were acquainted with the manual labor of the harvest deliberated Pliny's story about the Gallic reaper. It seemed unlikely. No archeological evidence had been found to support it, yet Palladius, writing around 400AD, said:

"In the more level plains of the Gallic Provinces they employ the following short cut (or labor-saving) device (compendium) for harvesting. With the aid of a single ox the machine outstrips the efforts of laborers and cuts down the time of the entire harvesting operation. They construct a cart (vehiculum) carried on two small wheels. The square surface of the cart is made up of planks, which slope outwards from the bottom, and so provide a larger space at the top. The height of the planks is lower at the front of this container (carpentum); at this point a large number of teeth with spaces between

13

are set up in line to match the height of the ears; they are bent back at the tips. At the back of the vehicle are fastened two very small yoke-beams, like the poles of a litter; at this point an ox is attached by means of a yoke and chains, with his head pointing toward the cart; he must be docile so that he will not exceed the pace set by the driver. When the latter begins to drive the vehicle through the standing corn all the ears are seized by the teeth and piled up in the cart, leaving the straw cut off in the field, the varying height of the cut being controlled from time to time by the cowherd who walks behind. In this way, after a few journeys up and down the field the entire harvesting process is completed in the space of a few hours. This machine is useful on open plains or where the ground is level, and on areas where the straw has no economic value."—Palladius, in De re rustica, "On Harvesting Grain". Lib. 7 Tit. 2 (Cited in White 1968).

Was Palladius describing Pliny's Vallus? Hardly, for there is much more information here. These two passages constitute the whole of the surviving literary evidence for the existence of Gallic harvesting machines. A Frenchman, Comte De Lasteyrie, wrote in 1820 that no example of this machine or carriage "has been found among any ancient monuments (Lasteyrie 1820).

The region of Gallia Cisalpina was an important granary, as well as one of strategic importance. It was attractive enough for Caesar Constantine to build his

Upper Left: Interpretation of Palladius' carpentum by the Frenchman, Comte De Lasteyrie, 1820, prior to the discovery of any of the monuments; Above: Site of the discovery of the important Buzenol-Montauban stone depicting the Gallic stripper-harvester. A replica of this stone is at the headquarters of the American Society of Agricultural Engineers in St. Joseph, Michigan.

summer palace there, at Trier on the Moselle. In 1958, archeologist Edwin P. Fouss, at work on the site of a medieval fort in Southern Luxemburg at Buzenol-Montauban, quite by chance uncovered in the foundations of a fortification a stone relief showing the vallus (Fouss 1958).

This limestone block was one of several that had been vandalized from Roman monuments in the Middle Ages and dragged to Buzenol-Montauban to build the fort. It was the first clearly identifiable evidence of Pliny's

A vallus First Day Cover. Stamp issued in Belgium in 1970 to commemorate the discovery of the vallus on the Buzenol relief, the first link in harvest mechanization. The photo on the cover is of the actual relief which is now housed in a small museum at the site. A replica was placed back in the Medieval fort ruins.

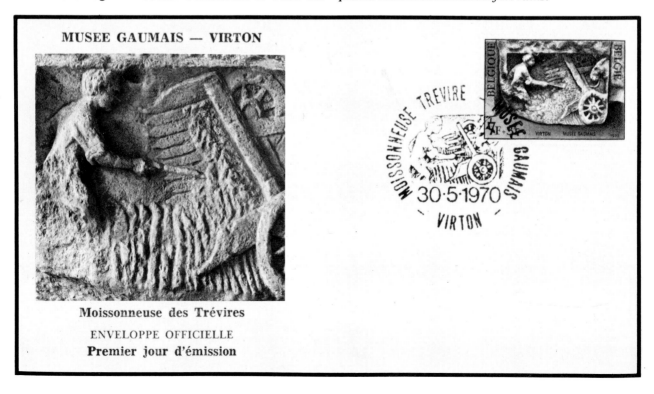

MUSEE GAUMAIS — VIRTON

Moissonneuse des Trévires

ENVELOPPE OFFICIELLE
Premier jour d'émission

Left: Reconstruction of the Trier harvester in the Landesmuseum, Trier (Cüppers 1964); Right: Vallus depicted on the Tableau of the Seasons at the Porte de Mars at Rhiems. Discovered earlier, but not identified as such until the 1958 Buzenol discovery. This monument has been dated as late second or early third century AD (White 1968).

vallus. This particular stone is well preserved. The discovery promptly led to the identification of three other stones from Gaul—all disfigured, but nonetheless giving partial views of the push-stripper harvester. Pliny was right. Each of the representations showed a box-like container on cart wheels, with teeth and dividers projecting ahead of the draft animal. There were two operators, one man walked slightly to the side of the machine wielding a rake-like tool to swipe off the heads

Below: Trier carpentum fragment. This mutilated relief was unearthed in 1890 behind the Landesmuseum in Trier, West Germany.

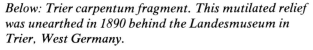

Above and Below: German engineers (DLG) reconstructed a full size model of their interpretation of the vallus. The simulated harvesting scene shows the DLG vallus in operation. The teeth were made of wood. The space between the teeth was tapered, not parallel (E.P. Fouss, Viton); Left: The Arlon relief-fragment of a funerary monument discovered in 1854 at Arlon, a Roman site near Luxemburg. The dimensions of this fragment do not match the Trier fragment as some writers have intimated.

trapped between parallel teeth and to thrust them into the catch-box. The other man drove the machine and used draft poles to maintain the correct stripping height.

This provided enough evidence for members of the German Society of Agricultural Engineers (DLG) to build a full size model and test it. The machine depicted on the monument was far more elegant than the slab-sided solid-wheeled dung cart that they illustrated. Form followed function. The wheels were spoked, the box was well proportioned and the dividers appeared to have been curved to deflect the stalks away from the wheels. The modern German replica was incorrect, having been built with tapered wooden teeth, which joined each other at the box. This design was bound to cause the crop to jam within a very short distance and was the root of the problems they described when the replica was field-tested.

The Buzenol vallus had parallel teeth and those who have worked with steel stripper harvesters in Australia have learned to appreciate the importance of the parallel teeth configuration. In all probability, the vallus had iron teeth which would have rapidly worn smooth so the headless straw could slide through.

Contemporary Gallic funerary carts were built with wheels similar to the vallus. A comparison of wheel rut spacing on Roman roads and the available evidence on the monuments was the basis of an estimate of the size of the Buzenol vallus and the machines capacity. The Buzenol vallus could have had a gathering width of about 4-4½ feet, with a productivity around 1/5 acre per hour (White 1967).

The crop heads collected in the box were partially threshed, depending on the dexterity and enthusiasm of the man wielding the beater. After bringing the material in from the field, a secondary threshing operation by treading and a final winnowing was all that was needed to finish the harvest. Productivity was indeed "completed after a few journeys, . . . ". Why did this bold and innovative step in mechanization not achieve wider popularity? Were not the drier lands of Northern Africa in Roman hands suited to harvesting by this method? Evidently freeing slaves and bondmen from drudgery was not a Roman incentive. It is impossible to lump the whole of ancient Western society into one generalization but, according to Finley (1965) "from the Homeric world to Justinian, great wealth was landed wealth. New wealth in Rome came from war and politics, not from enterprise"—or labor saving inventions. In Western antiquity "only the tongue was inspired by the gods, never the hand".

The ASAE headquarters building replica of the Buzenol-Montauban stone is displayed in the foyer. The replica was a gift of former ASAE national president Wayne Worthington.

"Premium List, Item 183. Machine for Reaping and Mowing Corn: For inventing a machine to mow or reap wheat by which it may be done more expeditiously and cheaper than by any method now practiced, provided it does not shed the Corn and that it lays the straw in such a manner as may be easily gathered up for binding; the Gold Medal, or Thirty Pounds. The machine with the certificates that at least ten acres of wheat have been cut by it, to be produced to the Society on or before the second Tuesday in December 1775." (Premiums offered by the Society for the Encouragement of Arts, Manufactures and Commerce, London 1774).

3
THE SOCIETY *for the encouragement* OF ARTS *Manufactures and Commerce*

Medieval writers divided society into three classes, those who work, those who guard, and those who pray; omitting altogether the merchants and townsmen. But it was these men without status—the anonymous middle class—who eventually changed the entire system.

In 1754, a group of such men gathered in London to found The Society for Encouragement of Arts, Manufactures and Commerce. The instigator, William Shipley, defined one object of the Society as simply, "that industry should be stimulated by means of prizes drawn from a fund contributed by public-spirited people."

England already had the Royal Society, devoted to pure science. To the new Society of Arts went the application of science. It concerned itself with matters as diverse as "shoes and ships and sealing wax . . . cabbages and kings"; nothing seemed too unimportant for it's attention.

The Society of Arts was a prime mover in the development of agriculture in England until the formation of the Royal Agricultural Society in 1838. In the first 22 years of existence the Society awarded £16,625 in prizes. These awards played no small part in the invention of many important mechanical devices and contributed to the progress of the Industrial Revolution. According to a later president of the Society, Sir Henry Trueman Wood:

"It may safely be asserted that of all the implements used by the farmer during the fifty or sixty years from 1760 onwards . . . there was not one which was not either introduced or improved in consequence of the Society's exertions and influence". (Wood 1913).

Britain of 1750-1850 was astir. It was an age of reformation, of creativity, of insatiable curiosity. It was a time when, as Dr. Trevelyan said

". . . our fathers conquered Canada and half of India, rediscovered and began to settle Australia, and traded on an everincreasing scale all over the inhabited globe; reorganized British agriculture on modern methods, began the Industrial Revolution in our Island, thence in later times to spread over the whole world."

Innovators were at work in Scotland too. James Watt's steam engine was developed there in 1784, and the world's first agricultural journal was published in Edinburgh by Arthur Young, Scottish agriculturalist, traveler and author. Capel Lofft, of Bury St. Edmunds, England, sent a letter to Young, advising him of the Society's offer of a gold medal for the development of a reaper. Young, in return, sent Lofft information on the Gallic reaper. Lofft replied with a translation of Pliny and Palladius' description of the vallus and sent notes suggesting improvements. These appeared in Young's *Annals of Agriculture* in 1785.

The Capel Lofft article inspired William Pitt of Pendeford, Stafford to propose a harvesting machine for "rippling (reaping) corn". Pitt wrote in Young's *Annals* in 1787:

"In pursuance of the idea contained in Mr. Lofft's translation from Pliny and Palladius . . . a machine for reaping, or rather, rippling corn is earnestly recommended to Mr. Winlaw; as preparatory to his rubbing (threshing) mill. . . were it brought to perfection, the business of rippling being cheaply and expeditiously done in the field, the corn is rendered fit for the rubbing mill, without farther preparation".

The rippling cylinder, with its iron combs, was supposed

to travel at twice the speed of the driving wheel and snatch off the ears, throwing them into the collecting box—

"the great car K, which being filled, the beast may be turned the contrary way, and the whole drawn to the store rooms".

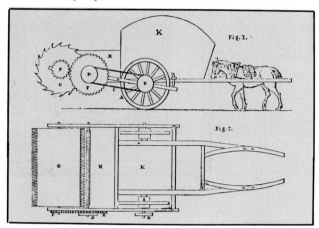

Machine for "rippling" or reaping corn, 1787. Proposed by William Pitt of Stafford, England. This was one of the earliest machine concepts with a ground-driven mechanism in England. M man with a shovel was supposed to transfer the rippled heads from the "small car" H to the container K. When this was full the horse was backed in and the cart pulled to storage. Pitt was not known to have built the machine (Young 1787).

Rotary Scythes

Arthur Young made no editorial comment on Pitt's proposal, but in 1799 commented editorially that he feared that it could be a long time before a practical reaper was designed.

His fears proved groundless, however, for in 1799 the world's first patent on a reaping machine was issued in England. Joseph Boyce of Mary-le-bone was awarded the first reaper patent, No. 2320, on July 4, 1799. This began a spate of inventions, the earliest of which were rotary devices that attempted to reproduce mechanically the action of the scythe.

James Dobbs of Birmingham has been labeled "the most mirth-provoking fellow of the whole craft" for the manner in which he promoted his rotary reaper. An actor by profession, he demonstrated his machine by cutting an artificial wheat crop 'planted' on stage. He advertised the "act" on playbills as if it was a theatrical performance (Fussell 1959).

The Avis Birmingham Gazette, Oct. 17, 1814 reported:

"Mr. Dobb's newly invented Patent Reaping Machine was shown at our Theatre on Friday evening, and appeared to be highly approved of. Mr. D's first experiment was completely successful and we have no doubt the other could have been equally so, had not the scenery obstructed the progress of the machine, which, causing a little embarrassment, prevented Mr. Dobbs from working it so effectively as he could have wished. Mr. D's explanation of the principles and properties of this invention was very satisfactory, and, we are inclined to think, it will prove of great public utility." (Woodcroft 1853).

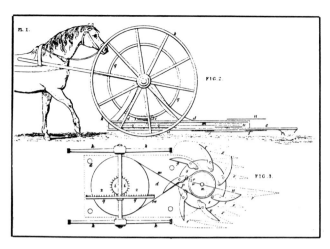

Above: Two-foot rotary scythe was illustrated in **Walker's Philosophy,** *published in 1799. Inventor's name was not stated, nor was there "any account of the machine having been made or worked". The rotating rod N was intended to sweep the "corn" aside after it was cut (Woodcroft 1853); Below: James Smith, manager of Deanston Cotton Works, Doune, Scotland, built and used this push-reaper in 1811. A cloth mesh enclosed the five-foot diameter drum to "increase the friction in carrying round the cut corn". The cutters were of German steel and projected 5 ins. beyond the drum. They required sharpening four times to the acre and could be sharpened with a scythe stone, after removal, in two minutes. His most successful model was built in 1815 and could cut an acre an hour. The Highland Society of Scotland awarded Smith a piece of plate valued at 50 guineas following successful demonstrations (Lee 1960).*

American innovators were not too far behind the British with rotary reaping devices. The first American reaper patent was awarded in 1803 to French and Hawkins. James Ten Eyck's 1825 design was a significant departure. Ten Eyck's machine had a drum on a horizontal axis and foreshadowed today's barrel-type lawn mowers.

Reciprocating Reapers

In 1800, Robert Meares of Frome, Somersetshire, England, patented a garden shears on wheels. There was no illustration accompanying Meare's patent, remarkable only for having been the earliest wheeled

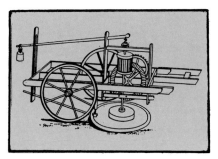

Jeremiah Bailey of Pennsylvania, 1822. Variation on the British theme of a rotary cutter.

Kerr of Edinburgh. The drum could be rotated in either direction to lay the swath on either side by use of another "bevil wheel". Kerr claimed priority over Smith and received a premium of 20 guineas from the Highland Society for his model which was field tested in 1811.

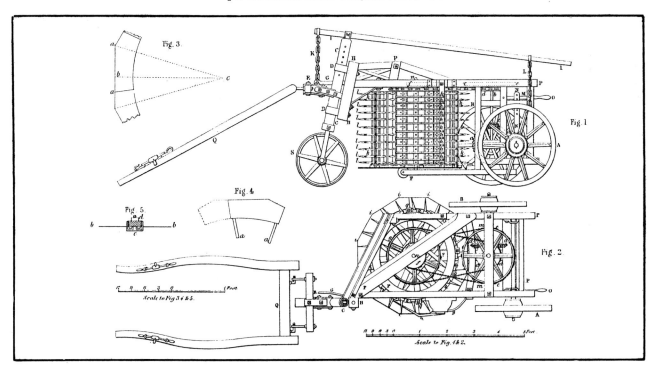

Above: Joseph Mann of Raby, 1820. The 1832 version is illustrated. The raking cylinder revolved at slower speed than the cutting drum. The machine performed for only a few hours. It was first to demonstrate the drive wheel in the line of draft (Woodcroft 1853); Below: Robert Meares of England: wheeled hand shears. British patent No. 2400, 1800.

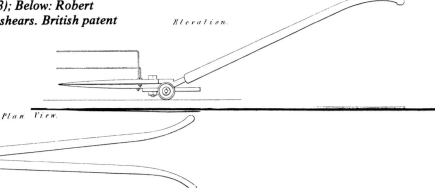

machine with a reciprocating cutting mechanism. Commissioner of Patents Bennet Woodcroft supplied illustrations of his own, based on Meares' patent specification.

Robert Salmon's 1807 design contains the first hint of an oscillating clipper with a row of smooth shear blades. This remarkable machine carried the first crop divider, and the first apparatus for collecting and delivering cut corn "in parcels like sheaves" ready for binding.

Salmon made significant contributions to farm mechanization. For 30 years he was the surveyor on the Duke of Bedford's estate. The "farmer Duke" held an annual sheep shearing at Woburn and used the occasion to exhibit agricultural implements, awarding prizes to the most meritorious. Salmon was a frequent exhibitor and won some prizes. He designed and developed a chaff-cutter, drill, plow, cultivator, the reaper illustrated, and hay-making machinery. He died at Woburn in 1821 at the age of 69, "well known and respected by the admirers of the fine arts and sciences, the inventor of many useful and valuable instruments of surgery,

Bennet Woodcroft's 1853 illustrative classification of rotary reapers was reproduced in Scientific American, *1888.*

agriculture and hydraulics". (*Gentlemen's Magazine 1821*).

The Napoleonic Wars in Europe of the early 1800's sorely depleted England's pool of harvest labor at a time when grain was in great demand. Migrant Irish reapers were available, but it was said that nothing could try a farmer's patience more severely than the management of a band of Irish "shearers". "To prevent altercations, the utmost watchfulness, diligence and determination were indispensable." There were causes enough for discontent. Living conditions for the transient help were often deplorable. Men and women were at times quartered together in sheds. There were also serious questions about the wisdom of providing reapers with alcohol as part of their keep, and the job was often so badly done that "a sheep could be lost in the stubble". The time was ripe for machines.

Living in the picturesque village of Denwick, Northumberland, was John Common, a millwright, who in 1811 hit upon the idea of using a "clipper-type" reaper. Some of the correspondence relating to Common's machine is of significance, both in its relation to the Society of Arts and for the documentation of Common's

CIRCULAR MOTION.					
CONTINUOUS AND ADVANCING.					
PITT PENDEFORD.1786.	SMITH DEANSTON.1811.	MANN CUMBERLAND.1820.	GIBSON NEWCASTLE.1846.	FRANCE.1851.	BURCH CHESHIRE.1852.
BOYCE LONDON.1799.	KERR SCOTLAND.1811.	BAILEY U.S.A.1822.	WHITWORTH MANCHESTER.1849.	MACKAY SWANSEA.1851.	CONTINUOUS WITH ALTERNATE.
FROM WALKER'S PHILOSY 1799.	CUMMING NORTHUMBERLAND.1811.	BUDDING GLOUCESTER.1830.	WHITWORTH MANCHESTER.1849.	MASON IPSWICH.1852.	SPRINGER AUSTRIA.1839.
PLUCKNETT KENT.1805.	DOBBS BIRMINGHAM.1814.	CHANDLER U.S.A.1835.	FAIRLESS NORTHUMBERLAND.1851.	SMITH LINCOLNSHIRE.1852.	TROTTER SUNDERLAND.1851.
GLADSTONE SCOTLAND.1805.	SCOTT ORMISTONE.1815.	U.S.A. DUNCAN LONDON.1840.	WINDER LONDON.1851.	COMPERTZ LONDON.1852.	
PLUCKNETT KENT.1807.	SCOTT ORMISTONE.1815.	PHILLIPS 1841.43 & 52.	BECKFORD & GOSLING MIDDLESEX.1851.	COMPERTZ LONDON.1852.	

design on subsequent American claims to priority for the invention of the reaper. Said Common:

"About the year 1803, I was at Fleetham setting up a new threshing machine for Mr. John Ostens. Mr. Ogle was then schoolmaster at Newham, and as a reaping machine had been mentioned in the newspapers as having been tried at the South of England some years before . . . he desired me . . . to help him contrive a machine. His was on the rotary plan at that time, and he could not find out one to cut like a scythe. I gave him my idea of a plan, which he thought well of, and we parted. I understand that he subsequently got Mr. Edward Gatis a cartwright in Newham, to make a model of his (rotary) machine . . . *My own machines were all clippers.* The first I made was a small one, and the shears were driven by a crank, but it had no apparatus on it for delivering the corn . . . it was made in the year 1811, eight years after Mr. Ogle's first, and tried in secret at night, in company with Mr. Thomas Appleby, among my own ripe corn, and it appeared to him and me to answer well. I don't know the night, but I recollect the moon was eclipsed. After this, Mr. Ogle came accidentally into my shop, and on seeing the machine he looked at it, and felt it with his hand and asked me if it was not for shearing corn, and I said it was, and after further conversation he advised me to let the Duke of Northumberland know about it, and I did so.

"Now my second machine was made on the same model, by desire of the Duke to send to a Society in London, I believe it was the Society of Arts. I delivered it to His Grace, according to my books, on the 5th of May, 1812; and after His Grace sent it to the society, they sent him a letter saying they could not give him a premium for it until a further trial was made, and no doubt the Society has the model yet. It was made to clip the corn, and *to deliver it with rollers into a swathe* . . . My third and last machine was made in 1812, the same year as my second, in full size, and to be drawn or pushed by men, by desire of the Duke. The shears were driven by a

Salmon of Woburn, 1807. This machine was a notable advance. It had the rudiments of the reciprocating sickle with crank and pitman drive, a crop divider rod and an oscillating rake to place the cut corn in "gavels" ready for binding (Woodcroft 1853).

sparrel (cam) instead of a crank and it delivered the corn into the swathe *by an endless sheet moving over two rollers, and it had a fan (reel) to bring the corn to the shears.* It was tried at Denwick, in Mr. John Thews field, called "The Havers", and did its work so well that it surprised the people who were present, but a little something at last happened with the shears that stopped it, as might likely happen with any new machine. A few days after I got it righted and tried it again at the Broomhouse, with like success and accident; so that I gave it over, as it had cost me a great deal of trouble and money. When I made this machine, I gave *Mr. Brown of Alnwick,* some patterns of it to make of cast iron, but as he was unable to cast them, he made them of wrought iron . . . " (Common 1907).

Unfortunately for posterity, neither the model, the drawings, nor the machines have been preserved. The Society of Arts' minutes of the Committee of Agriculture and Mechanics, April 30, 1812 attest to the accuracy of Common's letter (although he was 82 when he wrote in 1860):

"Read a certificate from John Thew and Thomas Appleby, dated Denwick, April 9th, 1812, stating that they accompanied John Common to a field at Denwick to make trials of his newly-invented reaping machine, where they saw it cut down a small patch of ripe oats with ease and dispatch, and that they are of opinion it will answer the purposes intended, and be of general benefit to the country.

"Examined the model sent. The apparatus is to be drawn by one or more horses, and is of considerable length; it appears that the two large wheels, from whence motion is given to the machine, are to be made of wood and iron spikes fixed in their periphyrae; they are fixed upon an axle of iron which moves round with them, and which axle also turns round a brass wheel fixed upon it, the teeth of which work in a pinion placed on a longitudinal axle, on the other end of the spindle or axle near to the shafts is fitted another cogd

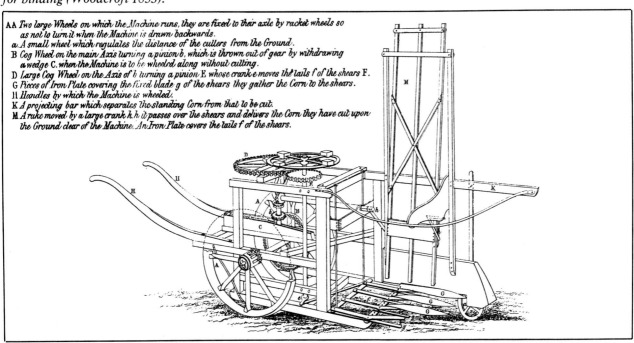

AA *Two large Wheels on which the Machine runs, they are fixed to their axle by racket wheels so as not to turn it when the Machine is drawn backwards.*
a *A small wheel which regulates the distance of the cutters from the Ground.*
B *Cog Wheel on the main Axis turning a pinion b, which is thrown out of gear by withdrawing a wedge C, when the Machine is to be wheeled along without cutting.*
D *Large Cog Wheel on the Axis of b turning a pinion E whose crank e moves the tails f of the shears F.*
G *Pieces of Iron Plate covering the fixed blade g of the shears they gather the Corn to the shears.*
H *Handles by which the Machine is wheeled.*
K *A projecting bar which separates the standing Corn from that to be cut.*
M *A rake moved by a large crank h, h it passes over the shears and delivers the Corn they have cut upon the Ground clear of the Machine. An Iron Plate covers the tails f of the shears.*

wheel, the teeth of which work in a pinion below, supported by a small wheel to prevent this part of the machine from touching the ground. The pulley and pinion last mentioned give an alternate backward and forward motion to *a set of angular knives,* so as to enable them to act upon a principle similar to the action of a number of scissors upon the blades of corn, directed to them by angular spikes of iron projecting before them as it falls upon a set of rollers, from whence it is delivered in a line upon the earth, as described in an account sent with it.

"Resolved,—It appears to this committee, upon inspecting the model of Mr. Common's reaping machine, and reading the account thereof with the certificate produced, that this invention is incomplete, and at present they cannot fairly judge of it, and therefore cannot recommend it to the further attention of the Society."

Apparently this Mr. Brown, to whom Common had given the patterns, took up the challenge of improving the machine. A trial of a machine made by Brown was reported in the Newcastle-upon-Tyne Chronicle, October 19, 1816 where it was stated that:

"... it was tried on October 3rd, 1816, in a field of wheat belonging to T. Dodds, Esq., South Side, and far exceeded the expectations of everyone who saw it at work, and bids fair to give satisfaction. It will cut six or seven acres a day, and much more even and low than by sickle, etc."

Schoolmaster Ogle then re-entered the picture. In the *Mechanic's Magazine,* November 12, 1825, he published a letter, and provided a facsimile of "Ogle's Reaping Machine".

Sketch and description of "Ogle's Reaping Machine" in the November 12, 1825 issue of Mechanic's Magazine, London. Note that Ogle had a straight blade, EE, where Common's had angular sections. Unfortunately there is no existing drawing of Common's sicklebar, but from the description it would seem to be the true prototype of the modern reciprocating cutterbar.

Mechanics' Magazine,

MUSEUM, REGISTER, JOURNAL, AND GAZETTE.

No. 116.]   SATURDAY, NOVEMBER 12, 1825.   [Price 3d.

" All science arises from observations on practice. Practice has always gone before method and rule ; but method and rule have afterwards improved and perfected practice in every art."—*Blair.*

OGLE'S REAPING MACHINE.

VOL. V.

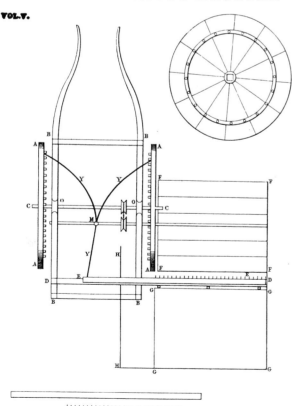

OGLE'S REAPING MACHINE.

Sir,—In the year 1822 I made a small model of a machine for reaping corn, but not being a workman in that business myself, and being on very friendly terms with one Thomas Brown, a founder, in Alnwick, and his son, Joseph Brown, I presented it to them ; they made a better model of it than I had made, of iron, and presented it to the public many market-days at Alnwick, thinking to carry it into execution by subscription, but were disappointed—the farmers considered it an impossibility. Thomas and Joseph Brown then made the machine at their own expense, and tried it first near Alnwick : it did not, however, altogether answer, the teeth of the frame, DD, where the knife cut upon (as hereafter described) was too long, and collected the dirt among the corn too much. They then made the teeth shorter, and tried it again at a place called South Side, near Warkworth, in a field of wheat : it then cut to great perfection, but still not laying the corn into sheaves, the farmer did not think that it lessened the expense much. Mr. Brown took it home again, and added the part for collecting the corn into a sheaf, GG, when he tried it again at Alnwick in a field of barley, which it cut and laid out in sheaves extremely well. Messrs. Brown then advertised, at the beginning of the year 1823, that they would furnish machines of this sort complete, for sheaving corn, at the beginning of harvest, but found none of the farmers that would go to the expense, though the machine was seen even to cut the lying corn where it was not bound down with new rising green corn. Some working people at last threatened to kill Mr. Brown if he persevered any farther in it, and it has never been more tried. It was estimated, from what was tried, that it would cut, at an easy rate, fourteen acres per day. For the encouragement of farmers and mechanics, I here give the following account of this machine.

Description.

DD, the frame which the knife acts upon.

EE, the knife that acts upon DD.

BBBB represent the frame on which the machinery is fixed, with the shafts for a horse.

CC, an axis which turns round in the frame at OO.

AA, &c. are the wheels which are fixed fast upon the axis, turn round with it, and give motion to all the other parts of the machine.

DD is a frame, of iron or wood, with teeth in it about three inches long.

EE is a knife which acts upon that frame a little upon the teeth.

YYY, an instrument fixed upon a centre upon a frame at M, turns upon it, and acts in the teeth of the wheels, AA, &c.; leaving the cog of the wheel on the one side and catching it on the other, keeps the knife sliding and cutting out and in, in a very quick motion.

FFFF is a rake that goes round by a belt or chain, upon two pulleys or wheels, and lashes the corn backward upon the knife ; it is just eight small iron bars or rods placed in a circle, in a cylindrical form.

GGGG, a platform, made of thin deal or tin, made to go on hinges on the back of the frame that the knife acts upon, and collects the corn as it is cut ; this frame is lifted till as much corn is collected as will be a sheaf, and then let fall by a lever, hh, over a fulcrum upon the frame, BB, &c. where the corn slides off, when it is a little raised again. It was found, however, to answer better when it was put off by a man and a fork towards the horse, as it is easier bound, and leaves the stubble clear for the horse to go upon.

I have only given a part of the framing, as most mechanics take their own way of fixing the main principles.

I am, Sir,

Your most respectful well-wisher,

HENRY OGLE,
Schoolmaster.

Renington, near Alnwick,
Northumberland.

Thomas S. Brown, a grandson of Thomas Brown of Alnwick, took exception to Ogle's claim of having invented the machine, and to the date claimed. In a letter to the editor of *Farm Implement News,* January 1886, he wrote:

"The fact is that Mr. Ogle invented very little of this machine. His model, as he admits, was a very crude thing, and doubtless very impracticable until the Browns gave it a mechanical organization. As he says, "Mr. Brown took it home and added the part for collecting the corn (grain) into a sheaf, and when he tried it again at Alnwick in a field of barley it cut and laid out the sheaves extremely well;" and for this success he received a mobbing; he had to flee for his life, while the machine was broken up. This took place, Ogle says, in 1823. I have one of the advertisements mentioned in your article, date 1820; so Ogle is out of the way three years in date. I will forward it to you, but request that you return it, as I value it very highly.

"Mr. Brown built afterward an entirely different machine from the one represented by the cut in your article. It was a geared machine with crank and pitman, similar to the most popular machines of the present day. I did possess an illustrated pamphlet of the same, but loaned it to a friend, and, of course, according to Shakespeare, lost "both it and friend". Mr. Brown emigrated to America in 1824 and settled at Stirling Center, Cayuga Co., N.Y., where he died at the advanced age of eighty eight. Yours truly, Thomas S. Brown."

Here was reference to the world's first reaper advertisement, in which Thos. Brown, "Smith and Founder" of Alnwick offered his customers:

"The situation which T. B. occupies enables him, in a peculiar manner, to execute orders in machinery on the lowest terms, and with the greatest accuracy . . . The following are a few of the many machines which T. B. has for a series of years been in the habit of making, and which have afforded general satisfaction:

... "all instruments of Husbandry, from the Threshing Machine (whose principal centers are raised a considerable distance from the wood beams on iron carriages to prevent friction and taking fire), to the "newly invented reaping machine . . . —Alnwick Foundry 1820"."

A mottled and stained copy of the full advertisement was reproduced in *Farm Implement News* 7(2):22. 1886. Dated 1820, it was the world's first reaper advertisement.

Thus thorough documentation exists of successful reaper activity in Northumberland around 1820-23 until threats to the innovator's life and the smashing of his machine curtailed development. In 1824, Brown and his family sailed for the United States, taking with him the patterns and ideas of the Alnwick reapers.

In subsequent chapters it will be seen that the Society was influential in other aspects of the harvest, but when it came to mechanized reaping, the Society's efforts were not so creditable. It offered prizes for a reaping machine for 46 years (until 1820) *without ever making an award,* in spite of the large number of machines brought out during this period.

Another important machine, which also failed to impress the Society, was Bell's reaper. This reaper was exhibited in Scotland in 1828 and 1829 and awarded a £50 prize by the Highland and Agricultural Society in 1829. But when it was submitted the following year to the Society of Arts it was turned down on the grounds that it was so well known that encouragement was unnecessary. Again, no prize!

Bell's Reaper

Patrick Bell was born in 1801 in Scotland, the son of a Forfarshire tenant farmer. Living frugally and working part time on the farm, including harvesting by hand, he managed his way through St. Andrews University, finally receiving his MA degree. He had a great interest in mechanics all his life, maintaining a workshop both on his father's farm and later at the parsonage of Carmyllie. He became interested in gas lighting and set up a gas plant to light his father's farm which was considered the wonder of the neighborhood. He grew sugarbeets and manufactured sugar for sale in Dundee. When he invented his reaper at age 25, he was still a divinity student at St. Andrews. In 1854, Bell wrote an account in the *Journal of Agriculture* (Edinburgh) describing his design. He related that he had carefully studied Smith's engraving of Deanston's rotary scythe and that he thought about the matter of machine-reaping for years, searching until it became an obsession.

The Reverend Patrick Bell, LLD. From the **Illustrated London News,** *March 7, 1868.*

Bell's reaper, as illustrated in **Transactions** *of the Highland and Agricultural Society, January, 1852. The machine had a five-foot width and, pushed by a single horse, could harvest an imperial acre an hour to keep six persons busy tying up the corn "reaped by this force".*

One evening in 1826, Bell caught sight of a pair of gardener's shears poking out of a hedge. Inspiration! He saw at one glance the principle he wanted. He seized the shears, crept through the hedge and began trying them on green oats:

> "It was well no neighboring gossip saw me, else the rumor might have been easily circulated that the poor student had gone crazed". (Hendrick 1928).

He built a model in 1827, followed in 1828 by a full-sized mockup with wooden facsimilies of metal parts which he took to the local blacksmith for fabrication in steel. He finished the work on the cutting shears by hand. This gave him:

> " . . . a world of trouble and vexation, when they came into my hand they were in a very rude state, and required my grinding, filing and fitting".

When he had assembled the first machine, he and his brother tested it in the outbuilding which was his workshop. For this purpose they stealthily conveyed to the shed a 6 in. layer of earth into which they planted, stalk by stalk, a sheaf of oats. When they tried his machine in the simulated crop it cut well enough but the straw

> "was lying higgeldy-piggeldy, in such a mess as would have utterly disgraced me in the harvest field."

Encouraged that it would at least cut, he set to work on a travelling canvas, or draper, to deliver the corn to the side and he developed a reel. Again he tested the machine in an artificial crop. It cut and delivered the corn to his satisfaction. He recorded that he congratulated himself audibly on his success

> "being convinced that I had converted the implement from a cutting to a reaping machine."

Completed in the summer of 1828, he could barely wait for the crop to ripen. The first field trial was held in secret, late at night, when everybody was in bed " . . . the good horse Jock was yoked to it", and they slipped away to a field of ripening wheat—"my brother and I speaking the meantime to one another in whispers." In their anxiety they forgot the reel and at first they were disappointed with the result, for while "the wheat was well enough cut, it was lying in a bundle before the machine".

They hurriedly fetched the reel and installed it in the dark. Success: now "the wheat was lying by the side of the machine as prettily as any that has ever been cut by it since". They congratulated each other, then took the machine back to the shed as quietly as possible to await public trial (Hendrick 1928).

The method of driving Bell's machine in front of the horses, with the canvas delivering to one side or the other, was considered one of the principal advantages, for the machine could cut the crop without the need for a man to "open up" the field with a scythe. The horses followed in the track of the cut crop without treading down the swath.

A report in the *Quarterly Journal* followed in 1828, stating that two machines were in use in the trial, one made by Bell constructed on the plan of the model transmitted to the Highland Society the previous December, and the other made by a mechanic from Dundee. It was estimated that the machines could be built for £30 each, and were said to have cut oats, barley and wheat at a rate of about an imperial acre per hour. They were each pushed by one horse.

Left: cutaway of an improved Bell reaper; Below: Original Bell reaper from the Science Museum in London. The apron driving gear was reversible so that the machine could deliver on either side (Science Museum, London).

"—the horse was wholly untrained—the people without experience in such work—the corn, also, reserved on purpose for this exhibition, being too ripe, was broken down and not in good state for reaping; and though we were well pleased with the operation of both machines we anticipate still greater facility—and superior management by additional practice . . . we beg leave to express our conviction that Mr. Bell's reaping machine will come immediately into general use—that it will confer a signal benefit on agriculture—that his invention is of national importance—and that he deserves the highest encouragement for his active and strenuous exertions for the public good." (*Quarterly Journal*, Report. November 1828).

The Highland Society awarded him £50 in 1828. Reports gave considerable publicity to his machine and a number were built, though a major manufacturer had yet to be found.

Bell reported in 1832 that 10 machines were at work and had cut 320 Scottish acres among them. Experience showed they could cut 12 acres a day and that the cutters required sharpening after 50 acres. The cost of reaping grain was reduced by about one third.

Bell did not attempt to patent his design, a decision he was to regret later. In 1854, he wrote in the *Journal of Agriculture*:

"Out of some twelve or eighteen machines made by various parties about this time, I am not sure that any one of them continued to work above a few years . . . Had I patented the machine all this bungling in machine making would have been avoided; and the issue perhaps proves that, for the public benefit even, this was the prudent course to have been adopted. But I was always averse to this step being taken. I wished the implement to go into the agricultural world free of any extra expense . . . "

Four Bell reapers were exported to the U.S.; others after 1833 to Australia and Europe. Unfortunately, Bell remained aloof from industry. Instead, he continued to pursue his calling to the Ministry. After completing his divinity course at St. Andrews he went to Canada in 1833 where he taught, tutored and preached on occasion, and recorded the four years of his life at Fergus, Ontario, in a detailed journal. (Byerly 1928).

Left: Bell's reaper as illustrated by Woodcroft, 1853. Two of Bell's reapers harvested 87 acres in 1830, and seven Bell reapers harvested 219 acres in 1831. Bell's machines were probably the world's first reaper exports, finding their way into Europe, America and Australia before 1854; Below: An original reaping machine made by Patrick Bell was given to the Science Museum in London by Bell in 1868.

The original machine that Bell built in 1828 continued in use on his father's farm, then on his brother's farm. In 1852 it was repaired and displayed in operation at the Highland Society's show, finally arriving at the London Science Museum. The only concession to change was a modified cutterbar, installed before 1868, and chain drives in place of some of the original bevel gears.

Bell spent the later years of his life quietly at his parish in Carmyllie, where he died in 1869. A window in the little church there is dedicated to his memory. He had never lost interest in mechanics and though he worked sporadically at various instruments in his workshop in Carmyllie, he gained little public recognition until the American reaping machine came into general use after the Great Exhibition of 1851. The part which Bell had played in the development of the reaper was belatedly recognized and various honors were bestowed upon him, including the degree of LLD., from his alma mater, St. Andrews, and a £1,000 award by the Highland Society.

Between 1786, when Pitt's design was published, and 1831, over 50 different reapers had been reported in England, Scotland, Europe and the US. The number, though impressive, reflects perhaps the persistence of the inventors rather than the progress made in adoption of machine reaping in Europe. The Alnwick reapers and Bell's embodied the essential principles of workable machines, but the social climate, combined with the economic downturn in Britain during the 1830's, retarded adoption. It remained for the combined genius of the American entrepreneurs and publicity given by the Great Exhibition in London 1851, to bring about the change from the scythe to reaper. Before 1851, few British farmers had ever set eye on a reaping machine.

Cyrus H. McCormick, inventor of the reaper, was born on this farm Feb. 15, 1809. Here he completed the first practical reaper in 1831. V.P.I. Student Branch, American Society of Agricultural Engineers. 1928. (Inscription in granite marker near the McCormick homestead at Raphine, Virginia.)

4

Who Invented The Reaper?

If the reaper was ever to come of age, a man was needed who would make the reaping machine his life's aim: a man who would take the concept, develop and manufacture the machine in quantity and sell it at a price that farmers could afford. Cyrus Hall McCormick was the man. Virginia, cradle of the nation and home of Presidents, was the place.

McCormick developed a reaper and built a business foundation on which a whole new American enterprise— the farm equipment industry—could grow.

Origin of The McCormick Reaper

Robert McCormick, prosperous farmer and mill operator, of Rockbridge County, Virginia, and father of three sons, experimented with a device for reaping grain as early as 1809. The oldest son, Cyrus, born Feb. 15, 1809, was seven when in 1816 he saw his father humiliated in front of skeptical onlookers when his rotary scything machine failed dismally. Robert McCormick was more successful with his patented hemp-breaks, blacksmith's bellows— which he made at the Walnut Grove smithy and sold for $12.50 apiece— and his threshers, five of these were sold in 1834 for $70 apiece. This environment was conducive to young Cyrus McCormick's mechanical abilities. At 15, he had built to suit his stature, his own cradle scythe and at 21, he patented and built an excellent self-sharpening plow. Father and son forged a close partnership at home and in business. They worked together manufacturing plows and later expanded into iron smelting, partly as an extension of their plow business.

In 1831, Robert McCormick finally abandoned his reaper, even though he had made improvements to it. Cyrus McCormick began where his father had left off. His faithful black servant, Joe Anderson, later recalled:

"Old Master Robert gave up working on the reaper when Master Cyrus he thought it could be done. It is like the good Lord who sent his son to save sinners. He began the work but his son did the work and finished it".

The interim May-July 1831 was an important time in the life of Cyrus McCormick. In this period McCormick assembled and tested his first reaper. He also went to Washington to patent his plow. He may have become familiar with reaper patent art while he was there, but later vowed he never knew of Patrick Bell's reaper at that time. It is believed that Bell's reaper did not arrive in the US until 1835.

Cyrus Hall McCormick was so absorbed in the reaper business that he did not marry until age 49. This 1860 photo was taken by Matthew Brady, famed for his coverage of the Civil War (International Harvester).

The Walnut Grove, Virginia, farm smithy, near the house where Cyrus McCormick was born. This building became the factory of "Rob't McCormick & Son" in the 1830's (International Harvester).

McCormick's first reaper had a straight, smooth-edged, sharpened blade driven by a crank and pitman gearing from the larger drive wheel. There was a crude form of crop divider. A reel was added after the first trial—in a patch of ripe wheat held over for the purpose by his father. Six acres of oats were also harvested in a field near Steele's tavern. McCormick himself commented that the job was less than perfect. To improve the cutting action, he had a blacksmith notch out the smooth blade with teeth, or serrations, all running in the same direction.

Though critical of his first machine, McCormick largely ignored his father's advice to give it up and devote attention to "more productive matters". Early in 1832, he turned up in Kentucky where he spent some time setting up his father's hemp-breaking equipment. He received this note from his father:

"I think the building of a grain machine in that country
  (Kentucky) might be attended with difficulty, as it will
  require a good deal of new modeling which when done at
  home is free from the watchful and jealous eye of
  strangers." Robt. McCormick letter of March 27, 1832 to
  his son, Cyrus. (First written documentation of the C. H.
  McCormick reaper and later used in establishing the patent
  claims priority of 1831.)

Above: McCormick's first public demonstration, 1831. The machine cut a four foot swath (International Harvester). There were enough differences between this reaper and Patrick Bell's 1828 model to lend credibility to McCormick's statement that he never knew of Bell's machine—in spite of later English claims that he copied Bell's reaper; Below: McCormick's second cutterbar, 1831. The original straight one-inch wide flat blade was modified with serrations to improve the cutting action. The blade vibrated in a slot through recurved steel fingerguards. The guards were pinned to wooden pegs (Swift 1897).

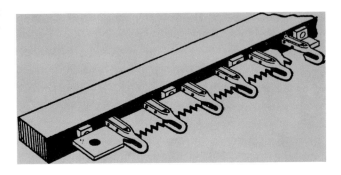

McCormick's reaper required only two operators—a man or boy to ride the horse and a man walking alongside to rake the cut grain off the platform. With this machine, the two men could cut as much grain in a day as four or five men with cradles (International Harvester).

In 1832, McCormick returned to Walnut Grove to add improvements to the reaper. Fifty acres of wheat were harvested successfully on the home farm and two public trials were held.

In 1833, the machine was modified again and equipped with a 4½-foot serrated sickle. Its performance was favorably commented upon in the local press. Then in the April 1834 issue of *Mechanic's Magazine* a cut and a description of Obed Hussey's reaper patent appeared. Though Cyrus had patented his hillside plow in short order, he waited three years before filing on the reaper, because he wanted to improve on it. Now he was goaded into action. He felt that Hussey's machine bore too close a resemblence. Necessity called for protection of his interests, although he knew Hussey's machine was not a finished product. He secured a patent for his reaper on June 21, 1834.

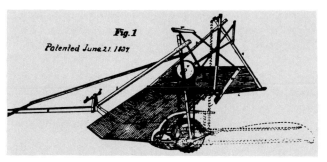

McCormick's first reaper patent, secured for a fee of $30 and issued June 21, 1834 extended to him "the full and exclusive right and liberty of making, constructing, using, and vending to others to be used, the said "improvement". There is evidence that McCormick moved ahead faster than he may have wished to go, since the patent emphasized the push mode rather than the side-pulled method which he had in fact used exclusively until then. Apparently, McCormick had felt that a larger machine might have too much side draft if pulled, and therefore, would need to be pushed. His fears proved groundless and he soon returned to the side draft hitch.

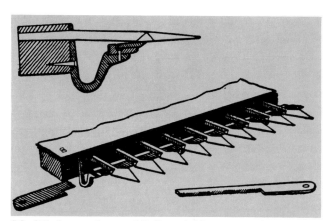

Modified cutterbar on McCormick's reaper, 1841. The saw-tooth pitch was alternated every 1½ ins. and the guards changed to a spear-shaped arrangement (Swift 1897).

Obed Hussey, 1792-1860. The black patch over his left eye and the loss of an arm—neither of which were ever explained—were as mysterious as the reasons that led this seaman to turn, at age 40, to the problems of reaping.

Actually the issuance of a patent in those days was no guarantee of originality. Before the 1836 Patent Office fire and subsequent reorganization, the Office was not empowered to reject an application. This proved a fruitful field for future litigation.

Competition

McCormick had been stung by the publicity given Hussey's patent. With characteristic boldness he issued this challenge in a letter to the editor of *Mechanic's Magazine*, 1834:

"Dear Sir,
Having seen in the April number of your "Mechanic's Magazine" a cut and description of a reaping machine, said to have been invented by a Mr. Obed Hussey, of Ohio, last summer, I would ask the favor of you to inform Mr. Hussey and the public, through your columns, that the principle, viz. cutting grain by means of a toothed instrument, receiving a rotary motion from a crank, with iron teeth projecting before the edge of the cutter for the purpose of preventing the grain from partaking of its motion, is a part of the principle of my machine and was invented by me, and operated on wheat and oats in July 1831. This can be attested to the entire satisfaction of the public and Mr. Hussey, as it was witnessed by many persons: consequently I would warn all persons against the use of the aforesaid principle, as I regard and treat the use of it, in any way, as an infringement of my right."

"Since the first experiment was made . . . I have been laboring to bring it to as much perfection as the principle admitted of, before offering it to the public . . . "

This was the first shot in the American reaper war. Hussey, who had sold his first machine in Ohio in 1834,

The Hussey reaper was said to work best when the horses were driven at a trot. This may have been due to insufficient sickle speed or to the absence of a reel. The machine eventually proved more useful as a mower than as a reaper. The cut grain had to be removed before the machine made its next pass, since it was discharged to the rear and would be trampled by the horses (USDA).

did not have a machine in Virginia and did not reply. There followed a lull for McCormick. Although his machine continued to cut each season on their farms, there were no further public showings until 1839. Financial failures and other pressures forced McCormick to postpone work on the reaper. He resolutely worked off his indebtedness and in 1839 returned to work on the reaper. The first two commercial units were sold to farmers late in 1839, in time for the 1840 harvest. These two machines did not work well. The following year McCormick modified the cutterbar and was able to sell seven units for the 1842 harvest.

The modified cutterbar worked so much better that McCormick was able to warrant that:

"Purchasers would run no risk since, if the reapers for 1842 were not strong and endurable, and would not cut fifteen acres a day and save one bushel per acre, ordinarily lost by shelling when the cradle was used, they could be returned." (Hutchinson 1930).

Obed Hussey

Little is known of Obed Hussey during the first 40 years of his life. Born in Maine in 1792 of Quaker stock, he went to sea at an early age and rowed more than once after whales in the Pacific Ocean. (Greeno 1912).

Hussey always contended that *he* invented the first practical reaper. He certainly had obtained a patent six months before his chief rival. But McCormick, as Hussey acknowledged, had used his reaper two years before Hussey's first trial on July 2, 1833 at Carthage, Ohio. Nevertheless, wasting no time, Hussey, had a manufacturer and agents in New York by 1834. He sold his first machine the same year and several more in 1835. In 1836, he was acclaimed in Maryland when he successfully harvested 180 acres of oats on the

estate of his sponsor Tench Tilghman. He received awards for public exhibitions there and established a "factory" in a barn to make the machines he sold in Maryland in 1836.

In 1837, the Hussey machine was produced with a single driving wheel in place of the two of equal size and was used for the first time as a mower. Hussey moved, for the last time, to Baltimore in 1838 and made about 12 machines for the 1839 harvest. The machines were so convincing in public trials that Hussey wrote:

" . . . Among those (attempts to produce a workable reaping machine) . . . I consider myself alone successful . . . every previous attempt has totally failed . . . and . . . gone into oblivion . . . My next year's machine will be much superior to any which I have before made and to which I apprehend but little improvement can be subsequently added."

Hussey overstated his case. The "little improvement" he made (to the gearing) caused the 36 machines sold in 1840 to "suffer a retrograde". Other circumstances combined to cut into his market so that by 1843, he could sell only two reapers. Then came the public competitions.

The First Public Reaper Competitions in the US

The lines for the next round of the reaper war were drawn on the pages of America's oldest farm journal, *The Southern Planter*. Perturbed by the growing popularity of his rival's reaper, Hussey wrote in the Jan. 20, 1843 issue that he was "apprehensive of an impression amongst the farmers of lower Virginia unfavorable to my Reaping Machine."

Hussey then defended his machine, citing successes, and closed with the following challenge:

"I see from your 'Planter' an account of another reaper in your State which is attracting some attention. It shall be

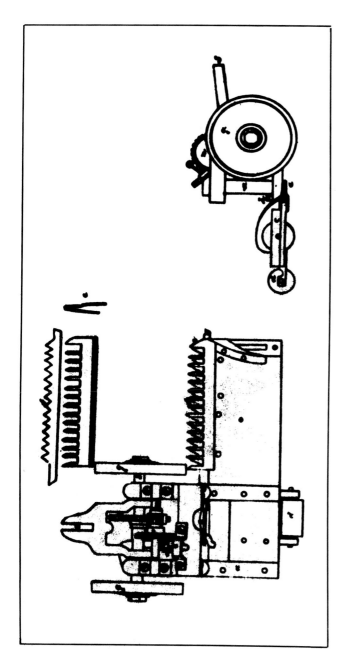

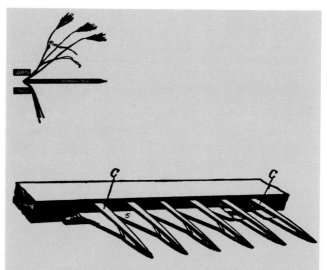

Left: Hussey's reaper patented December 31, 1833. The machine did not have a reel. The severed grain was pushed off the platform to the ground in gavels by the man riding on the machine who used a special rake for the purpose. He was expected to lean forward, gently dip into the grain and sweep the straw across the cutter. He was, in short, a "human reel". (*Mechanic's Magazine*, April, 1834); Above: The Hussey cutterbar. Double-bevelled triangular steel blades were attached to the knife back or "rod" and were reciprocated through "strong iron spikes . . . with . . . a space for the blades to play in". Thus Hussey's knife provided a "draw cut" or double shear action, but was inclined to plug in wet grain. Blade width was 3 in. . . . "causing all the blades to move at once, each blade vibrating between two spikes, passing out of one spike into the other, cutting the grain as it comes in between the spikes as the machine progresses; each double spike forming a bearing to resist the straw both below and above the edge of the blades" (From Patent drawing and **The Farmers Magazine**, England, where Hussey wrote a letter offering the machine for $150 in 1841).

Shortly after 1840, George Rugg, who operated a repair shop in Ottawa, Illinois, conceived the idea of serrating the cutting edges of Hussey's cutterbar. The serrations were a great improvement for cutting clean grain free of weeds. When McCormick's machine appeared, Rugg took out the original straight sickles and "applied his improved design with considerable approbation in the neighborhood" (Deering 1900).

Hussey's open-backed guard design of 1846, patented on August 17, 1847. This modification practically removed the problems he had experienced with clogging in wet grain and assured the success of Hussey's design as a mowing machine. This 3 in. pitch cutterbar is the ancestor of today's designs. Hussey realized $200,000 from the sale of the patent rights even when there were only two years left before the patent expired (Deering 1900).

my endeavour to meet this machine in the field in the next harvest. . . I think it but justice to give this public notice that the parties concerned *may not be taken unawares* but have the opportunity to prepare themselves for such a contest, that no advantage may be taken. These gentlemen, who have become prudently cautious, by being often deceived *by humbugs* will then have the opportunity to judge for themselves.''

The insinuation "humbug" stirred the Virginian to quickly accept the challenge. Richmond, Virginia, was selected as the site. The outcome of the contest was destined to give Hussey many a headache. Said the judges:

"The undersigned were called upon at the farm of Mr. A. Hutcheson, to witness the performance of the wheat reaping machines of Cyrus H. McCormick and Obed Hussey, and to decide upon the merits of the same. We are unanimously of the opinion that both of them are valuable inventions, and richly merit the encouragement of the farming community. They both performed most admirably. The committee feels great reluctance in deciding between them. But upon the whole, prefers McCormick's. . .''

The *Southern Planter* continued:

"Mr. Hussey contended however, that he had not had a fair chance, as much as the field had been selected by his adversary, and was not calculated to test those qualities in the machines in which he excelled; moreover, he said that circumstances compelled him to come to trial with a low-priced inferior machine (built under contract) which was not at all the one generally known as Hussey's reaper. He, therefore, invited Mr. McCormick to meet him again at Mr. Roane's on the following Wednesday.''

Editor Botts continued:

"At this exhibition we were present and were much pleased with the operation of both machines. . . The company, consisting of fifteen or twenty gentlemen, seemed pretty equally divided between the two. For our own part, we thought there were some advantages pertaining to the one that did not belong to the other, and vice versa. For instance, McCormick's is the lightest draught, being worked by two horses whilst Hussey's required four. . .

"From all we could learn and judge from the construction of the two, we should infer McCormick's would cut best in damp grain; but on the other hand in lodged or tangled grain Hussey's certainly possesses great superiority. It is also a heavier and stronger, and more efficient machine, cutting we should suppose, if well attended, from a fourth to a third more a day. The price of Hussey's machine is $160, that of McCormick's $100.

"The greatest advantage we conceive about these machines is the extreme cleanness with which they both cut. They shatter too, infinitely less than the cradle. Either is worth more than its cost to any farmer who cultivates a large tract of smooth level land in wheat or oats; still we would advise no one to go into his harvest relying on his machine alone; what with wet wheat, gullies and hillsides, he will find that the cradles cannot be dispensed with, and it will be necessary to have them ready to take the place of the machine where such circumstances oppose its operation." (*Southern Planter.* August, 1843).

It was a dramatic moment when the judges returned to proclaim the victor. The spectators were "as tense as the contenders". The decision left Hussey stunned. The editor's observation that the Hussey machine would not

Hussey's side-delivery reaper illustrated in England.

cut wet grain seemed justified when Hussey refused to meet McCormick in a second contest at the Roane farm because of damp grain conditions.

Hussey kept sniping at his triumphant rival through the *Southern Planter* for several years with McCormick, on occasion, returning a wrathful reply.

Editor Botts eventually tired of the verbal sparring and announced in March 1845 that this thing must end. McCormick sent in another invective, in spite of the ultimatum. Botts refused to publish the letter, saying that the argument had degenerated into a personal matter. McCormick wrote to his brother William:

"I have written an answer to his (Hussey's) piece at Botts, but doubt whether he will insert in the June no. Think him a little fishy at any rate, neither flesh nor fowl."—Letter from C. H. to W. S. McCormick, May 15, 1845. (Hutchinson, 1930).

The publicity gained from the reaper competitions were grist for the mill for agricultural writers who were, in turn, influential in increasing reaper sales, exactly as each inventor had hoped. McCormick sold 29 reapers in 1843 and 50 in 1844 at $100 apiece. Orders came from his home state, from New York, Tennessee, Ohio, Indiana, Illinois, Missouri and from as far away as Wisconsin. Hussey sold only 11 machines in 1843 but his business gradually increased so that by 1846, when he had developed his "open-backed" guard, 50 Husseys were sold. This level was maintained for the next few years, with Hussey, in the meanwhile, contesting McCormick's popularity. He complained of the "disastrous effects on my interests" and maintained that the McCormick reaper was being wrongly esteemed. In 1848, he said "it had not proved a useful invention to the public." That same year McCormick sales topped the 800 mark.

McCormick Looks at the Midwest

McCormick decided to personally investigate the potential of the newly-opened prairies of the Middle West, a source of a trickle of new business. He traveled North and West in late 1844, a trip that fired his imagination. He wrote to his family that the reaper, while a luxury in Virginia, was a necessity on the great Western plains. He began plans to move Westward and while on the trip contracted with others to build reapers to meet the demand that was impossible to fill from Walnut Grove. These included:

- Fitch & Co., Brockport, N.Y. (Cyrus was to receive a patent fee of $20 for each machine).
- A. C. Brown, of Cincinnatti, Ohio.
- Seymour and Morgan, Brockport, N.Y. (The contracts were to be valid until 1848, when the 14 years of the original patent were due to expire).

By 1845, Walnut Grove was stretched to the limit to build 48 reapers out of the total of 123 machines sold that year. In 1846, 190 machines were sold. McCormick spent much of the time away from home attempting to improve and maintain the quality of construction of those reapers built under contract.

A New Industry for Chicago

When Cyrus McCormick built his first reaper, the wheat center of the US was just south of Rochester, New York. Chicago was still a frontier trading post for furtrappers and Indians. Improved transportation, particularly the completion of the Erie Canal in 1825, sent Americans surging Westward. European immigrants poured into the country and settlers reached for the wide expanses of unplowed prairie. Political and economic factors, combined with the favorable climate on the rich prairies, moved the breadbasket a thousand miles West. The Southern States soon found they could not compete with the quality, quantity and price of grain grown in the Mid-West.

In 1848 the first load of wheat moved into Chicago by rail, the same year that the first load of freight was "locked" through to Chicago from England. Chicago, "gateway to the West", was now a port as well. New settlers who arrived there clamored for free land. The railroads had the land and began rapid expansion in anticipation of larger harvests—partly because of the reaper. Cause and effect worked hand in glove. The decade 1840-1850 has been called "the first revolution" in agriculture. (Rasmussen 1974). The Mid-West of the 1850's was about to leap from a home economy to mechanized production.

The Reaper's Profit Potential

McCormick was able to make a reaper for $50 at Walnut Grove and sell it for more than twice that

Robert McCormick, father, counselor, best friend and business partner of his son, Cyrus. The death of the elder McCormick on July 4, 1846, overshadowed everything else on his son's mind that year.

amount. This factor and the rapid growth of the country made it imperative that McCormick build his own machines in a factory close to these markets. McCormick chose Chicago. He formed a partnership with C. M. Gray on August 30, 1847. They purchased lots on the North bank of the Chicago River and immediately began construction of a factory that could produce 500 reapers for the 1848 harvest. That same year the original patent was due to expire. McCormick began a protracted and unsuccessful battle to seek renewal of his patent. He foresaw that if the original patent was not renewed, profitable agreements could no longer be made with other firms. McCormick then took out a new patent on improvements to the reaper, including the addition of a raker's seat, and inaugurated a vigorous advertising campaign. An era came to an end when production ceased at Walnut Grove. In fact, the poor

The McCormick Reaper Works was established in 1847 on the north bank of the Chicago River, just east of the present Michigan Avenue bridge. Today, the corporate headquarters of International Harvester are situated on two of the original lots purchased by McCormick for his factory (International Harvester).

McCORMICK'S
PATENT
VIRGINIA REAPER.

The above cut represents one of McCORMICK'S PATENT VIRGINIA REAPERS, as built for the harvest of 1849. It has been greatly improved. Lower cost lower, by the addition of a seat for the driver. By a change in the position of the crank, some further advantages gained from the cutting operation it: (thereby very much lessening the friction and wear of the machinery, by dispensing altogether with the lever and resistance rod) and also on the heel wheel which operates more gently, on the grain than the round ones; by a direct of care on the platform, which very much reduces the draft of the platform; taking it by an increase of the wire weight and strength of the wheels of the machine; and by improvement made in the cutting apparatus.

D. W. BROWN,
OF ASHLAND, OHIO,

Having been duly appointed Agent for the sale of the above valuable labor-saving machine (manufactured by C. H. McCormick & Co., in Chicago, Ill.,) for the Counties of Seneca, Sandusky, Erie, Huron, Richland, Ashland and Wayne, would respectfully inform the farmers of those counties, that he is prepared to furnish them with the above Reapers on very liberal terms.

The Wheat portions of the above territory will be visited, and the Agent will be ready to give any information relative to said Reaper, by addressing him at Ashland, Ashland County, Ohio.

Ashland, March, 1850.

Copy of an 1850 handbill advertising McCormick's "Patent Virginia Reaper". The agent, D. W. Brown, claimed to be selling "in all Northern Ohio" in 1850 (International Harvester).

quality of the last 35 machines built there under the youngest McCormick brother's supervision was the only set-back in 1847.

The "Virginia Reaper" Manufactured in Chicago

McCormick showed shrewd sales psychology in choosing to associate his machine with a State name. Virginia, "home of Presidents and cradle of the nation", probably evoked the most emotional appeal of all the States.

Production at the Chicago factory placed 120 men on a primitive assembly line. The 30 hp steam power plant became one of the wonders of Chicago. At the close of the year, McCormick had reached a personal milestone—he had bought out his partners, including Mayor Ogden—and had discontinued subcontracting. Now all the units could be built in one facility under his personal supervision. McCormick's younger brothers eventually joined him in Chicago. This gave him latitude to attend personally to business away from the factory, with the assurance that the company would be in trustworthy hands.

McCormick introduced credit sales and an ambitious advertising program, both novel at the time. His advertising, while tedious and lacking in conscious humor, left little to the imagination:

"I warrant my reapers superior to Hussey's and all others. I have a reputation to maintain. Let a farmer take both and keep the one which he likes best."

The 1849 McCormick design finally took the driver off the horse and seated him on the reaper. The raker's seat, however, was not much of an improvement. One English writer, on seeing the machine in operation described it this way:

" . . . (the raker) a short rose-faced, muscular man who rode backwards astride a sort of rail, with a stout piece of board attached to keep him from falling off: His legs were so short that he could only touch one toe; and he hung thus in the air,

McCormick's American Reaping Machine.

doing his work much as a man lifts a basket over a fence across the top of which he lies balancing himself. If it is not hard work for him, it is hard work to look at him." (Lewis 1941).

Although McCormick had already produced 2800 machines through 1849 and his competition only several hundred, that situation was soon to change.

A Raft of Reapers

In the East, the Baltimore Quaker, Hussey, had been McCormick's earliest and most serious rival. But there had been other practical minds at work.

The first US Patent on a reaping machine was issued in 1803 to Messrs. French and Hawkins of New Jersey for a rotary scythe design. The Schnebleys of Maryland built four machines along the style of Patrick Bell's reaper some time after 1825, but had faded from the scene for "lack of patronage and the notice of the press". (Marsh 1886). William Manning of New Jersey patented and built a reciprocating cutterbar with a "sway-bar" drive and skid shoes in 1831. (Ardrey 1894). A successful cutterbar which was first to have been belt-driven was developed by Enoch Ambler of New York and patented in 1834.

William Randall was said to have built and used a reaper in the State of New York by 1833. His machine appears to have met nearly all the requirements of a successful hand-raking reaper. A remonstrance was filed in the Patent Office early in 1848 by a New York citizen's group on behalf of this local inventor and opposing the extension of McCormick's original patent. (Deering 1900).

Ferdinand Woodward of New Jersey manufactured several reapers in 1843 and after. A Woodward reaper of special interest was the one in Illinois acquired on a trade-in by Marcus Steward, a farmer, and J. F. Hollister, a mechanic. They rebuilt the machine and had George Rugg's serrated sickle installed in place of

Top Left: William Manning's patent of May 3, 1831 featured a "scalloped" sicklebar. Note the long driving link or "sway bar" which was pivoted at H. At the drive end, the sway bar ran in a zigzag groove in the wheel hub (Butterworth 1888); Left: F. Woodward's reaper was a pusher design similar to Bell's. The driver directed the machine with a tiller. Some were manufactured in 1843 and a patent awarded the New Jersey inventor on September 30, 1845. The description mentions that the cut grain was accumulated behind the cutterbar in a shallow box that could be tilted to unload the gavel (Deering 1900); Above: This 1833 Randall reaper built in New York lacked little in the way of meeting all the requirements of a successful hand-rake reaper (The Smithsonian Institution); Below: Seymour & Morgan introduced their own improvements, such as a raker's stand and driver's seat and Hussey's scalloped sickle on the "New Yorker" reaper after the McCormick patent expired in 1848. Five hundred "New Yorkers" were built in 1851. In 1852, Morgan licensed a manufacturer to build "New Yorkers" for sale in Ohio.

the original 10-ft cutterbar. The modified machine was said to be capable of 60 percent more work than any machine made in 1845. Cyrus McCormick inspected this machine and promptly threatened legal proceedings if these mechanics did not desist. He was ignored and the machine, one of several they modified, was said to have operated for 30 harvests before it was retired. This same farm near Plano later figured in the development of the celebrated Marsh "Harvester". McCormick threatened suits against practically all users of machines sold that were not under license to him. The

CUTTERS WORKED BY HAND.	RECTILINEAR MOTION.				
	ADVANCING ONLY.	RECIPROCATING AND ADVANCING.			
	ANCIENT ROMAN.	SALMON BEDFORDSHIRE. 1807.	RUNDELL U.S.A. 1835.	Mc.CORMICK. U.S.A. BROOMAN. LONDON. 1852.	WRAY & SON YORKSHIRE. 1852.
	CLADSTONE SCOTLAND. 1826.	OGLE ALNWICK. 1822.	Mc.CORMICK U.S.A. BROOMAN. LONDON. 1850.	POOLE LONDON. 1852.	HARKES CHESHIRE. 1852.
ISLAND OF JAVA. ANI ANI	ESTERLY U.S.A. 1844.	BELL SCOTLAND. 1826.	STACEY MIDDLESEX. 1852.	CROSSKILL YORKSHIRE. 1852.	HUSSEY MANCHESTER. 1852.
MEARES SOMERSETSHIRE. 1800.	BLAIKIE GLASGOW. 1851.	MANNING U.S.A. 1831.	DRAY LONDON. 1852.	DRAY LONDON. 1852.	JOHNSON LONDON. 1852.
TAYLOR LANCASHIRE. 1851.	SIDELONG AND ADVANCING. LILLIE MIDDLESEX. 1847.	HUSSEY U.S.A. 1833.	RIDLEY NORTHUMBERLAND. 1852.	FOWLER BRISTOL. 1852.	GOMPERTZ LONDON. 1852.
	EXALL READING. 1851.	Mc.CORMICK U.S.A. 1834.	RANDELL CORNWALL. 1852.	NEWTON LONDON. 1852.	

"Rectilinear Motion" class of cutting mechanisms. Bennet Woodcroft's classification of 1853 was extremely thorough. Woodcroft, as England's Commissioner of Patents, was instrumental in arranging for Bell's reaper to go to the Science Museum in London, where it can be seen today.

inclusion of such warning threats in his advertisements probably made many a farmer hesitate to purchase a rival machine when there seemed to be a risk of being dispossessed of it after expensive court action. (Hutchinson 1930).

The Great Exhibition, London, 1851

Probably the most important work ever undertaken by the Society of Arts, Manufacturers and Commerce, and one that certainly had far-reaching effects on the arts, trade and industry, and on the adoption in Europe of the reaping machine was the world's first international exhibition (Wood 1913). The Great Exhibition was initially proposed in 1828 as a national exhibition of industry. As enthusiasm mounted, Royal Patronage followed and it was made international in scope. A Royal Commission was established and a magnificent "Crystal Palace" was built especially for the Exhibition.

In the United States, the year 1851 practically marked the end of the "experimental phase" of mechanical grain reaping. McCormick and Hussey were both ready for new fields to conquer. Hussey's reapers were actually being built by Dray in England. McCormick had opened negotiations with English licensees and had taken the precaution of patenting his design in England through his agent Richard A. Brooman. A specially-finished reaper was prepared for England's Royal Agricultural Society, replete with the painted Eagle-"talons, bolts and all."

McCormick was selected as representative for the United States farm industry in the American exhibit. Over confidence had led the American Commissioners to reserve more floor space than they could fill. Their exhibit, with the reaper as center piece, and with emphasis on the functional rather than the beautiful, was ridiculed by the English press for the first few weeks of the Exhibition. The "Prairie Ground", as the scoffers called it, seemed even more barren and ridiculous flanked by the magnificent displays of art from Russia, Austria and France. As for the reaper, The *London Times* of May 1, 1851, said it was "a cross between a flying machine, a wheelbarrow and an Astley chariot."

Neither McCormick nor Hussey found it expedient to sail for England until the harvest season of 1851 was well advanced. McCormick planned well, and sent a skilled mechanic and operator with his machine. When the first competitive field trial was held prior to the Exhibition, at the farm of wealthy manufacturer J. J. Mecchi, the weather was foul but McCormick's man had the machine working reasonably well—considering the "sour, dark, drenching day", as observer Horace Greeley described it. The jury estimated that the machine was capable of 20 acres per day. Hussey's reaper clogged, making raking-off impossible. Perhaps if the inventor had been present it may have given a more creditable showing. The only English machine, a Tollemache, failed to start. The crowd gave four cheers for McCormick but a final decision on an award was postponed until fairer weather and a second trial. The point had been made. The English press carried long accounts of Mecchi's Tiptree Farm trial. McCormick and his reaper were momentarily the most talked-about topics associated with the Exhibition:

"The "Prairie Ground" of America is now thronged. McCormick's machine is put back in its place and I believe yesterday more visited it than the Kohinoor diamond itself." (Hutchinson 1930 citing agent B. P. Johnson's report from the Exhibition, July 1851).

The deciding trial for the award of a Royal Council Medal was next held at the farm of Philip Pusey M.P., who was described by Disraeli as "one of the most distinguished country gentlemen who ever sat in the House of Commons."

Pusey cited the following reasons for failure of the reaper to make inroads in English agriculture prior to 1851:

- More difficult cutting conditions, (unreliable climate, damp grain and tangled straw) compared with the lighter, free standing, drier crops of the US.
- ridges and drainage furrows common to English fields made soil surface uneven.
- small fields, hedged in, without gateways wide enough for machines.
- hired labor was paid less than in the US. (Pusey 1851).

At exactly the right moment, at the time of the second trial, McCormick arrived; while Hussey was in France on "a profitless excursion". Both machines were thoroughly tested, Hussey's again failed to work well, while McCormick's operated to the admiration of farmers and the Jury. McCormick was awarded the Council Medal.

How the tone of the press changed!

The previously scornful *London Times*, conceding earlier misjudgement, was now convinced that
"the reaping machine from the United States is the most valuable contribution from abroad, to the stock of our previous knowledge, that we have yet discovered".

And by August
"that the McCormick reaper alone was doubtless worth more than the cost of the entire Exhibition".

McCormick and his machine made a brief triumphal tour of the English countryside; he arranged for 500 reapers to be built in England, then returned to the plaudits of the press in the United States.

McCormick's successes in England were shortlived, for when Hussey returned from France he solved his problems and actually tipped the scales in his favor during the next trials. But he was too late for the Grand Council Medal. Agricultural opinion was divided when the harvest of 1852 arrived. In Scotland, Bell's reaper was favored and was being accorded belated recognition. The consensus was that:

Left: London's Crystal Palace built for the Great Exhibition, 1851; Below: Burgess and Key's McCormick reaper of 1854. Conditions in England were different from those in the US. If "raking off" by hand was a tough job in the US, in England, the job required super strength (Slight 1854). The Burgess and Key patent improvement used an auger discharge for side delivery of the swath (Neilsen 1970).

Cause and effect. The production of McCormick reapers 1839-1856 and wheat shipments from Chicago during the same period.

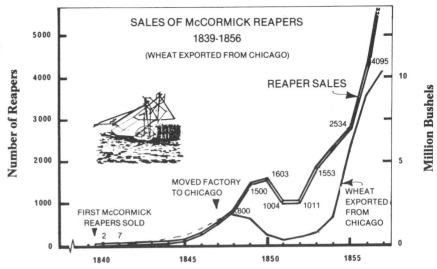

SALES OF McCORMICK REAPERS
1839-1856
(WHEAT EXPORTED FROM CHICAGO)

Number of Reapers

Million Bushels

REAPER SALES

4095

2534

1603   1553
1500
800   1004   1011

MOVED FACTORY
TO CHICAGO

WHEAT
EXPORTED
FROM
CHICAGO

FIRST McCORMICK
REAPERS SOLD
2 7

5000

4000

3000

2000

1000

0

10

5

0

1840   1845   1850   1855

(1) Hussey machines were "cheap, handy implements which work well if the corn is dry and has little undergrowth." The lack of the reel, the heavy draft and need for the horses to drive at a trot were serious drawbacks.
(2) McCormick's cutting principle "as now adopted" was simple and more efficient, "not liable to be stopped by the weather," draft was light, two horses might work at the ordinary gait all day with little distress." But the machine was more elaborate and expensive.
(3) "the mode of delivery. . .of. . .Bell's reaper is the best (of all) in principle." "None of the reapers can be considered completely satisfactory, but, . . .no long time will elapse before the great bulk of both corn and hay crops will be cut by machine." (Thompson 1852).

Certain men with amazing hindsight began to criticize, seeing in the American machines clear proof of the return to the old country of Ogle's, Common's and Bell's reapers. But the reapers of 1851 were not the reapers of the 1830's. For example, almost all the elements of McCormick's reaper had been revised. A host of secondary inventions had improved the machine in performance and reliability: the draft, the seats, the cutter-bar, etc.

The Great Exhibition was the turning point in British—and indeed European—harvest mechanization and there was a confidence that British machinists would doubtless improve on the American machines. The press now vigorously admonished the British farmer

to mechanize for his economic survival. Furthermore, the Irish potato famine had now made cheap labor scarce. Manufacturers and licensees rose to the occasion to move from practically no reapers in the England of 1850 to almost one thousand machines in 1852. (Hutchinson 1930).

The problem of adapting the McCormick reaper to British conditions was essentially solved by Burgess & Key's 1854 patent auger table and side-discharge. Hussey's was likewise equipped with a reel and side-delivery, and pierced knife sections to prevent clogging. Bell's Crosskill machine, in turn, was equipped with the McCormick style of sicklebar.

Around 1855, the high price of grain in England gave added impetus to mechanization.

The reaper spread rapidly to the Continent, even as far as Russia. The circle was completed when the successful introduction of American reapers in Europe in turn heightened the interest of farmers at home.

Had McCormick stayed on his Virginia farm instead of moving to Illinois in 1847, his name may have forever stayed on the lists of anonymity. McCormick possessed the American secret of making things work and simultaneously exploiting them.

"One such practical man as McCormick, does more for his kind than a thousand theorizers on Guano."—J. S. Wright, Editor of *The Prairie Farmer*, Oct., 1849.

"It may be proper to add, besides the loss of grain, thrashing with the flail is a very unwholesome occupation, (from the quantity of dust perpetually flying about) and that those who make thrashing a constant occupation, are never healthy, and seldom live long. Of what consequence therefore is it, not only to increase the quantity of grain, but to render an unhealthy occupation unnecessary".— Wm. Lester 1811.

5

Tribulum, Roller and Drum

Threshing is the next process after gathering and reaping; reaping being understood to mean the process of simultaneously cutting and bundling grain.

The word "thresh" (or "thrash" in earlier days) by its very origin implies beating. But friction and shearing forces alone without impact may do the job. The ancients frequently used cattle to tread grain out of the ear or to drag threshing implements which rubbed the grain from the crop spread out on a threshing floor.

The Threshing Sledge

The threshing sledge may possibly be the oldest farm implement in recorded history. A clay tablet uncovered at Uruk, Southern Iraq, of around 3000 BC has a primitive representation which archeologists have identified as a threshing sledge.

The ancient Egyptians used wooden sleds or stone drags for threshing. They purposely roughened the underside of the sled to create the desired shearing and straw-chopping action. Around 720 BC Isaiah wrote:

"Lo I will make you a new threshing sledge furnished with sharp teeth."—Isaiah 41:15 (Moffatt Translation)

Later, in Roman times, Marcus Varro (116-27 BC) recorded the following observations on threshing methods:

"The grain should be threshed on the floor. This is done in some districts by means of a yoke of oxen and a threshing sledge (*tribulo*). The latter is constructed either of a board roughened with stones or pieces of iron embedded in it, which separates the grain from the ear when it is dragged by a yoke of oxen with the driver or a heavy weight on it; or else of a toothed axle running on low wheels, called a Punic cart (*plostellum poenicum*); the driver sits on it and drives the oxen which pull it: this is the method in use in Eastern Spain and elsewhere." (cited in White 1968).

The modern word "tribulation" comes from the name the Romans gave the threshing sledge—*tribulum*.

Threshing floors of considerable antiquity have been discovered in Israel, Syria, Greece, Spain and elsewhere. Many bore the distinctive markings of the teeth of threshing implements. (White 1967).

Threshing sledges are still widely used. For example, it has been estimated that there were over two million threshing sledges on Turkish farms in the 1960's. (Bordaz 1965). The sledge is known there as *doven* and it is sold in hardware shops, along with spare flints. The Turkish farmer brings his crop in to the threshing floor by ox-cart, spreads it in layers up to one foot deep in a circle 40 feet in diameter. The sledge is hitched behind a pair of draft animals and driven round and round over the crop. The task, although a simple one with docile oxen, is hard on the animals if the weather is hot. The Biblical injunctions against muzzling and unequal yoking were evidently

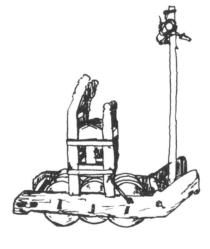

Above: An excellent example of a tribulum seen on Cyprus and reported in 1935. The underside of the two wooden balks were studded with triangular bits of flint. Still in use today, such implements are typically 6 feet long, 2½ feet wide and 2½ in. thick; 600 to 800 knapped flint inserts might be used (Crawford 1935).

written with the well-being of the valuable draft animals in view:

"Thou shalt not muzzle the ox when he treads out the corn"— Deuteronomy 25:4.

"Thou shalt not plow with an ox and an ass together"— Deuteronomy 22:10.

As the sledge is drawn around, attendants constantly stir the straw with forks until practically all the grain in the batch has been separated and the straw has been finely chopped. Then the crop is shovelled aside, heaped up, and winnowed.

The threshing task is laborious, taking as long as two months, and the *doven* causes even more damage than animal treading alone. Grain losses as high as 30 percent have been noted. In other parts of the Middle

Top: Tribulum in use on Cyprus. The children riding on the implement add weight to increase its effectiveness. The flint teeth chop and thresh the straw and ears (Crawford 1935); Left Center: Illustration of an old Egyptian threshing drag or traha (Leser 1929); Right Center: Lean cattle draw this threshing cart in contemporary Sudan (photo courtesy of Abdul Hafiz, FAO-UN); Bottom: Centuries old threshing cart design from Egypt. The discs were made of stone. (Leser 1929). The traditional name of this machine in Egypt was "noreg", in Hebrew "Charuz". In the days of Roman Carthage, it was "Plostellum Poenicum" or Punic cart.

East, basalt or even iron teeth have been used. (Lerche 1968). A reason for the continued popularity of the *doven* is that it is inexpensive and it provides the farmer with two vital commodities: first there is the grain,

Left: Olpad disc thresher made in India. The 10 gauge steel discs are 18 in. in diameter and set 6 in. apart by cast iron spacers. Available in 8-, 11-, 14-, and 20-disc sizes. (Kline et al. 1969); Above: Illustration of an Olpad thresher at work. The Olpad has proven about five times as productive as ox-treading alone (Hopfen 1960).

THRESHING OF WHEAT IN THE ETHIOPEAN CENTRAL HIGHLANDS, 1968
(Data adapted from Kline et al. 1969)

Method Used	Oxen Required (worked in shifts)	Men Required	Man Hours/Day	Quantity of Grain Threshed/Day (not cleaned)	Relative Costs/Unit of Grain
Ox Treading					
Old varieties	6	3	9	440-770 lb.	100
Improved Varieties (higher yielding, more disease-resistant)	6	3	12	(200-350) kg	106
Threshing Sled (Tribulum)					
Old varieties	4	3	7.5	660-1100 lb.	54
Improved varieties	4	3	9	(300-500) kg	60
Olpad Disc Thresher					
Old varieties	4	2	4	Up to 1875 lb.	41
Improved varieties (productivity lower with Olpad in improved varieties).	4	2	5	(Up to 850) kg	46

then there is the chopped straw which is used for fuel, fodder and for structural brick-binding.

The Threshing Cart—Disc Threshers

The threshing cart or *noreg* has a long history in the Middle East and Asia. Variants can still be seen in use today. At one end of the scale, there have been ox-drawn machines with stone discs; at the other, tractor-drawn steel disc cultivators have been used on threshing floors.

The 'Olpad' Disc Thresher

The Olpad serrated disc thresher was developed in India in the past two decades and has proven about five times as productive as ox-treading alone. The results of a survey comparing the productivity of the Olpad with two ancient methods are tabulated above.

Threshing Rollers

Threshing rollers have been known for centuries in Europe and Asia. In parts of Asia the method is still in use. Some threshing rollers were quarried from stone, others were made of wood. Some had fluted surfaces, still others were smooth, truncated cones, so built to better accommodate to the rolling circle and reduce draft. Steensberg (1971) reported his observations on the preparation of a threshing floor and on the use of threshing rollers in India. The surface of the threshing floor was smoothed with liquefied cowdung to produce a hard smooth finish. A smooth roller, quarried from local stone, was used for surface preparation as well as for threshing. The driver sat on a wooden frame fitted to the roller axle and draft pole. He guided the ox team by means of reins passing through the nose cartilage of the oxen.

A muzzle was used to prevent oxen eating the grain during threshing.

From Iran, Lerche observed the Ghardji peg-tooth threshing roller in use (1968). This implement con-

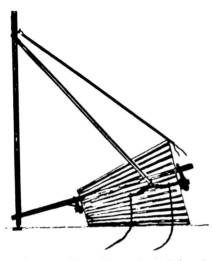

Friesland, 1808. Cone-shaped fluted threshing roller for use in the barn (Van der Pool 1967).

China, 1637, as illustrated in Liu.

China, 1937. Drawn from photo of a roller used for preparing threshing floor in Anhwei Province (Hommel 1937).

Straw-chopping roller thresher in India, 1974 (W. F. Buchele photo).

Grain harvest in Iran, 1967. Ghardji thresher with iron paddles on wooden rollers 35 in. wide and 5 in. diameter (Lerch 1968).

sisted of two wooden rollers equipped with iron "scutchers" mounted between a morticed wooden frame. Two oxen were used to draw the Ghardji over the cereal crop.

In the US during the early 19th century, a roller thresher, appropriately called a "porcupine", was used in the Atlantic States for cereal threshing. Reminiscent of the rolling harrows used in Europe in the Middle Ages for breaking down soil or clods, this roller thresher was used inside a barn. The "porcupine" was made from a stout oaken log, up to 12 feet long, with about 240 oak pegs driven into it. It was pivoted to a center post in the barn and drawn around the threshing floor by horses. (Van Wagenen 1927).

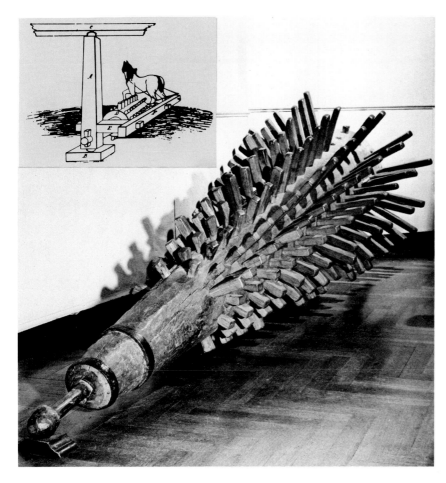

A US patent on a peg-tooth threshing roller was issued to W. Loomis on March 6, 1828.

This porcupine thresher was used in upstate New York in the early 1800's. It was made from an oak log 12 feet long fitted with 240 oak pegs and was pulled by horses round and round the barn pivot post (Henry Ford Museum).

"Anthropomorphic" Threshers—
Mechanized Flails and Hooves

Wheat was notoriously tough to thresh with a flail. An output of 10 bushels a day was exceptional, seven was average. Some varieties needed persistent pounding to get the last grains from the ear. Robbie Burns called flailing "the most degrading of occupations". The first attempts to harness horse or other power in place of manual power to thresh grain were directed at imitating the action of animals' hooves or simulating the human arm and flail. Nevertheless, the flail continued in use for threshing rye for decades after it had been superseded in wheat because rye was easier to thresh and the rye straw had a high value if it was left straight and unbroken.

The first patent on a mechanized thresher was the 1636 English patent No. 92A awarded to the displaced Moravian knight Sir John Christopher Van Berg. Van Berg's patent petition to Charles "king of England, Scotland, France and Ireland and Defender of the Fayeth" makes quaint reading but imparts no details as to how the "Invencon to bee agitated by winde, water or horses, for the cleane threshing of corne, whereby much corne that is nowe left may bee saved, and the strawe made neere as good as haye."

The next patent on a thresher, petitioning George II, 98 years later, was No. 544 issued in 1734 to Michael Meinzies of East Lothian, Scotland. Meinzies was much more specific. His machine consisted of a series of mechanical flails capable of giving 1320 strokes per minute, or "as many as 33 men threshing briskly" and "as moved by a great water wheel and triddles." A committee of the Scottish Society of Improvers was impressed enough to publish an account in 1735 and judged that the Meinzies thresher would . . . "be of universal advantage". It was abandoned shortly afterwards. (Fussell 1952).

When the celebrated Jethro Tull, who died in 1740, endeavored to banish the flail from the barn, his neighbors denounced him. "The tradition of the neighborhood" still is (in 1843), that he was "wicked enough to construct a machine, which, by working a set of sticks, beat out the corn without manual labour". (Ransome 1843).

The first agricultural machine ever officially tested and awarded a prize by the Society of Arts in London was a 'threshing mill', a model of which was presented to the Society in 1761. The Society was sufficiently impressed with the design to have a full-sized machine built for trial. Threshing was done by a gate weighing "about half a hundredweight" which rose and fell like a sash window. Crop was fed under the gate from a sloping hopper. It never worked successfully despite numerous improvements suggested by the committee.

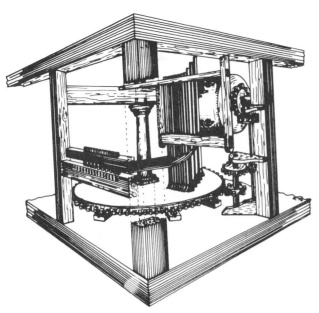

Above: Mechanical hooves. "Windmill for grinding and threshing corn" by a treading action was patented by William Evers of Yorkshire in 1768. The corn was laid on the revolving floor to be fed under the "stampers." The whole machine was driven by gearing from a windmill; Below: Philadelphia, 1778. Machine consisting of multiple jointed flails on a horizontal axle is reminiscent of modern flail forage cutter.

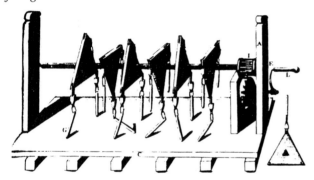

The designer, John Loyd of Hereford, was reimbursed for his trouble. Four years later another £15 bounty was awarded to a Mr. Harvey who, it was said, threshed grain in the Connecticut colony with a large revolving ribbed cone. Then in 1769, William Evers of Swillington, Yorkshire was awarded 50 guineas for his mechanical treading machine with moveable threshing floor. The hooves or "stampers", were of wood shod with iron. (Bailey 1772). The last bounty for a threshing machine was awarded in 1810 for a machine similar to Meinzies.' (Hudson and Luckhurst 1954). Meinzies efforts also inspired a Mr. Alderton to construct a "wringer" type of threshing machine in 1772. He used two rollers, one with fluted surfaces and the other cellular, to squeeze the ears, forcing the grain to be 'squirted' out into conveniently located cells and discharged by the further motion of the rollers.

In 1785, a William Winlaw was encouraged by his employer to construct a thresher. He had a degree of

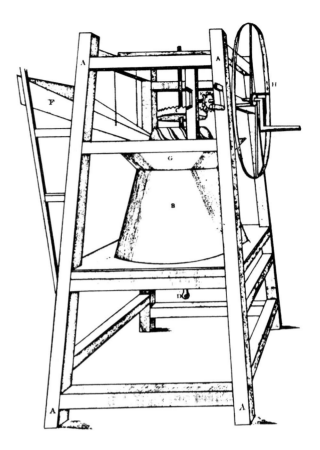

Above: William Winlaw's axial thresher, 1785. "Meat grinder" action rubbed the grain from the detached ears. The straw was not fed into the machine. This was the ancestor of axial threshers; Below: French thresher with rotating clubs, 1765 (photo courtesy Tim Fogden, Manns of Saxham).

success with his conical rubbing mill. First the heads were combed free of straw by hand, then fed down a chute axially into the "grinder" where a spiral rotor rubbed the grain from the ears. This apparatus was the forerunner of axial threshing machines.

The Threshing Drum Or Cylinder

Possibly the earliest thresher using a rotating drum in a casing was conceived by a Scot named Leckie in

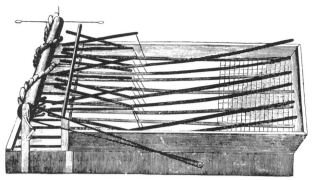

Wardrop of New Jersey invented this oscillating flailer in 1794. He attempted to imitate mechanically the action of the human arm and flail (Butterworth 1888).

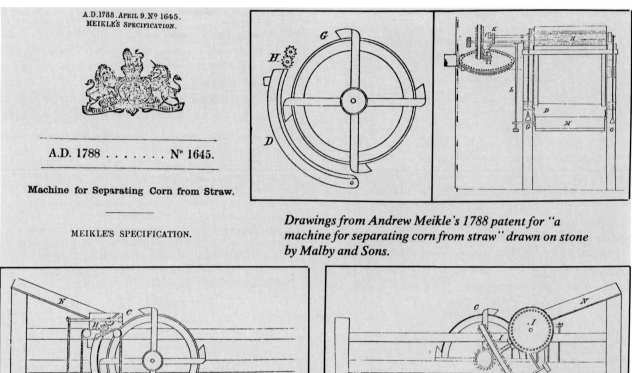

A.D. 1788. APRIL 9. Nº 1645.
MEIKLE'S SPECIFICATION.

A.D. 1788 . . . . Nº 1645.

Machine for Separating Corn from Straw.

MEIKLE'S SPECIFICATION.

Drawings from Andrew Meikle's 1788 patent for "a machine for separating corn from straw" drawn on stone by Malby and Sons.

The result is now an important milestone in farming history.

Andrew Meikle's Thresher

Mention has been made of some of the Scottish implement makers who poured forth a stream of inventions between the rebellion of 1745 and the early 19th Century to contribute to the agrarian revolution. Andrew Meikle's threshing machines were probably the most important.

Letters Patent No. 1645 were granted to Andrew Meikle in 1788, pre-dating the first British Patent Act by some 64 years. Provision of Patents prior to

1758. He invented a "rotary machine which consisted of a set of cross arms attached to a horizontal shaft and the whole thing enclosed in a cylindrical case." *(Farm Implement News* 1886). In 1772, Ilderton and Stuart developed a machine in Scotland in which the crop was threshed by "the sheaves being carried round between an indented drum and a number of fluted rollers."

Sir Francis Kinlock of Gilmerton, East Lothian, took a model of this machine to Andrew Meikle of Houston Mill for testing and in the course of the trial "in a few minutes the model was torn to pieces." The incident stimulated Meikle to tackle the problem himself in 1785.

1852 was on a private monopoly basis, granted by individual petition to the monarch in council and subject to considerably more rigamarole than today. The patent was not issued as such until 1856. Meikle's patent specification is enlightening. It read in part:

"Machine for separating all kinds of corn from straw:
When the mill is set a going, the sheaf of corn is taken up and spread upon the board marked (N), when the two fluted rollers marked (H) take hold of it and feed it on gradually, so that the scutchers marked (C) coming round scutches the corn off from the straw, and the breast marked (D), moveable upon a centre below, moves back and forward when the corn is put in thicks (sic) or thin. When the corn comes from the scutchers, it falls into the harp marked (M), by which the corn is separated from the straw, and below the harp a pair of fawners may be placed so as to separate the corn from the chaff."

The first Meikle "threshing mill" was erected in 1786 for a Mr. Stein at Kilbargie, Clackmannanshire, England. It could be powered by waterwheel or horse power and the undershot cylinder was geared up to run at 200 revolutions per minute. Meikle used four iron shod "scutchers" or beaters rotating within about 5/8 in. of the feed rollers. The crop was metered diametrically into the path of the scutchers to be swiped off and carried over the pressure-loaded concave or "breasting". The jogging screen or "harp" sorted the grain from the straw after the initial separation at the concave.

Meikle went on to improve his threshing mills, and make further contributions to the art. He added rotary-rake grain separators, separating grates or concaves, and the fanning mill. In 1805, it was reported that Meikle's threshers had proliferated to the extent that they were in use

"upon most of the principal farms in the country and were wrought in different ways, by steam, by wind, by water, and by the strength of horses."

Windmill powered threshers at first posed the difficulty of governing drum speed, but that defect was overcome by a sail-furl designed by Meikle. Several sizes of thresher were manufactured, the smallest costing £40 and the largest £250. (Sommerville 1805).

Most of Meikle's ideas, although patented, were pirated and thus he profited little. It was Sir John Sinclair who was to make Meikle's old age more comfortable by raising a public subscription of £1,500 and investing it for his declining years. Meikle lived to a ripe old age and was still plying the pipes at 90. When he died, he was buried near the scene of his labors at Preston-kirk. On his grave appears a fitting epitaph:

"Andrew Meikle: Beneath this stone are deposited the remains of the late Andrew Meikle, Civil engineer at Houston Mill, who died in the year 1811, aged 92 years. Descended from a race of ingenious mechanics to whom the country for ages had been greatly indebted, he steadily followed the example of his ancestors, and, by inventing and bringing to perfection a machine for separating corn from straw (constructed upon the principle of velocity, and furnished with beaters or scutchers), rendered to the agriculturalists of Britain and of other nations more beneficial service than any hitherto recorded in the annals of ancient or modern science." (Poole 1928).

"See I will make of you a sharp threshing-sledge, new and studded with teeth; You shall thresh the mountains and crush them and reduce the hills to chaff; you shall winnow them, the wind shall carry them away . . . "—Isaiah 41:15, 16.

6

Pitch Forks and Devil's Wind

Threshing tends to throw together detached grain, husks, chaff, broken ears and pieces of straw, necessitating further separation. When threshing was a manual operation, men wielding forks frequently accomplished two stages of separation at once—they divided the straw from the finer material and, using the natural breezes, winnowed away chaff and dust. Some very beautiful grain forks can be seen in museums today that were literally grown by one generation for the next.

Whether the threshing of grain was accomplished by animal trampling, by sledges, or by men flailing, getting the grain out of the ear was only half the task, for there was always the winnowing. Two thousand years have passed since the ingenious hand of the Chinese made a tremendous technological advance: they harnessed the wind.

In 1958, a tomb of the Han Dynasty (206 BC-221 AD) was uncovered in the San Chou region in the Northwest Honan on the Yellow River. Archeologists unearthed a trove of delicate pottery models. Among the pieces was a pottery farmyard diorama, with a hand cranked winnowing fan in one corner. The model presents a lively scene, showing three primary operations connected with the processing of rice. There is a pounder or huller, a grist mill, and the first winnower. The first crank handle and the first centrifugal fluid impeller in history! (Needham 1965). A worker is seen purposefully treading the lever of the tilt-hammer and a dog, possibly a greyhound, sits erect and alert, perhaps in the act of detecting a rat in the grain.

Operation of the First Winnower

The hulled grain was doubtless put into the hopper at the top; the plug in the wall directly below the hopper was eased out to meter the flow while the operator turned the crank. The wind from the centrifugal fan blades blew the chaff to the far side to fall out the left-hand opening. The heavier kernels of grain sifted through the airstream into the opening below the hopper and were scooped up and dumped into the grinding mill on the opposite side of the yard.

This is not the only evidence of the Chinese rotary

Funerary model of a rotary mill with pedal tilt hammer and rotary winnowing fan from Han Dynasty 206BC-221AD (Courtesy William Rockhill Nelson Gallery of Art, Atkins Museum of Fine Arts, Kansas City, Missouri).

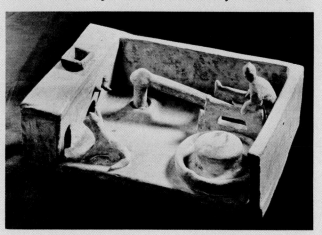

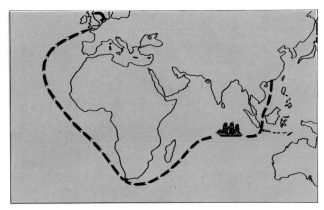

Trade routes of the Dutch and Swedish East India Companies from Macao and Canton and the path of technology diffusion of the fanning mill into Renaissance Europe.

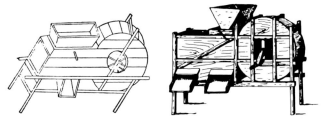

Left: Shrouded Chinese fanning mill illustrated in Nung Shu, a Chinese book of agriculture, in 1313 (Wang 1956); Right: The Chinese winnowing fan travels abroad. A Swedish fanner of the late 1600's (Berg 1976).

fan. Joseph Needham, doyen of historians of technology, in his monumental "Science and Civilization in China", observed that while China has been to European eyes like the moon—"always showing the same face", it had "attained an early scientific and technological superiority over the West, so that until the Renaissance it had more to give than to receive."

In Volume 4, Part 2, Needham sifts through evidence of devices from hand-fans to the punkah to unshrouded rotary winnowers, concluding that the use of fan blades on a continuously revolving axle was certainly developed by the time of the Han Dynasty. No longer need men be dependent on the caprice of the wind. Other documentary evidence exists to establish the Chinese provenance of the winnower. The earliest rotary winnowers in Europe appeared in the 16th Century and showed undisputable resemblance to the classical Chinese form.

The path of diffusion of this Chinese invention has been traced to Dutch and Swedish East Indian Company traders and scholars traveling with the traders who studied and brought Eastern technology to Renaissance Europe. (Berg 1976).

A writer of 1737 reported on a machine of the Chinese style "for tossing corn" in which

"the air was driven with such force that, when Judge Lars
  Ehrenmalm went to look in the opening from which the
  air came out, the blast lifted the wig off his head and carried
  it a good distance along the floor". (Berg 1976).

Fanning mills subsequently appeared further afield, in Germany, Carinthia and Transylvania, although with variation in design.

A striking feature of the shrouded Chinese rotary fans was that the air intakes were always shown in the middle—for axial air entry—and the discharge was circumferential, as would befit the centrifugal action. By contrast, mineshaft ventilators illustrated in the 1556 Latin manuscript *De re metallica* by Georgius Agricola show both inlet and outlet on the outer casing.

The blades of Agricola's ventilators were tipped with goose feathers, another early Chinese innovation. Agricola's peripheral fan ducts are faintly reminiscent

of the cross-flow fan which appeared in the 19th Century and which was dependent on the development of a forced vortex within the rotor (Quick 1971). Agricola's fan could not have generated much of a cross-rotor air flow.

Canny Scot Sees The Fan's Potential

During the religio-political upheaval in Scotland that followed the ascension of Charles II to the English throne, a Lord Salton, traveling under the family name

Chinese winnower illustrated in a 17th century manuscript shows how attendant meters grain flow with left hand while cranking with right hand (Liu 1963).

48

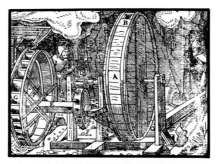

Far Left: An example of a fanner from Schlesian in SW Germany, 1717, where it was known as a "tossing machine" (Feldhaus 1914); Far Left and Above Right: Mineshaft ventilation fans illustrated in Agricola's **De re mettallica**; *Lower Right: Chinese open winnowing fan, the rotary motion produced by foot-treadle action to drive the elevated fan. Both shrouded and unshrouded rotary fans are mentioned in the Nung Shu, 1313.*

of Andrew Fletcher, sought refuge in Holland. There he became impressed with Dutch agricultural methods. In 1710, having resumed his title and lands in East Lothian, he authorized his brother to commission James Meikle, millwright of Wester Keith to

" . . . go to Holland with the first fleet that sails thither after the date of these presents, and there learn the perfect art of sheeling (milling) barley, both that which is called French barley, and that which is called pearl barley, and how to accommodate, order and erect mills for that purpose, in so far as he can with his uttermost industry, and the recommendations given him.

That as soon as the said James Meikle shall find himself sufficiently instructed in said art, he shall return with the first fleet or man of war, he can have safe passage in for Scotland or Newcastle; that in the meantime, if he shall be forced to wait, he shall endeavour to instruct himself in any other useful trade or manufactory.

That when he returns to Scotland, he shall be obliged to communicate the arts he has learned to Salton or to any whom he shall appoint, and shall communicate them to no other person but by Salton's permission . . . But if it shall happen, that the said James Meikle shall be taken prisoner either going or coming, Salton shall be obliged to relieve him and pay all his expenses." (Somerville 1805).

For this effort Meikle was allowed two shillings sterling every day, one shilling for his entertainment and one for his work.

Meikle went beyond the call and brought back with the barley mill a more enduring gift to his country— the fanner for winnowing. However, under the articles of agreement with Salton, Meikle was prevented "from making use of the art or of teaching it to another", and it remained for his sons and others to capitalize on the device.

The pot barley mill that James Meikle erected for Lord Salton was very successful, although for many years it was the only one of its kind in Scotland. The

dressed barley from it was known as "Salton barley" in the Lowlands. Elsewhere, barley continued to be prepared on the primitive "knockin-stane" and winnowed by hand. (Handley 1853). The Salton mill buildings were still standing in 1928. (Poole 1928).

A strong factor that militated against Meikle's Dutch fan was local prejudice. Some of the clergy labeled it "Devil's wind", on the grounds that the newfangled machine would be "impiously thwarting the will of Divine Providence, by raising wind . . . by human art, instead of soliciting it by prayer . . . " (Handley 1953).

The old method cleaned the grain from the chaff "by

English open fan. This machine had blades of loose cloth that assumed their fanning action when rotated by the flexible handle attached to the four-arm flywheel. An original is in the museum of English Rural Life, Reading, from which this was drawn.

riddling the corn in the draft between open barn doors made opposite each other for that purpose." "This is productive of great inconvenience" said Lester in 1811, and "In a long track of calm weather the corn is often unavoidably spoiled for want of wind."

Andrew Rogers, a mechanically-minded farmer of Roxburghshire, made a fanner for his own use in 1737 and—prejudice notwithstanding—later began selling fanning mills locally, on both sides of the border. In 1768, Andrew and George Meikle, of threshing mill fame, and the sons of James, obtained a patent and began manufacturing fanning mills. The task of winnowing was reduced from about half an hour by hand to several minutes per bushel, even allowing for the need to pass the "corn through the fanning mill two or three times for thorough cleaning." (Ransome 1843).

Origin of the Name "Tailings"

Prior to the introduction of sieves and screens, the grading of grain was carried out in crude fashion. A wide shovel was used to pitch the grain across the barn floor into a draft so that the crop fell in an elongated heap, the heaviest grains traveling furthest and the lightest forming a tail on the near side of the heap. In spite of the modern refinement of the process the name "tailings" is still retained for shrivelled and broken grains. (Spencer and Passmore 1930).

Open-Shrouded Rotary Fans

A fan of a different design using a rotor but lacking a housing also originated in China. This type, less sophisticated and less effective, may be older than the shrouded Chinese type and must consequently "be at least as ancient as the Han period." (Needham 1965). It is mentioned in the Nung Shu 1313:

" . . . some people raise the fan high up (without enclosing it) and so winnow: this is called the 'shan chhe'." (Wang, 1956, cited in Needham 1965).

The open fan was also used in England as early as 1677, where . . .

"Those who have only small quantities of corn produce a draft of air with the aid of a cloth . . . but the wheel fan saves a man's labor, makes a better wind, and does it with more expedition . . . Implements of this type were in use until the end of the 19th Century, particularly in remote country districts."

The implement was

"a sort of reel upon which was fastened, by one edge, pieces of canvas these, as they revolve, cause considerable draught, the corn during this time being sifted through widemeshed sieves called riddlers, and during its passage to the ground the lighter particles are blown away." (Andrews 1853).

The fan was equipped with linen or woollen cloths on cross bars mounted on a horizontal axle between upright supports. In operation, the blades straightened out and created a draft. Grain was cleaned by an operator who "shakes the corn on the sieve to and fro".

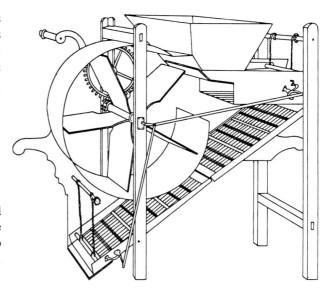

William Evers' 1761 English patent showed a combined fanner and grain cleaning "riddle" or sieve (Baily 1772).

The same open fan was recorded in Sweden in other applications, for example, as a blower in the primitive smelting of bog ore at Dalarna and for the boiling of wood tar. (Berg 1976). Agricola also illlustrates an open fan for mine ventilation (1556).

Addition of Cleaning Sieves to The Fanner

It is tempting to suppose that the Chinese may also have been first to have linked mechanical sieving with the winnowing machine. Certainly the Nung Shu of 1313 describes a system for coupling oscillating sifters and tilt-hammers to water wheels, thus converting rotary to oscillating motion. However, the evidence is lacking to show whether the early Chinese winnowers were integrated with mechanical grain cleaning sieves. Possibly the earliest *documented* combined fanner and sieves, or "riddles", was William Evers' British patent of 1761.

By 1880, fanners were widely known in Europe. The fan had been improved and combined with shaking sieves for systematic seed sorting as well as for chaff removal. Close-fitting circular fan housings were illustrated in 1858 catalogs, but the more efficient scroll-shaped housing was shown in Cooch's 1800 patent in England.

The fanning mill used on Lord Ducie's estate at Whitfield in the 1830's had a scroll-shaped housing and backward curved blades. It was designed by R. Clyburn of Uley, who, recognizing that radial blades in a tight circular housing "tended to keep the air whirling constantly within the casing", overcame the problem with backward-curved blades and scroll-housing.

Early American Fanners

The winnower or fanning machine made its way to

Below Left: Fanning mill with end shake, opposed action sieves and cast iron spur gear (Henry Ford Museum); Below Right: John Springer's horizontal fan winnowing mill, New England, 1831 (Henry Ford Museum).

Opposite Page and Right: Two early American wooden winnowing fans on display at the Henry Ford Museum in Dearborn, Michigan. The fan on the opposite page is a vertical axis type, ca 1800. Note the wooden gear teeth on the right angle drive. The grain was fed into a hole half-way along the tunnel and the clean grain collected underneath. The fan on the right is a simple box winnower of all wood construction, ca 1840.

Above: Advertisement in 1928 promoting Racine fanning mill. A good fanning mill was touted as a first class investment. Seed cleaning eliminated weed seeds and dockage and ensured a better market price as well as better crops; Right: The Chatham fanning mill sold by Massey Mfg. Co., in Ontario in 1894.

the colonies in North America prior to the Revolutionary War. (Rogin 1930). But it was slow in adoption and for several decades winnowing was generally performed by means of a basket and sheet, or wicker fan. Around 1800, Governor Reynolds of Illinois said that winnowing with the sheet was "about the hardest work I ever performed."

With the rapid expansion in wheat production in the period before 1850, winnowers became so widespsread that in 1842 the editor of *The Cultivator* wrote that

"of fanning mills there are an almost endless variety . . . all of which will perform good work, principally differing in rapidity of execution. Wherever the threshing machine is being used, winnowing by the old method was so slow as to be out of the question."

The necessity for seed cleaning was appreciated by farmers who recognized that a single wild oat seed could produce 200 others, a mustard plant 10,000 of its kind, and that crops would produce better when graded seed was sown (graded using different mesh sieves). They also realized a better market price for cleaned and graded seed. A great variety of designs, many having great intrinsic beauty were manufactured, and today "wooden box-winnowers" can often bring good prices as antique collector's items.

Old Fanners Never Die

Even after the threshing machine was adopted for separating the grain from the chaff, and not many years ago, the fanning mill was still a common farm tool.

Farmers often ran their seed grain through a winnower before putting it in the grain drill. This prevented stoppages of the grain drill mechanism and ensured a more uniform crop stand. Today, this operation has been practically taken over by the specialized seed-grain producers who use factory-style seed grading, cleaning and treatment equipment. But the basic principles of cleaning are still the same.

> "Probably when a thrashing machine is invented, that will separate the corn from the straw, with the same ease and facility, as the winnowing machine does it from the chaff, they will become universal . . . I am confident that the adhesion betwixt corn and its straw is not so great, as to require the strength of horses to separate them, if the power made use of were properly applied." (Lester 1811).

7

Getting It All Together: "Combined Thrashers"

William Lester of Paddington, England in 1811. Innovator, writer and consultant, he was one of the first of his profession to call himself an agricultural engineer.

IN 1788, the Scottish millwright Andrew Meikle patented his celebrated threshing machine. Meikle included the forerunner of the reciprocating straw rack in his design of a jogging screen or "harp". Twenty years earlier, the same enterprising Scot and his son George had built and patented a "double blast" grain cleaning machine. The first fanning stage of their machine was intended to "dress" the grain or remove chaff, after which the partially-cleaned grain was elevated for a "finishing" pass through the blast from the second fan (Ransome 1843). The Meikles' became the first threshing machine manufacturers when they began selling combined threshing and winnowing machines in the closing years of the 18th century. Others pirated Meikle's threshing and cleaning principles, added straw separators, and began selling them as "combined thrashing machines".

These early straw separators were of the rotary-rake type and claims for priority of the invention were made independently by Gladstone of Castle Douglas (1794), Bailey of Northumberland (1798), and Palmer (1799). There is, furthermore,

"every reason to believe that rotary straw shakers were used (by Meikle) in some of his earliest machines." (Pidgeon 1892)

Use of Threshing Machines Spreads Rapidly In The North Country

Threshing machines were used extensively on the larger farms in the South of Scotland and border districts of Northern England. By 1796, most incorporated winnowers of some description (Partridge 1969). However, an agricultural survey by Arthur Young in 1800 contained this complaint about threshing machines:

"That farmers were spending their money without any prospect of meeting with a return. In consequence of the price of thrashing machines having been by the first inventors fixed too high, almost every mechanical knave has been tempted to set up the trade of making them; there are swarms of them, therefore, not worth a shilling..."

In 1805, John Ball of Norfolk patented a thresher with an open concave made of iron, positioned under the drum. This was the forerunner of the "English" style of thresher. Ball's machines were exhibited at early Royal Agricultural Society meetings and became popular in the Eastern and Northern counties of England (Pidgeon 1892). William Lester was another inventor who felt that the "scutching" principle of the Scottish threshers left something to be desired, noting that most of the energy went into beating the straw to pieces. He patented several threshers between 1802 and 1805 in which the thresher

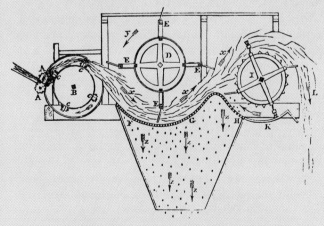

Early Scottish threshing mill with "rotary rakes" for separating grain from straw. This type of machine with overshot thresher operating on Meikle's "scutching" principle found widespread service before 1800 (Ransome 1843).

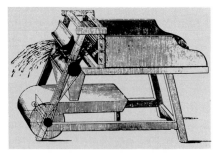
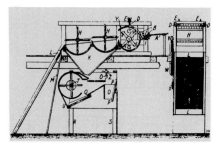

Clockwise: William Lester's threshers were patented between 1802 and 1805 and were claimed to have worked on a rubbing principle rather than by "scutching" or impact (Lester 1811); Jeremiah Bailey's thresher, set up in Chillingham, England, could thresh 35 bushels of wheat an hour when driven by a water wheel; In 1795, Jubb of Lewes, England, patented this combined "Thrashing-machine". The straw was fed up to the intermeshing beaters by infeed rollers (Ransome 1843); 1805 Scottish threshing mill with single-blast cleaning system. The cylinder of the Scottish machines beat the straw as it was metered through the infeed rolls.

"works efficiently by sleight . . . (it) rubbed the grain out of the straw, and was particularly beneficial for smutty wheat because it did not break the (smut) balls." (Lester 1811).

The Royal Society of Arts offered a gold medal in 1801 for an improved threshing machine and awarded the prize in 1810 to H. P. Lee of Maidenhead Thicket, England. Lee's machine consisted of four "vanes" working over a concave, with motion given to the machine spur gearing by a horse-gear. The special merit of his machine seems to have been "the high speed at which it could be driven" (Wood 1913). Men such as Lee, Ball and Lester led the movement away from the Scottish "scutching" principle, with its feed rolls, to the "English" style of thresher and open concave.

"Scotch" vs. "English" Threshers

Meikle's "scutching" principle of threshing had a lasting influence on the development of early threshing machines, particularly in Scotland. The grain was "scutched" (beat) from the ears on the Meikle thresher as the crop was metered forward by the infeed rolls. In order to be effective, this action of metering the crop diametrically into the path of the revolving sharp-edged beaters demanded very even feeding and "heads first" orientation. The crop sheaves had to be of uniform straw length, with all the heads at the top. Such uniform sheaves were customary where the grain was cut with the sickle, but not so from the scythe. Any ears or straw sheaf-bands that were not fed in correctly tended to be broken off unthreshed or were otherwise incompletely threshed, as the threshing process was "all over in the space of a few inches" from the feed rolls. Furthermore, the closed concave or "breastwork", which wrapped around one-third of

the cylinder circumference, was as much as 3 inches away from the beaters, and contributed little, except to stop the grain from flying all over the barn. These factors, Ransome said, necessitated a "seconds" elevator to return unthreshed heads:

"whose business it is continually remind the scutchers of any defect in their first performance. I have some fear that the same cause which allows the ear to pass unthreshed the first time may admit of its doing so the second..." (Ransome 1843).

Thus it was important when reaping...

"...to have the corn laid down...regularly, with the heads in one direction; ...when the thrashing machine is to be used. For in consequence of the ears of crops gathered in untidy sheaves not being... so regularly presented to the rollers of the threshing mill, the thrashing was not done so perfectly." (Slight and Scott-Burn 1858).

The tip-speed of the Scottish thresher drum was usually set no higher than 3000 feet per minute and the feed-rolls were geared so that the crop would be struck about 19 times for each foot of advance through the machine.

The "English" machines, on the other hand, were supposed to work on more of a "rubbing principle". Feedrollers were not essential and had been dispensed with. The crop was fed in from the feed board tangentially to the cylinder. The cylinder bars mounted on the open or "skeleton" cylinder were thicker and blunt-edged and separated the grain from the ears as the crop advanced over the open concave or "mesh." The concave wrapped around three quarters of the cylinder circumference, drum speed was around 4000 feet/minute; threshing took place over a distance of as much as four feet and there was less straw break up.

"In an apparatus of this kind it is impossible that an ear of corn, enter how it may, can emerge unthreshed, or rather unrubbed." (Slight and Scott-Burn 1858).

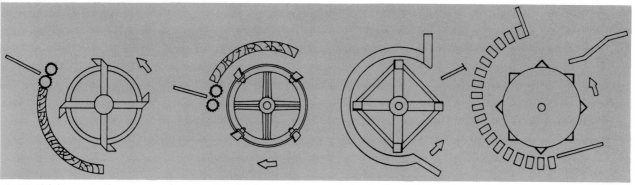

1. *Meikle's original undershot design, patented in 1788.*

2. *Later Scottish threshers. Overshot "scutching" action, 1800-1840's.*

3. *Lee's English thresher with four beaters, 1810.*

4. *English open concave design, c.f. Garrett's 1840.*

Development of Scottish and English Threshers.

The ultimate supremacy of the "English" thresher was assured because in England, and later in Scotland, the crop was more commonly being reaped by scythes or cradles rather than by sickles, and the sheaves were more irregular. Again, there was the matter of the straw which was:

> "more particularly coveted by farmers residing near large towns, to whom the production of clean unbroken straw is frequently an object of more importance than the thrashing out of the greatest possible quantity of grain in a given time." (Ransomes and Sims Catalogue 1856).

Finally, there were the matters of relative power and capacity. Ransome (1843) observed that, given the same four-horse team to drive the horseworks, the English machine could thresh 36 bushels of wheat in an hour whereas a similar sized "Scotch" machine could manage only 26 bushels an hour.

The Threshing Machine Riots of 1830

The spread of threshing machines in England coincided with a period of considerable unemployment. Farming had boomed during the Napoleonic Wars, but with the coming of peace following the Battle of Waterloo (1815), many of the 200,000 returning veter-

Garrett's open drum two-horsepower threshing machine set down for work. A man atop the sweep rotated along with the horses. Each time the horses made a circle, they had to step over the "tumbling rod" which drove the machine (Garrett's Catalogue 1852); Below: A Ransomes and Sims' illustration of a "one-horse-power thrashing machine with winnower driven through an intermediate motion (gear) by a strap". The caption states that "If any accident occur through careless or over feeding, or from a stone or a horse-shoe being fed in with the straw, the strap flies off the pulley and everything stops", (Ransomes Catalogue 1858).

RANSOMES & SIMS' ONE HORSE-POWER THRASHING MACHINE,
WITH WINNOWER DRIVEN THROUGH AN INTERMEDIATE MOTION BY A STRAP.

ans, most of them farm laborers, were without work. The threshing machine became a target of the unemployed. Unrest erupted at the beginning of the 1830 harvest as bands of farm laborers roamed the countryside burning hay-ricks and destroying machines. While the machine was a symbol, the main issue was work and decent wages. In two years of violence 400 machines were destroyed. Many of the rioters were arrested and brought to trial. Nineteen were executed, and 481 transported to Australia as convicts (Hobsbawm & Rude 1969). As a result of the bitter episode of the

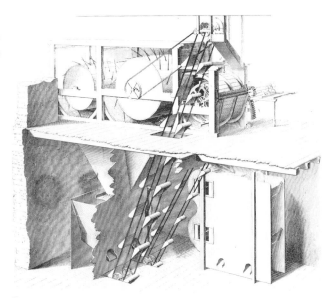

Right: Double blast "Scotch threshing mill" for "finishing" grain. This type of machine necessitated two floors in the barn. The sheaves were carefully pitched into the machine at the upper floor. The finished grain was elevated back up from the ground floor for bagging (Morton 1856).

Above: J. A. Ransome's "hand thrashing machine", ca 1843. To drive the cylinder, two men operated the lever action on one side to produce motion through a ratchet mechanism, while two others on the other side rotated the crank, geared up 8 to 1. Another man was needed to feed the machine and still others to bring up the sheaves. Capacity was about 10 bu/hr for wheat. Hand-powered threshers met with only limited acceptance (Ransome 1843); Center: A "Swing" letter. Captain Swing was a pseudonym for one of the organizers of gangs of unemployed laborers who rioted to draw attention to the injustices of the rural poor (Hobsbawm & Rude 1969); Right: Even segments of the law abetted rioters of 1830 by encouraging farmers to dismantle their own threshing machines.

Sir

This is to acquaint you that if your thrashing Machines are not destroyed by you directly we shall commence our labours

signed on behalf of the whole

Swing

PUBLIC NOTICE.

THE *Magistrates* in the Hundreds of *Tunstead* and *Happing*, in the County of Norfolk, having taken into consideration the disturbed state of the said Hundreds and the Country in general, wish to make it publicly known that *it is their* opinion that such disturbances principally arise from the use of Threshing Machines, and to the insufficient Wages of the Labourers. The Magistrates therefore beg to *recommend* to the Owners and Occupiers of Land in these Hundreds, to *discontinue the use of Threshing Machines, and to increase the Wages of Labour* to Ten Shillings a week for able bodied men, and that when task work is preferred, that it should be put out at such a rate as to enable an industrious man to earn Two Shillings per day.

The Magistrates are determined to enforce the Laws against all tumultuous Rioters and Incendiaries, and they look for support to all the respectable and well disposed part of the Community; at the same time they feel a full Conviction, that *no severe measures will be necessary*, if the proprietors of Land will give proper employment to the Poor on their own Occupations, and encourage their Tenants to do the same.

SIGNED,

JOHN WODEHOUSE.
W. R. ROUS.
J. PETRE.
GEORGE CUBITT.
WILLIAM GUNN.
W. F. WILKINSON.
BENJAMIN CUBITT.
H. ATKINSON.

North Walsham,
24th November 1830.

J. PLUMBLY, PRINTER, NORTH WALSHAM.

Notice issued by Norfolk magistrates, November 1830

threshing machine riots, the pace of innovation in British farm machinery was temporarily halted. Farmers were not about to risk buying threshing machines and inciting a renewal of mob violence. A few farmers went so far as to voluntarily dismantle their threshers and revert to the flail crew.

Raspbars and Straw Walkers

The first "rasping bar" cylinder was patented by John Goucher of County York, England, in 1848, but it was some years ahead of its time. Even in the 1860's British agricultural writers were still debating the merits of the "Scotch" and "English" threshers, both of which had essentially flat beater bars.

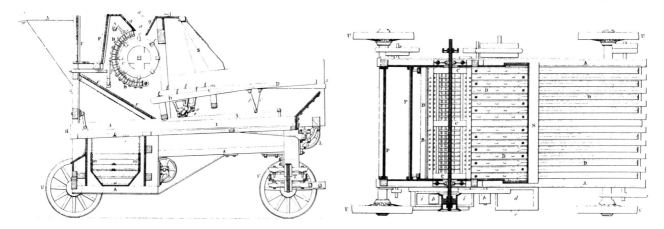

English threshing machine, ca 1845 with earliest form of straw walkers, known then as "harps" (Morton 1856).

Goucher's raspbar design was intended to reduce the crushing of grain and straw breakage by employing grooves or channels in the surface of the rounded-profile beaters. He accomplished this by winding iron wire around the rolled steel barstock, "care being taken to secure the wire at each end to prevent the wire becoming loose"—a far cry from the drop-forged steel bars in use today.

On open concave machines, most of the grain threshed from the ears is separated through the concave. The remainder must be removed by the separator section. On the early English machines, this was accomplished by an operator known as "the forker" who quite naturally was equipped with a pitchfork. The forker was dispensed with when straw walkers were perfected. The earliest individually oscillating walkers, known then as "harps", were probably those used by John Morton on Lord Ducie's thresher at Whitfield. His machine consisted of 30 "frames" made of 2 x 3/4 in. timber, 6 feet long, spaced 3/4 in. apart, mounted on a double crank linkage and so designed that as one set of "frames" was rising, those adjacent were falling, but always being maintained parallel to one another. The straw motion under the action of the 3½ in. stroke cranks was such that not only was the straw vigorously shaken to separate the grain, but the material was progressively "walked" away from the cylinder and out the back of the thresher (Ransome 1843).

By the 1850's English threshers were being built with open-type straw walkers. Hornsby's machine, advertised around 1856, was even equipped with an auger-type grain pan under the concave and walkers (Morton 1856).

Steam Power for Threshing

James Watt, "father of steam power", added to his patents in 1784, a "portable" engine mounted on a wheeled carriage. An early application for his machine was grain threshing. The first **portable** threshing machines appeared in East Anglia in 1800. Richard Trevithick, who produced one of the first practicable steam vehicles in 1803 (Gray 1975), built a fixed steam engine to successfully power the threshing machine on Sir Christopher Hawkins' estate in 1811 (Pidgeon 1892). In 1814, William Lester patented a self-propelled portable steam engine specifically for driving and transporting a threshing machine (Pidgeon 1892). This **may be the earliest patent for an "agricultural tractor"**. Lester was an engineer of great vision—it was he who in 1811 proposed that a "General Depot of Improved Implements in Husbandry 'be set up' to test and sell by commission the most improved implements of the **British Empire"**.

By the 1830's, it was claimed that portable threshing machines operated by contractors were the rule rather than the exception in the Eastern counties of England, although they were often rudimentary machines, lacking fanners or separating rakes. Steam power and even combined threshing machines up to 1850 were generally used only on large estates. The simple portable threshers, horse powered, were economically justified when they guaranteed the farmer the opportunity to receive higher prices for grain threshed early in the season—before the usual rapid fall in prices as the season progressed.

Adoption of the Threshing Machine in the US

The first threshing machine imported into the US was a Scottish unit—probably one of Meikle's—brought to New York by Baron Pollnitz in 1788. Operated by just one man and a boy, this machine had a capacity of 70 bushels a day, a tenfold increase over the flail (Rogin 1931). The first US threshing machine patent was issued to Samuel Mulliken of Philadelphia on March 11, 1791. It was the seventh patent on record (Kennedy 1860). On the other hand General George Washington, who was said to "be in the vanguard of agricultural improvement," was experimenting with a "treading floor" for animal threshing, in place of bare ground used previously. He lamented that on one of his estates in 1795,

"instead of using proper flails for threshing grain, I have found my people at this task with hoop poles". (Rogin 1931).

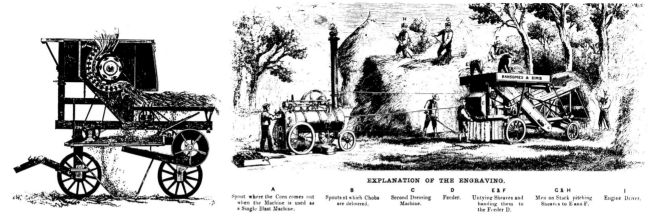

EXPLANATION OF THE ENGRAVING.

A	B	C	D	E & F	G & H	I
Spout where the Corn comes out when the Machine is used as a Single Blast Machine.	Spouts at which Chobs are delivered.	Second Dressing Machine.	Feeder.	Untying Sheaves and handing them to the Feeder D.	Men on Stack pitching Sheaves to E and F.	Engine Driver.

Top Left: Garrett's improved threshing machine, 1851. Garrett & Son, Leiston Works; Top Right: Ransomes' Model A portable threshing machine of 1862 required an eight horsepower steam engine. Ransomes & Sims, Ipswich, England; Bottom Left: A peg-tooth thresher with combined rotary straw separator and cleaner was patented by A. Savage in the US on March 28, 1822; Bottom Right: Howe's novel threshing machine, patented in the US on December 3, 1822, was designed with a vertical cylinder and driven by belt from the horse works, with the horse walking on a circular platform.

 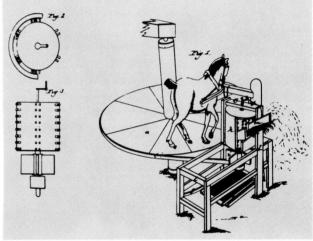

A more progressive Thomas Jefferson imported a threshing machine which, in 1796, was capable of 120 bushels a day. (Wilson 1943).

Although there was a burst of inventive activity in the US, local machinery agents were not impressed with their designs. Advertisements in the US farm press around 1822 more often promoted imported Scottish and English machines. A simple, portable thresher was listed at $375, a stationary thresher sold for $325, while a four-horse portable "combined thresher" was listed at $500—an "immense and unattainable sum for the ordinary farmer of the day". Nevertheless, threshing machines had "ceased to be a curiosity" in the grain-growing regions of the middle (Eastern) States around 1820, and rapid strides were made so that there were 700 different machines patented or on the US market. (Rogin 1931). Steam power was still the province of the contractor.

Jacob Pope of Massachusetts was an early threshing machine manufacturer. He received his first thresher patent in 1802. His 1826 models, still hand-powered, were similar to Meikle's but it was said of them that

it was "harder work to turn the crank than to swing the flail." His machine was then adapted to horse power drive and proved quite popular. It was capable of 150 to 200 bushels of rice a day when attended by three men and a boy, with the horse walking on a circular platform. By 1843, the *Southern Planter* reported that "the use of the threshing machine is universal in Virginia". Notable among the earliest Virginia thresher manufacturers was Robert McCormick, father of Cyrus.

Threshing machines were introduced into Ohio in 1831, Indiana in 1839 and into Illinois sometime around 1847. In Racine County, Wisconsin, there were 500 to 600 threshers produced in 1851 alone (Rogin 1931). The adoption of mechanical threshing was rapid in the Western States as the region rose to dominance in wheat production. Notable departures from the British designs soon appeared. The peg drum and concave, first patented by A. Savage in 1822, and improved by A. Douglass in 1826 and by numerous other inventors, rapidly gained acceptance when farmers found that they were easier to drive and feed. They were also inexpensive

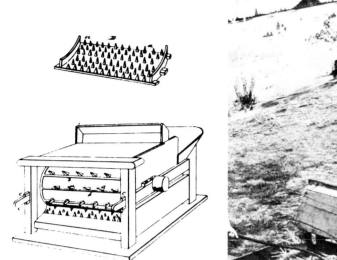

Left: Peg-toothed threshing machine patented by A. Douglass on September 18, 1823, and improved in a September 8, 1826 patent. The concave was adjustable and removable for service and repair; Right: Early American "Ground Hog" thresher. The machine was staked into the ground, hence the name "Ground Hog"; Below: H. A. Pitts "Thresher-Cleaner" of 1847 was a ground hog style machine. It combined the thresher and fanning mill.

to manufacture, in some cases pegs were simply driven into a log. Since the straw was frequently burned after threshing, the problem of straw breakup was of little consequence. Early machines were staked to the ground rather than being mounted on wheels and as a result were known as "ground hogs".

Wherever threshing machines were adopted, winnowing by the old rod and sheet method could not compete with the threshers' output and became obsolete. The adoption of winnowing machines was hastened and the idea of combining the machines as "thresher-cleaners" soon followed.

Left: Hiram A. Pitts (1779-1860), principle inventor of the Pitts Thresher (Farm Implement News 1886); Right: An 1877 Pitts endless apron thresher.

An improved Pitts separator of the 1840's.

The "Sweepstakes" thresher with Pitts separator, ca 1850.

An early Westinghouse thresher (Westinghouse Co.).

An early Wemple thresher and horse sweep power.

Westinghouse ground hog thresher with vibrator and separator and tread-or railway-horse power, ca 1850 (Farm Implement News, 1886).

Enter the Brothers Pitts

The brothers Pitts, Hiram and John, began work in Maine as itinerent threshermen. With their "ground hog" they could thresh 100 bushels of wheat in a day. In 1830, Hiram patented an improved horsepower design of the "railway tread" or endless-track type and the brothers began manufacturing the tread-power thresher. Hiram next turned to combining a thresher and fanner into a combined "thresher-cleaner" which the brothers began selling in 1834. In 1837, they patented a design with an endless "apron" conveyor and separator. In 1847, the Pitts moved to Illinois, first to Alton, then to Chicago. Production of the "Chicago Pitts" thresher began there in 1852. When driven by eight horsepower, this machine was capable of averaging 300 to 500 bushels in a day.

Simultaneously, thresher production was progressing rapidly in Ohio. Although threshers had only been introduced some 20 years earlier, by 1851 they were "being manufactured in every town of any consequence". John, who parted company with his brother to work in Ohio, returned to the East and Albany, New York.

Joseph Hall, of Rochester, New York was another pioneer thresher builder. In the 1830's, his claim to fame was the "bull" thresher, so called for the bellowing sound made by the cylinder when operating at full speed. The cylinder, turned in a lathe out of a solid log and then fitted with iron spikes, was all the noisier because Hall left the top of the cylinder exposed. John Pitts joined forces with Hall and they opened a factory in Buffalo to produce the famed "Buffalo Pitts" thresher. This machine was produced in large quantities in New York and later in Canada.

Wemple and Westinghouse

About 1830, Jacob Wemple, a blacksmith and wagon-maker of Mineyville, New York became interested in building threshers as a result of frequent repairs he made to the machines available at the time. He developed an improved tooth design for bull threshers so that they were less likely to loosen or become unbalanced. In 1840, he entered into a partnership with George Westinghouse, father of the celebrated inventor of the airbrake and founder of the company that bears the family name. They patented and manufactured improved threshers at Fonda, New York under the Wemple name. Westinghouse shortly after withdrew to go into business for himself. Wemple went to Chicago in 1848, sold out to Hiram Pitts and retired.

J. I. Case

Jerome Increase Case was born in Williamstown, New York and reared on the family farm. When he was 16, he was entrusted with the operation of the farm's one-horse treadpower "ground hog" thresher and

he did such a good job, that from his thresher he was able to pay for his education through to a business academy. In the spring of 1842, he procured on credit six "ground hog" threshers and transported them to Racine in what was then Wisconsin Territory, after carefully considering the potential of the newly-opened Middle West. He promptly sold five of the machines and used the remaining one to thresh grain throughout the territory. In the spring of 1843, he rebuilt the worn-out machine, modifying it into a model superior to those back East. Until then, there had been no separator or fanner. After the 1843 harvest, he began building a machine modeled after the Pitts' "sweepstake" pattern, and the result became the first combined thresher in the West. In the fall of 1844, he rented a small shop in Racine to build six machines. For the next three years he continued to improve the design

Clockwise from Lower Left: An apron-type thresher as it appeared in the 1870 Case Catalogue; Case two-horse tread power; Jerome Increase Case (1819-1891); Case treadmill-powered thresher, 1855; Case "Agitator" thresher which was announced in 1880 when the company decided to build walker-type separators. All illustrations, except that of J.I. Case, were taken from the "Case Wood Engravings" book published by Case in 1942.

and in 1847 built a substantial shop of his own, laying the foundation for what became the largest business of its kind in the world, the J. I. Case Threshing Machine Co., Inc. He was a public-spirited citizen of Racine, serving as mayor for three terms and as a state senator for one term.

"Vibrating" Separators

Possibly the first patents on straw walkers or "vibrating" types of separators were issued in England to Docker of Findon (1829) and Ritchie of Melrose (1837). There certainly was a "vibrator" in operation on Lord Ducie's demonstration farm around 1840. The first manufacturer of US vibrator threshers was Cyrus Roberts of Illinois, just before 1850. John Nichols of Nichols and Shepherd improved on the design and gave their machine the trademark "Vibrator". Controversy raged whether the Pitts' apron-type of separator was superior to the vibrator type, with the vibrator emerging the victor. When the J.I. Case Threshing Machine Company changed to the vibrator type in 1880, they named their machine the "Agitator".

Nichols & Shepherd's first "Vibrator" thresher, 1858.

A succession of names became associated with the threshing machine business that became bywords in the industry during the 19th Century: Geiser, Aultman & Taylor, C. M. Russell, Robinson, Avery, Rumely, Gaar-Scott, and many others.

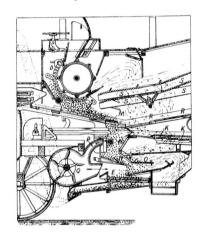

Far Left: Innovations in European threshing machines in the latter half of the 19th century included this 1880 dual-outlet fan under the separator from R. Wolf, Magdeburg-Buchau (Fischer 1910); Left: There was apparently still a small market for handpowered threshers, even in 1870. Belcher & Taylor's trade circular of that period offered Whittemore's patent grain thresher, a simple machine for $35 "well adapted for the threshing of rye . . . and being introduced into the Southern States for the threshing of rice. To any farmer who raises a few hundred bushels of grain, this Machine will pay for itself in one season".

Below: British threshing machine made by Wallis and Stevens of Basingstoke in 1876 (Scientific American, February 1876).

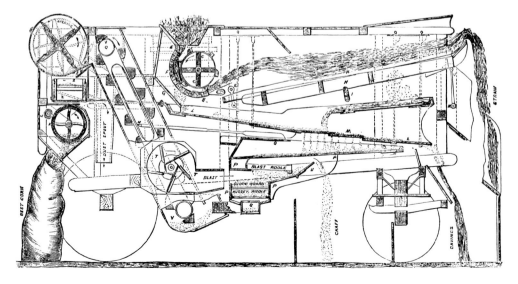

"It is a matter of arithmetical calculation, how much can, on the whole be saved by the use of the best machine yet invented. To cut fifteen acres will require on the one hand a machine worth $125, one or two span of horses and eight men. On the other hand it will require five cradles worth $25 and eleven men. We leave it to any man to say which he can the easiest command. If he be a small farmer having not over fifty acres to harvest, it would not be with our advice that he would buy a machine. If he has a larger harvest he can put horses in the place of men and drive his work through quicker than by reliance on cradles alone. This is not our opinion only, but that of a skillful and enterprising farmer who has tested more than one harvester."—J. S. Wright, Editor of The Prairie Farmer, September 1846.

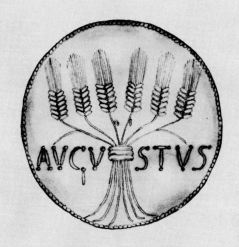

The sheaf, symbol of the harvest since antiquity.

8

Automatons, Headers and Barges

Grain crops that are cut and bundled into sheaves can be harvested earlier in the season. The grain is less likely to be harmed by rain than if the crop is left standing uncut and exposed to the mercy of the weather by harvest delays. The sheaf, an orderly collection of grain stalks, is a conveniently-sized bundle to carry or to pitch from point to point. The season can be shortened by cutting early and maturing the sheaves in the "shock".

For centuries, sheaves were bundled and tied by hand. But as machine reaping became increasingly practical inventors met the challenge of developing mechanisms for dividing the continuous ribbon of crop that fell from the reaper into sheaf-sized bunches or "gavels". Each gavel could then be tied with a few deft twists of a straw-band.

As early as 1807, Robert Salmon of Woburn had proposed a machine to cut and lay the gavel to one side, ready for binding. His concepts remained a curiosity until a reliable reaping mechanism was perfected. Salmon was half a century ahead of his time.

Self-Raking Reapers

The sale of McCormick and Hussey reapers in the US were swiftly followed by complaints about the difficulty of raking the cut grain from the platform.

Here and there, inventors brought out self rakers to overcome the problem. Foster's 1846 US patent eliminated the need for the "raker-off". Andrew J. Cook of Richmond, Indiana, was awarded a patent in November

1846 for improvements to the Hussey-type reaper. He added a bat reel and one arm equipped with tines which did the raking off. But there was still the same problem of dumping the gavel behind the machine, just as on Hussey's reaper. The sheaf-bandsters had to clear the way before the reaping machine could make the next round.

From Prairie Farmer Editor to Manufacturer

Farm inventiveness was spurred on by farm magazines that were particularly influential during the latter half of the 19th century. Farm papers sprang up everywhere, achieving considerable popularity and prosperity until they were curbed by the Great Depression of the 1930's. While their greatest service may have been in the cause of improving crops and livestock, their zeal for innovations, especially labor-saving devices, kindled intense reader interest. Certainly, the editors welcomed farm machinery for its advertising revenue. But it was

In 1811, Lester of Paddington, England advocated the use of this modified fork which enabled sheaves to be hand-tied with a hempen band. The bands cost "twopence a dozen."

Left: Grain binder's wheat rake. The operator did not have to stoop. When the bundle was collected, the operator lowered the handles and thereby raised the crop bundle above the stubble to a convenient position for tying. In the 1850's the device was sold in the east by Emery & Co. of New York for $4 (Thomas 1858); Right: Cook's self-raking reaper, patented November 1846. One arm of the reel did the raking off. Sheaf- bandsters were required to clear the way before the next pass of the machine around the field. A few of these rear-delivery reapers were manufactured under license (Deering 1900).

not until the turn of the century that machinery and purchased supplies began to edge out livestock advertising as the principle source of magazine income (Johnson 1976).

John Stephen Wright, founding editor of what is now North America's oldest surviving farm magazine, *The Prairie Farmer,* saw the need to dispense information on farming the prairie lands that were unlike the farmlands of the East. He helped organize the Union Agricultural Society in 1840, then sold Society members on the idea of a farm magazine, *The Union Agriculturalist.* The magazine was a financial disaster, so the Society gave Wright permission to produce the paper on his own. Wright changed the name to *The Western Prairie Farmer* in January 1841. A year later, it became simply *The Prairie Farmer,* and as Illinois agriculture prospered, so did the new magazine. Wright successfully used *The Prairie Farmer* in a campaign to establish schools in rural Illinois. He also encouraged the University of Illinois, now a land-grant college, to expand the curriculum to include mechanics and agriculture rather than only the classics which he had received there (Lewis 1949).

Wright was a keen advocate of reaping machines. He lavished columns of type and woodcut illustrations on them. He predicted that reapers would soon displace the cradle everywhere, although he cautioned against the "mania" for building home-made reapers. Manufacturers curried Wright's favor for the publicity he gave their machines in the Midwest's most powerful farm paper. They purchased advertising space in the magazine, published testimonials from satisfied users and wrote bitter denunciations of other reapers—even demanding the correction of unwarranted slights. All this was grist for the mill for readers. In November 1849, Jonathan Haines drove his new reaper all the way from

Whiteside County, Illinois, into Chicago so that he could put his machine through its clanking paces mowing imaginary grain up and down Lake Street in front of the windows of *The Prairie Farmer* office.

Around 1850, the impartiality of editorial comment became an issue. Wright had been dabbling in the construction of Hussey reapers under license. He was unsuccessful in persuading Hussey to move to Chicago. In 1853, Wright felt compelled to publish a disclaimer in *The Prairie Farmer.*

"It is proper for me to say that the editor of this paper has no connection with any Store, Warehouse, Wool Depot or Cabinet Shop; nor has he any interest in any Mower, Reaper, Thresher, Dog or Cat Churn, Seed Drill, Rat Trap, Cucumber Washer or Patent Jewharp, or any other machines or inventions of any sort. He is solely an editor. He is therefore without any undue bias to any one of these things above another."

Almost simultaneously, however, Wright became interested in challenging the McCormick reaper by backing the Atkins' "Automation" self-raker.

Atkins' Automaton

Jearum Atkins, a millwright by trade, had become permanently bedridden by injuries suffered during an accident in 1842. A neighborly farmer, knowing Atkins' mechanical skills, showed him a reaper and remarked that if he "would only attach a rake to it he would make his fortune". That farmer's remark awakened Atkins' inventive spirit. In need of a means to sketch designs, friends gave him a drawing board which they suspended above him so he could work lying on his back. He began by saying "if it will not work in my brain, it will not work anywhere". After weeks of intense effort, he drew and modeled a new principle—a self-rake which imitated the action of the human arm and hand. The "Atkins movement" became an immediate success when sold in 1852 as an attachment to the Hussey reaper.

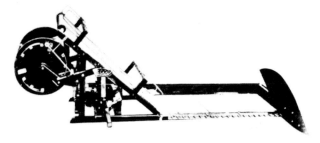

*Left: Advertisement from the **Iowa Northern Farmer**, 1857 for the Wright-Atkins self-raking reaper and mower. Note the name of the company and factory (Johnson 1976); Above: Atkins ''Automaton'' self-raker in 1852. The rake would reach out, sweep the cut grain from the platform, withdraw and then repeat the cycle. (**The Prairie Farmer** June 1853); Below: Hurlbut's reaper, February 4, 1851. This mechanism, for automatically providing uniform parcels of grain without the need to stop the machine, had an elevating canvas to bring the crop up over the drive wheel. The grain fell into a three-lobed receptacle poised beneath the elevator. When sufficient crop had accumulated in the cell to effect counterpoise, the cell's contents were dumped and the receptacle revolved one-third turn (Deering 1900).*

Wright began to manufacture the Atkins', notwithstanding his editorial disclaimer.

Sales of the Wright-Atkins Automaton enjoyed a meteoric rise. This stirred up McCormick, who launched bitter tirades against it in *The Prairie Farmer*. He even threatened to start an opposition farm paper. Wright, in reply, only taunted the McCormicks by saying in 1855:

"You have been twenty-two years in the business and have got out 4,000 machines; I have been at it four years, and shall get out for next harvest at least 3,000 . . . Truly it does seem as though there was a good deal of reason for your fear of rivalry of this machine."

The McCormicks were silent. Reaper competitions had by then deteriorated into theatrics, with some competitors going to the extreme of chaining their reapers together and then whipping their horses to see which machine could pull the other apart!

Wright expanded his "Prairie Farmer Works" in 1856, determined to overtake the McCormicks' annual production of 4,000 reapers. He advertised his intention to have 5,000 machines ready for the 1857 harvest.

He had gone too far! A financial panic struck the US in 1857. Wright not only "lost his shirt", but control of *The Prairie Farmer* as well. In a year that saw wide spread drought, farmers could not afford the magazine, let alone the reaper. The final straw was a complete loss of customer confidence in his machine when green wood that was used warped out of shape. Unseasoned wood was all that had been available under the pressure of Wright's boast to build 5,000 machines in 1857.

The Prairie Farmer rode out the depression, minus Wright. Today, the Farm Progress Shows, founded by *The Prairie Farmer,* are possibly the largest annual farm machinery field days in the world. These shows testify to the survival of comparative demonstrations—and of America's oldest farm magazine.

Above: Continental Bank Note Company engraving of 1865; Top Right: Aaron Palmer and S.G. Williams' self-raking reaper, patented February 4, and July 1, 1851. The centrally-pivotted raking arm was geared to the drive wheel to sweep across the quadrant platform located directly behind the cutterbar. The finger on the rake moved over a slider-rod, so arranged to fall into the crop at the forward position and drag the parcel of grain to the side. The design was so effective that the Seymour and Morgan Company, which purchased the rights, promptly put it into production, selling their "New Yorkers" in large numbers by 1854; Center: An early "New Yorker," first produced in 1851 by Seymour & Morgan after the Palmer & Williams design; Lower Right: The later "New Yorker," the first commercially successful self-raker with quadrant platform, was produced in large quantities after 1854 by Seymour & Morgan of Brockport, New York.

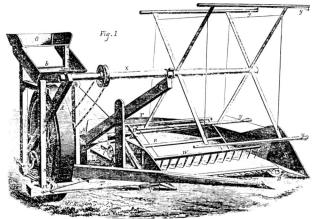

William Seymour

Side-Delivery Self-Rakers

F. S. Pease of New York was one of the first US inventors to attempt to mechanically deliver the bundle to the side of his reaper. The gavel-delivery mechanism of the 1848 Pease patent was not automatic; the operator had to stop the machine, then use a lever-operated linkage with rake-teeth sliding across the slotted platform to push off the bundle. Hurlbut's reaper patented in 1851, had an elevating canvas to bring the crop up over the main wheel and drop it into a measuring receptacle. When this receptacle was full, it would automatically dump the contents to the side of the machine.

The Quadrant Platform and the "New Yorker"

A quadrant-shaped platform for reapers, first disclosed in an 1849 patent, resulted in one of the most valuable monopolies in the development of the self-raker class of reapers. Nelson Platt's quadrant platform reaper attachment provided a rake to sweep the grain around the arc of the quadrant and onto the ground. Aaron Palmer and S. G. Williams, quick to realize the

Top Left: J. H. Manny's hand-rake reaper, 1855. Proponents of hand-raking cited the unreliability of the self-rakers and the lack of compactness of the gavel as major drawbacks; Lower Left: Ketchum's machine in use as a reaper, 1855; Below: Adriance Platt and Company's self-raker at work. Cam followers ran on a curvilinear track to produce the raking arm motion peculiar to the vertical-axis family of reapers. There was now space for the driver on the reaper, although the rakes whistled by close to his ear (Johnson 1976).

advantages of the quadrant principle, devised a way to incorporate the rake and quadrant immediately behind the cutterbar instead of merely attaching it to the side of the platform as Platt envisioned. Williams, in bad health, and Palmer, in dire need of money, sold their rights to Seymour and Morgan of Brockport, New York. The Palmer and Williams patent, and Seymour's own 1851 patent became the foundations for the celebrated "New Yorker" self-rake reaper which Seymour and Morgan began manufacturing in quantity in 1854. A number of other companies were then licensed to manufacture the quadrant design. As a measure of success, it was still being sold 80 years later.

Self-Rakers Sweep the Field

By 1853, reapers without a raker's seat, or some form of automatic raking device, were becoming outmoded. At the end of the decade, there were over 20,000 reapers with self-rake delivery in use. Hand-rakers, however, continued to be produced in small quantities—they were preferred by those who cited unreliability of the mechanisms and the lack of compactness of the gavel as the major drawbacks of the automatic rakers.

Vertical-Axis Self-Rakers

In 1856, Owen Dorsey was granted a patent for an "improved harvester rake" which dispensed with the need for the reaper's bat reel. Based on the unpatented Hoffhein design, Dorsey's mechanism consisted of raking arms mounted on a vertical shaft geared to the drivewheel. A universal drive mechanism coupled to

the rake arms caused them to vertically enter the crop one at a time, gently bring the crop back to the cutterbar, sweep the cut material across the platform quadrant and then deposit the gavel on the ground, before withdrawing to repeat the cycle.

On the first Dorsey machines, the sweep of the raking arms left no room for the driver, forcing him to ride the drafthorse. This detail was cleared up by an operator's seat provided in Whitenack's 1861 design and others that followed. The improved vertical-axis self-rakers, sometimes known as "pidgeon-wing" reapers, secured a firm place in the industry for the next half century.

The War of Emancipation

The Civil War of 1861-1865 created a shortage of harvest labor by simultaneously drawing men into the Union Army and stimulating grain production. One million strong-armed wheat cradlers rallied to Lincoln's call to arms. During the conflict the automatic self-raker was refined in design and surpassed all other reapers in sales. Each reaper equalled the labor of four or five men.

By 1864, harvesting machinery was big business. There were now 203 manufacturers, producing 87,000 reapers, harvesters and mowers annually (Hutchinson 1935). Some of the contemporary manufacturers who "cut their teeth" in the reaper or mower business of that period are profiled on the following pages.

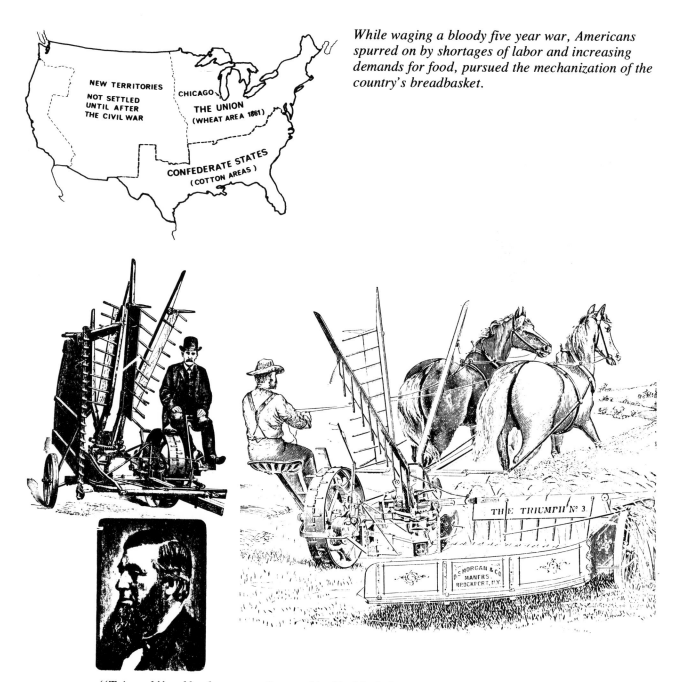

While waging a bloody five year war, Americans spurred on by shortages of labor and increasing demands for food, pursued the mechanization of the country's breadbasket.

"Triumph" self-rakers manufactured by D. M. Osborne & Co., Auburn, New York.

As the geographical center of US grain production moved steadily westward, the eastern region became more of a market for mowers, less for reapers. **Obed Hussey** concentrated on improving the mower. In his latter years, he also worked on steam power. He died in a railroad accident in 1860. As long as his patents on the knife and "double-finger" guard were in force, virtually every reaper and mower manufacturer was obliged to pay royalties to Hussey or his heirs.

In 1854, **Cyrenus Wheeler,** an astute Yankee inventor from Cayuga County, New York, was granted the first of his 17 patents for a jointed-bar combined mower and reaper. This flexible design permitted the machine to better follow ground irregularities. In the next 12 years, he purchased 67 other patents, and in the process acquired

the name "Patent King". In 1859, he formed the Wheeler Association to prosecute infringers and to license other manufacturers to build his "Cayuga Chief" machines. Between 1853 and 1888, over one million machines containing improvements embraced in the Wheeler Association patents were built. In 1874 the Wheeler Association consolidated with D. M. Osborne & Co., of Auburn, New York.

At the age of 22, **Cornelius Aultman** began building threshers and Hussey reapers under license in Canton, Ohio. In 1849, he and his partner, Ball, directed their co-worker Lewis Miller to build their first front-cut mower. Their two-wheeled "Buckeye" was an overnight success. The "Buckeye" design reduced draft and eliminated the need to place outrigger wheels on mowers.

The "Buckeye" self-raker
from the 1893 Aultman, Miller
Company catalogue.

John H. Manny

Abraham Lincoln

The purchase of an Esterly header in the late 1840's started farmer **Pells Manny** in the machinery business when he began manufacturing this header with his son John as his partner. They advertised their "Gem of the Prairie" header between 1847 and 1850, but found only a limited market for headers in the Midwest. So, in 1850, the Mannys turned their attention to the development of mower/reapers similar to McCormicks; extremely similar, as it turned out. By 1852, the Manny hand-rake reaper commanded prizes at the Geneva trials in New York, while the McCormicks' failed to place. In 1854, Manny reaper sales exceeded 1100 units

and were considered a sufficient enough threat that Cyrus McCormick initiated legal action to constrain competitive reaper manufacturers.

Abraham Lincoln and Manny vs. McCormick

The McCormick vs. Manny legal battle, involving a claim of a monopoly over most reapers, was tried in Illinois in 1855. It proved of great economic importance to farmers and reaper manufacturers. The case had an even more far-reaching effect—it's bearing on the career of Abraham Lincoln, hired as counsel for Manny. Lincoln was sent by his employers, an Eastern law firm, to Rockford, Illinois, to see how the Manny reaper was constructed and to prepare an extensive argument on his client's behalf.

Complainant McCormick held that Manny had infringed on his patents and that his patent improvements gave him a monopoly on all reaping machines, including the large number then being made by Manny.

Much was at stake in the court's decision. If McCormick won, he could place most of the reaper manufacturers of the land under his tribute.

Lincoln in turn was excited about the case because of its importance. It was his biggest case to date. Harding and Stanton, senior legal counsel for Manny, had hired Lincoln in the first place so that they would have an Illinois lawyer pleading their case before an Illinois judge. Lincoln received a $500 retainer, and subsequently, a $1000 fee which helped nurture him through a period of extensive political activity, including his forthcoming debates with Stephen Douglass, which he waged in a losing effort for a Senate seat.

Lincoln's plain, back-woods appearance created a poor impression on Harding and Stanton. They did not let the gaunt-appearing Lincoln utter a word in court. Lincoln later recorded that he had been roughly treated, but much to his credit, he did not permit this treatment to cloud his recognition of the great abilities of Manny's senior representatives. In fact, he was determined to become the equal of these gifted college-bred lawyers. Emerson wrote that Lincoln's style, manner and speech markedly improved after this case. When

Lincoln became President, he appointed Edwin Stanton Secretary of State.

The retainer permitted Mary Lincoln to proceed with her plan to build a home in Springfield, today a historic landmark, visited by thousands of tourists each year. As for the case, the court declared non-infringement. John Manny received the good news, but died of consumption shortly after. The bitterness of McCormick's defeat and heavy legal costs he bore were hardly relieved by the consolation which Justice McLean handed down in his January 16, 1856 decision which showed considerable interest in the welfare of the patentee:

"He is left in full possession of his invention which has so justly secured to him, at home and in foreign countries, a renown honorable to him and to his country, a renown which can never fade from the memory, so long as the harvest home shall be gathered. The bill is dismissed at the cost of the complainant".

The Adriance Group

John P. Adriance of Poughkeepsie, New York, founder of Adriance, Platt and Company, began making Manny mower/reapers for the New England market in the 1850's. He also obtained a license from Aultman and Miller to make the "Buckeye". Later these companies merged and sales territories were assigned to each. By 1868, Adriance had expanded their territory, and had sold over 120,000 harvesting machines. They also received more than $500,000 in license fees from over 25 manufacturers.

Above: Johnston's self-raker, 1876; Right: William Whitely and the "Champion" reaper ca 1867.

Samuel Johnston of Syracuse, New York became interested in improving the action of self-rake reapers. The Dorsey vertical-axis type swept a gavel from the platform with each revolution of the reel, regardless of crop density. Johnston's improvement consisted of a foot-operated mechanism that allowed the driver to determine the size of the gavel. Each of the four arms of the reel became a rake and the operator could use one, two or all four arms to discharge a bundle at each pass. Patented February 7, 1865, sales of his auto-delivery models eventually surpassed the "vibrating" rakes of the Palmer and William style. The Johnston was adopted by Adriance in 1867.

William Whitely became a Johnston licensee in 1867, and soon became the largest manufacturer of the Johnston style of self rakes. By 1880, Whitely, Fassler and Kelly of Springfield, Ohio claimed to have the largest agricultural machinery factory in the world. In 1867 alone, they manufactured 40,000 machines under Whitely's "Champion" logo.

Walter A. Wood, of Hoosick Falls, New York began manufacturing the Manny mower/reaper in 1853. Shortly before the Civil War, he developed his own "light draft" two-wheeled mower/reapers and was enjoying an annual sales volume equal to the McCormicks'. By 1865, he was the leading Eastern manufacturer of mower/reapers. His English licensee, W. H. Cranston of London, sold over 200 of his "chain rake" reapers in Britain between 1858 and 1862. Wood's company became one of the earliest farm machinery firms to adopt parts standardization and interchangeability, and to distribute parts catalogs (Giedion 1949).

The McCormick's Late Entry
into the Self-Raker Business

Cyrus McCormick's singular unremitting and deepening involvement in the business of producing and/or selling reapers blocked out his own interests in further machinery invention. He had "plowed deep, but in a single furrow" and the business had made him a millionaire by 1851. His marriage, at the age of 50, in 1858, led to increasing social and political involvement. He came to regard Europe as the "last outpost" for the reaper and, in 1862, embarked on a promotional drive in Europe, leaving the business in the hands of his brothers. He travelled Britain, France and the German

Left: Walter A. Wood and the Wood Company logo. Wood's company was the first in the farm machinery business with interchangeable replacement parts; Right: Wood's "chain-rake" reaper. This version was produced at the Hoosick Falls Works in New York in 1861.

Left: McCormick's "Reliable" self-rake reaper, 1862; Right: McCormick's "Advance" reaper for 1871. It was lighter in draft than the "Reliable," had sickle height adjustment and a controllable rake, features which led to the company's discontinuing production of the Reliable in 1870.

States for two years. Although he received numerous awards for his reapers, they did not sell as well as his competitors.

In Chicago, it was becoming increasingly apparent that the McCormicks' campaign to drive the self-rakers from the field by ridicule was not succeeding. By 1860, the Company's share of the market had slipped to barely 10 percent. Between 1858 and 1861, the Company negotiated for patents on reel rakes using the quadrant platform. For the vertical axis principle they selected the McClintock Young reel-rake design. These concepts were incorporated into the first McCormick self-raker marketed in 1862 under the name "Reliable". The McCormicks grudgingly paid over $60,000 to Seymour and Morgan between 1862 and 1872 for the privilege of using the self-rake and quadrant platform combination.

The McCormick self-raker was cumbersome, heavy and lacked controllable delivery. As the Civil War drew to a close it became increasingly obvious that the company was badly in need of "a machine that will mow well, and reap a few acres each year indifferently well". (Letter from Office Superintendant C. A. Spring Jr., to C. H. McCormick).

In the Summer of 1868, Company craftsmen divided their attention between a machine called the Sieberling "Dropper", which proved unsuccessful, and a two-wheeled mower-reaper with controllable auto-delivery. This effort was the "Advance". Three thousand "Advance" combined mower/reapers were made for the 1869 harvest. A wet season "drowned" the heavy pulling "Reliable". Modified for 1870, the "Advance" was offered at the same price as the "Reliable"—$190 cash.

In 1870, 6,000 "Advances" could not keep up with the demand and the decision was made to concentrate on the "Advance" and discontinue production of the "Reliable".

The Chicago Fire of 1871

The 1871 Fire which made Chicago "a howling wilderness" on October 8 and 9 included the McCormick Works in its path of destruction. The brothers had been contemplating a move to more spacious quarters when the consequences of the fire in O'Leary's cow barn forced the matter to a head. Within days, a decision was made to rebuild a few miles away.

Fortunately there was a surplus of "Old Reliables" built in 1870 to meet part of the market demand. By a remarkable effort on the part of the McCormicks the new factory was completed in January 1873. The McCormicks were able to sell 8445 machines for the 1874 harvest.

The period 1873-1879, however became the most precarious in the history of the firm. Disagreements between Cyrus and brother Leander (who subsequently resigned), poor climatic and discouraging economic conditions, and bickering over patents, prevented the company from effectively entering the race for a rapidly emerging innovation: the self-binder.

Who Needs Sheaves? Esterly's Header

George Esterly of Whitewater, Wisconsin, thought: why not deliver the crop to the thresherman in bulk, as loose ears on the straw? Esterly, a farmer, realized in 1844 that wheat could not be grown profitably in Wisconsin as long as harvesting required extensive manual labor. He also concluded that the new Hussey reaper took too much draft, McCormick's was unproven, while Moore's "Michigan Combine", then drawing some notice, was too expensive. On October 3, 1844, Esterly patented a harvester which was essentially a two-wheeled container with a cutting drum on the front. Esterly's push "header" arrangement was reminiscent of the Gallic stripper, but it featured a spirally-formed reel which cut against a straight knife edge and impelled the severed heads into the wagon box.

The development of the Esterly headers went through several stages, including the addition of a canvas conveyor to move the crop back into the container. They sold well from the outset. Then Esterly decided to revert to more conventional cutting and gathering means by adding a side-discharge canvas that would deliver the crop up and into wagons, an idea he borrowed from Jonathan Haines of Pekin, Illinois. This Esterly header version sold well in the late 1840's, so much so that Esterly felt confident enough to write derisively of McCormick in an 1849 issue of *The Prairie Farmer*:

> "What is all this frothy and long-winded fuss about?—I wonder whether I could sell Mr. McCormick a few Virginia reapers, as I am frequently offered them at half-price in exchange for my harvester".

George Esterly

George Esterly's first push harvester or "header," patented October 2, 1844. One farmer wrote in an 1846 issue of **The Prairie Farmer** *that this machine "can cut a field all to destructive smash – it walks over the ground like an elephant." Esterly's machine improved considerably during the next two decades.*

McCormick had cause to be indignant at the time. There had been a competition and Esterly had just been given the accolade for the best harvesting machine by a judgment of the Chicago Mechanics' Institute, although the committee never saw the machines work in a comparative field trial. As a final straw, one of its members was none other than senior editor J. S. Wright, then Esterly's agent.

Esterly devoted his efforts to the manufacture of reapers, mowers and self-rakers. After a series of almost-disastrous experiences with licensees of his self-raker in 1859, he rebuilt his business and soon made his "Light-running" machines an important factor in American agriculture.

In 1849, Haines built a push-header using the Hussey-type cutterbar. It had a side discharge canvas to convey the clipped heads into a wagon, called a "barge", which was pulled alongside.

The early headers, like the combine, did not find a ready market in the Midwest. Grain grown in the Midwest of the 1850's needed to be dried before threshing. Some farmers preferred to leave the crop to "sweat" in the shock or rick for some days, while others wanted to harvest all the straw at once. Uneven ripening and weeds were other factors working against the commercial success of the header in the newly-opened Prairies. A decade later, when the bonanza farms of California were opened up, "Haines' Illinois Harvester" became a fore-runner in the header business, even though it had to be shipped around Cape Horn to reach customers.

9

Harvesters and Binders

The first step in the transformation of the reaper into a "harvester" and binder was the discovery of a way to lift the cut grain over the driving wheel. The Bell, Esterly and Haines machines accomplished this by using a moving canvas. In 1849, Jacob J. Mann and his son Henry, of Clinton, Indiana, developed a reaper which lifted the crop into a receptacle outside the drive wheel. A boy also rode the machine; his task was to actuate a rake that swept the unbound gavel out of the container onto the ground. The Manns and their licensees sold well over a thousand of these harvesters in the next 15 years (Hutchinson 1935). Sheaf "bandsters" still had to follow the machine on foot.

The second step in the development of the harvester/binder—the placing of men on the machine to do the binding—was solved in the summer of 1850 by Augustus Adams and J. T. Gifford of Elgin, Illinois. Using one of Mann's machines, they replaced the receptacle with a platform to catch the falling grain. Bandsters rode on the platform tying each sheaf with a handful of straw. When Gifford died, Adams formed a partnership with the picturesquely named Philo Sylla to develop a machine which they patented in 1853. The cumbersome five-foot cut machine carried three bandsters and a fourth man to fork the grain off the plat-

form across to the bandsters. Balancing the machine was a problem, as was the tendency to inundate the bandsters with inflowing crop.

Adams and French harvester (Farm Implement News, 1888).

The Marsh "Harvester"

Among the licensees of the Mann reaper was one with an agent in Illinois, Charles W. Marsh, a young farmer blessed with mechanical talents, but an ambition that leaned toward law. Marsh, who lived 45 miles from Elgin, probably knew about Adams and Sylla's riding binder. One day early in the 1857 season, according to the story in the January 1891 issue of *Farm Implement News*, Marsh was driving a Mann harvester and thinking about the problems of binding sheaves in the field when he conceived a way to arrange the known elements into a better design. He and his brother set about to build a machine, figuring that two men on a "Harvester", as they called it, could do as much as four on the ground. Their first "Harvester", built in 1858, was based on a Mann machine, to which they added parts scavenged from a reaper and a corn sheller. In 1860, they built 12 machines, exhibiting one near Sycamore, Illinois.

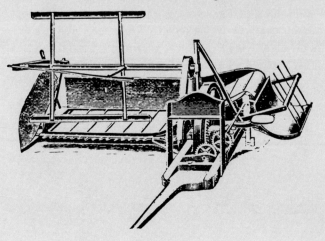

The Mann reaper patented in 1849. Actuation of a lever-operated rake swept the gavel out of the container onto the ground, ready for the "bandsters" (Scientific American, July, 1896).

Among those who saw it was Philo Sylla, who was disdainful, while McCormick's representatives said the Marsh harvester resembled "a cross between a windmill and a threshing machine" (Hutchinson 1935).

In truth, the machines did not work well, but by now the brothers had mortgaged their home and had involved family and friends in their costly venture. Yet help was at hand, in the person of Lewis Steward, a farmer and promoter from Plano, Illinois. He told the brothers:

"Boys you are on the right track. If you can run your machine successfully ten rods it can be made to run ten miles, and there is a man in Plano who can make it do this".

*C. W. Marsh, inventor of the Marsh Harvester and later, for many years, editor of **Farm Implement News**.*

Steward's encouragement led the Marsh brothers to mechanic George Hollister. Together they built and tested machines in an abandoned Plano factory. By 1863, their design was considered to be sufficiently developed for "the test of sale". Twenty-four machines were built by the firm of Steward and Marsh for the 1864 harvest.

A strong demand for the labor-saving "Harvester" was created by the war, and other companies were soon licensed to build them. By 1870, production at the Plano factory had reached 1000 units. It was now under the management of Gammon and Steward, the Marsh brothers having sold their interest in 1869. In their place appeared William Deering, a wealthy drygoods merchant from Portland, Maine, who brought both business talent and $40,000 to the Plano enterprise. Probably 10,000 Marsh Harvesters were sold by the end of 1872, when the original patent was extended.

The "Harvester" established the Marsh brothers' reputation. Their's was a landmark design. Unfor-

1866 Marsh Harvester built by licensees Gammon and Prindle, Chicago. The names of Steward, Hollister, Deering, Eastern and others were associated with this landmark machine. The binding tables were raised on later machines to avoid excessive stooping (Rogin 1930).

tunately for them, their success was shortlived. Later enterprises turned sour and by 1884 they were bankrupt. Charles W. Marsh abandoned the building of machines to become the editor of *Farm Implement News.*

Deering prospered. He moved to Chicago and by 1875 his 6,000 harvesters a year exceeded the output of the McCormicks.

McCormick's Riding Binder of 1875

In 1873, Cyrus McCormick was told by his field agents:

. . ."the evidence accumulates that the Marsh Harvester or that class of machine, is fast becoming the most popular machine on the market—the Marsh has been the rage". (Hutchinson 1935).

Two men riding on the Marsh Harvester could do the binding, using straw bands—a job that used to take four or five men on foot. The Marsh machine, more-

The McCormick Harvester, a riding binder of the Marsh class, manufactured between 1875 and 1883. The two bandsters riding on the machine tied the sheaves with straw bands (International Harvester Archives).

74

over, later proved suitable for the incorporation of mechanical binding devices. McCormick found it necessary to take out licenses on the Marsh patents. He even set up a new factory to produce Marsh Harvesters in 1875.

Towards an Automatic Binder

John E. Heath of Warren, Ohio, was the first to build a machine to bind sheaves mechanically. His landmark patent of July 22, 1850, clarified the general problem involved: compressing the straw bundle, wrapping the twine or cord around to secure the sheaf, and preparing a knot in the twine, which in his patent was tied manually. He made a number of machines in the 1850's which worked well under favorable circumstances. Others saw sufficient potential in the machines for Heath to make a tidy sum on the sale of his patent rights.

Above: The 1853 Watson and Renwick patent for an automatic twine-wrapping grain binder. None of the various complex machines built by these inventors sold (Deering 1900); Below: McPhitridge's 1856 patent—the first issued on a wire-tying sheaf binder.

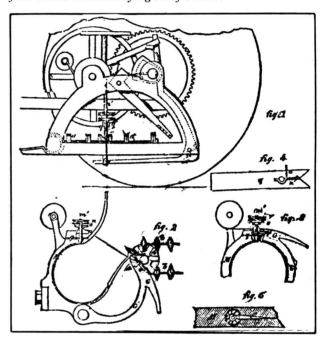

Among other twine binder inventions of the fifties were some exceedingly complex machines patented by Renwick and the Watsons who apparently wanted to patent every conceivable form of binding device. They failed to get beyond the experimental stage (Deering 1900). The time was not yet ripe for the twine binder and in the meantime, wire-tying came into vogue.

Wire Tying Binders

The stiffness of wire enables the ends of a loop to be secured together with a simple twist, rather than the more complex knot necessary for tying cord. Furthermore, the need for fences in the developing farmlands of America had given the wire industry an enormous boost. Wire was a relatively cheap and reliable product at the time and the first commercially successful mechanical binders tied with wire.

In 1856, C. A. McPhitridge of Missouri received the first patent for a wire tying binder. Alan Sherwood of New York, starting in 1858, and persevering for

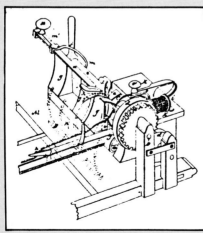

Above: John Heath, first patentee of a twine- or cord-binding machine. His 1850 patent was remarkable for its originality and the directness with which it aimed at the goal. It included the first use of spooled twine, a cord holder, sheaf compressor, knife and practically all the elements to complete the task except for making the bow (Farm Implement News, 1886); Below: Heath's 1850 patent binder illustrated in Farm Implement News in 1886).

The Bursons' wire binder at the "Great Reaper Trial", Dixon, Illinois, 1862 (Farm Implement News, 1886).

many years to improve on the wire tying mechanism, succeeded in having a number of his binders built and used in the East.

The Bursons of Illinois moved one step closer in 1861 with hand-operated wire and twine-knotter attachments for reapers. These were demonstrated at agricultural fairs to show how a sheaf could be bound mechanically. Approximately 1000 binders equipped with Burson hand-powered wire tying attachments were sold over the four harvests of 1862 to 1865. Over 50 patents were eventually granted to members of the Burson family for tying devices used with harvesting machinery and machine-knit hosiery.

The Burson's mechanism aroused the McCormicks' interest. It was also used on John Manny's six-foot cut reaper and by others. Talcott, Emerson & Co., built 1000 for sale in 1863. These worked "tolerably well", but they were hand-powered and thus required an extra attendant to ride on the machine.

The first automatic self-binder sale occurred in 1873. This was a "Packer" binder built by John H. Gordon to fit on a Marsh Harvester. His price: $300. Gordon and his brother James were soon hired by Deering to fit their wire binder to Deering's harvester. Gordon's wire binder was also subsequently licensed to Gammon, D. M. Osborne, and the "Buckeye" factories. Deering's

The Locke binder of 1875 was the first commercially successful automatic self-binding harvester. It tied with wire. (Scientific American, July, 1896).

principal rivals, the McCormick brothers, also attempted to exploit the Gordon binder patents, but were dissuaded after an unsuccessful legal battle.

Sylvanus D. Locke of Janesville, Wisconsin, took out many patents on wire-binders. After failing to reach an agreement with the McCormicks' he began production on his own, selling three machines in 1873. Locke then joined the Walter A. Wood Company and, by 1875, 250 Wood-Locke wire binders had been sold. Locke, a prolific inventor, took out 43 harvesting machinery patents between 1865 and 1879.

McCormick's Wire Binder

The McCormicks' reactions to wire-tying binders are worth noting. In 1864, when the company was busy promoting the "Old Reliable" self-rake reaper that they had only started producing two years earlier, McCormick representatives publicly labeled wire binders "a swindle", and "sheer humbug". But in fact, they were exploring this market and were soon enmeshed in legal battles over patent rights. Then in 1872, Charles B. Withington, a watchmaker from Janesville, Wisconsin, patented his first wire binder.

Above: McCormick's Hand Binding Harvester, a Marsh-type machine (Giedion 1949); Left: The Gordons' "Packer" binder on a Marsh-type harvester. In 1873, the binder became the first recorded fully automatic (wire) binder ever sold.

The McCormick Harvester and Wire Binder, 1876. Between 1877 and 1885, 50,000 of these machines were sold, the largest sales of wire binders in the industry (International Harvester Archives).

During an interview in Chicago he explained his device to McCormick. As Withington talked at length about his twin-spool design, the aging McCormick nodded off to sleep. Disappointed, the watchmaker gathered his invention and returned home, leaving the manufacturer to his rest. After Withington had left, McCormick awoke with a start to find the man gone. He remembered enough of what he had seen to dispatch a man to bring Withington back from Wisconsin. The outcome of the negotiations became the most successful wire binder on the market. Withington sold a half-interest in his design to McCormick in 1874 and the McCormicks built several of them, attaching them to the Marsh-type harvester for testing at Elgin, Illinois in July, 1875. About 40 were sold in 1876, with scheduled production beginning in 1877. By 1885, no less than 50,000 wire binders had been built at the McCormick plant in Chicago.

In spite of success, the wire binding principle was not without drawbacks. Cattlemen became increasingly vocal over the harm digested pieces of wire were inflicting on livestock ("hardware disease"). Opposition also came from threshing crews and flour millers upset by stray pieces of wire in their machinery.

Other opponents called attention to those occasions when cows were killed by wire mixed in the hay in the rumen, and played up the possibility of someone biting into a piece of binding wire in their bread. Straw and twine, being digestible and degradable, presented a far more attractive alternative. A practicable means of using either to automatically bind sheaves remained a sought-after goal.

Straw Binders

The earliest patent for a self-tying device to bind sheaves with straw was issued in the US in 1858. Various inventors tried unsuccessfully from time to time to produce an effective and reliable design. As late as 1905, an enormously long patent (37 pages of diagrams alone) was issued for an automatic straw-binder. This record-length patent stands as mute testimony to the persistence of William Douglass, a Nebraska blacksmith, who spent much of his life trying to perfect the self-tying straw-binder. He was firmly convinced that using straw instead of twine would reduce the costs of grain production, and that it would certainly prevent crickets from eating the twine. After 26 years of work he could not attract a licensee, although he advertised his machine and did receive token financial support at one time from Deering and Co. The machine was as complex as the patent was long.

Below Left: William Douglass, inventor and blacksmith, may have been issued the lengthiest farm machinery patent ever—US Patent No. 789,010, May 2, 1905 (Nebraska SHSM); Below: Automatic self-tying straw-binder developed by Douglass after 26 years of effort. It tied the sheaves with a "wisp" of straw twisted from the strip of crop cut and collected separately on the left of the machine (Nebraska SHSM).

Above: Some early twine knots. These were hand tied. They are shown loose for clarity; Below: Behel's billhook or knotting bill, patented February 16, 1864. The left cut shows Behel's first wooden model, the right, the first Behel metal bill. This device became the standard for all knotters (Farm Implement News 1886).

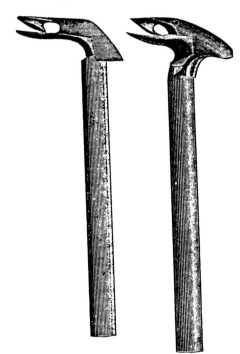

The Appleby and Behel Bills

The acceptance of the Marsh harvester and the wire binder paved the way, but a number of developments had to unfold before a reliable twine binder was developed. At first, progress was also hampered by the absence of an acceptable quality of twine—and at an acceptable price.

The story of the twine binder had its parallel in the career of J. Appleby—there was a vigorous start, a slump in interest, then a triumphant conclusion. In 1857, Appleby, a rugged 18 year-old farm boy was working for a farmer near Whitewater, Wisconsin, who had just purchased a new reaper. After watching it in operation for a while, he candidly remarked that he liked the way the machine performed, but that he believed it could be made to bind the sheaves as well. This suggestion brought only scorn in response. But before the end of the year, Appleby did indeed have the idea for a workable knot-tying device. According to Marsh, the vital lead came to Appleby while watching a little girl playing with her pup. When she accidentally dropped her skipping rope over the dog's neck, he shook himself, and backed free. As the rope slipped off his nose, a knot fell to the ground. The first model of Appleby's "bird's bill" knotter was made from applewood cut on the farm. In 1858, he made another out of steel at a Beloit, Wisconsin gunshop, but did

How the standard "round" knot is tied by a twine knotter (Diagrams courtesy of Geoffrey F. Cooper, Toronto, Ontario).

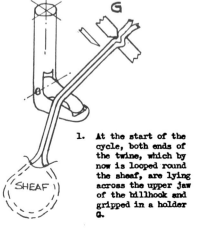

1. At the start of the cycle, both ends of the twine, which by now is looped round the sheaf, are lying across the upper jaw of the billhook and gripped in a holder G.

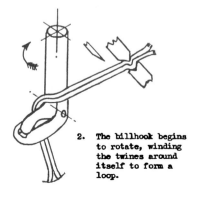

2. The billhook begins to rotate, winding the twines around itself to form a loop.

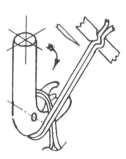

3. When the billhook has rotated about ¾ turn, its jaws open.

not feel that the idea was worth the expense of patenting. Then he laid the project aside for something more compelling—the Union Army.

When the Civil War ended, he turned again to the problem of mechanical sheaf binding, but this time used wire for the band. In 1869, he took out his first patent on a wire-tying binder. But more about Appleby later. It was the 1864 patent of Jacob Behel of Rockford, Illinois, that turned the key that unlocked the problem of twine binding. Behel's "hawk bill", or "billhook" as it was to become known, contained the essential elements for tying the "round" knot that was to become the standard, although other knots and devices such as Appleby's preceded it. The most important feature was a pair of jaws mounted crosswise at the end of a rotatable shaft. Just exactly how the knot is tied is shown in the simplified diagrams below.

W. W. Burson, who had moderate success with wire binders, obtained patents on twine tying mechanisms as well. His co-worker, Emerson, of Talcott, Emerson and Co., tried everywhere to procure suitable twine for Burson, finally building equipment to make twine himself. However, the project was terminated, when fire razed the Emerson twine works.

C. W. Lavalley has been credited by Marsh (1886) with the idea of paying out the cord to the knotter from the center of a round ball of twine. His patent was issued April 17, 1880, and it was an important idea that was to save miles of unravelled twine and soothe frayed nerves by eliminating snarled-up packages of twine.

In 1874, Appleby returned to twine tying. By 1877, he had several twine binders working on Marsh Harvesters. Then in 1877, Appleby found the sustaining support he needed in the form of William Deering, who this way:

"In William Deering, of Chicago, Illinois, formerly of the firm of Gammon and Deering, I found a man farsighted enough to see the importance of my invention. To him belongs the credit of forcing my binder onto the market with sufficient energy to convince the farmer of its practicability. His demonstration of the practicability of the invention soon led other manufacturers to adopt it." (Deering 1900).

In 1880, Deering startled the country with the release of 3000 twine binders and accompanied them with 10 carloads of suitable twine that he had secretly manufactured. (Church 1949).

The machines were sold outright, and Deering's "second fortune and Appleby's fame grew apace, the one because of the other". Appleby's twine knotter was to be licensed for use on other harvester-binders: Esterly (1880), Excelsior (1880), McCormick (1881), Buckeye (1882), Champion (1882), Osborne (1883) and Wood (1892).

Binder Twine—the Tie that Binds

The knot is central to the twine binder. The growth of the twine binder business was phenomenal—once a suitable supply of reliable quality twine became available. Even the best knotter would not tie every sheaf if the twine was defective. Indeed, some binders continued to be sold with a bandster's platform alongside the knotter so that the rider could catch the missed sheaves and hand-tie them.

Twine and cordage has been made from a variety of plant fibers—paper, wood, grass, and rushes, but the most suitable binder twine comes from a mixture of sisal and manila hemp fibers. Flax proved eminently suitable too, but was just too tasty for crickets and grasshoppers.

The chief supplier of sisal hemp has been Mexico. Sisal provides a short, stiff fiber that is cheap and only useful after mixing in about equal proportions with the stronger and more durable, rot-resistant manila hemp. The most even quality manila hemp fiber has come from the Philippines.

In the 1870's the cheapest twine to bind grain cost from about 75 cents to two dollars an acre. It was not until 1880 that twine became competitive with wire for grain binding, clearing the way for the twine binder. As competition increased and manufacturing techniques improved during the late 1880's, twine became much cheaper to use than wire.

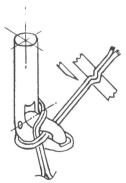

4. During the remaining ¼ turn, the jaws close, now gripping the twines.

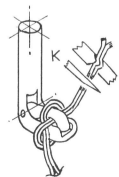

5. A knife K now moves across and cuts the twines, to free them.

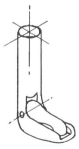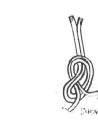

6. The loop around the billhook is now caused to slide off the hook; in doing so it passes over the free ends which are still held between the jaws. Further movement pulls the free ends out, completing the knot.

THE ESTERLY STEEL BINDER
AND
THE ESTERLY ENCLOSED-GEAR MOWER.

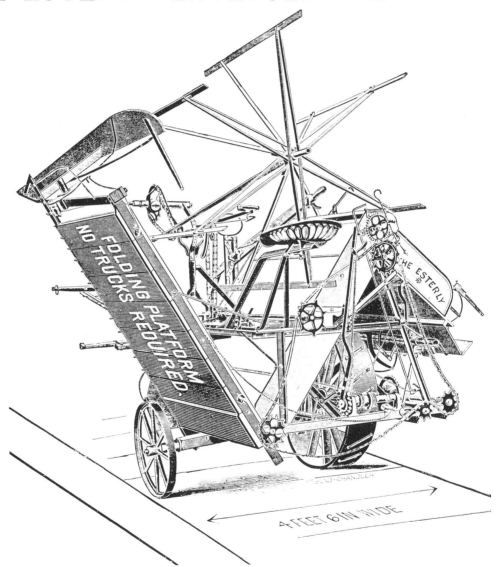

FOLDING PLATFORM NO TRUCKS REQUIRED.

THE ESTERLY

4 FEET 6 IN WIDE

To Our Friends and Patrons

Another circuit of the seasons has brought us to the threshold of another time of preparation for the one that is soon to come, and it is with pleasure we thank those who have favored us with words of commendation for our efforts to please them in furnishing the most

RELIABLE GRAIN AND GRASS CUTTING MACHINERY

that the inventive genius and skill of the best mechanics can produce. It is with no small degree of pride we can say that, notwithstanding we placed upon the market the past season **a much larger number of machines than ever before**, nearly **all of them have been sold**. Never before have we "carried over" so few unsold machines, even when we manufactured less than one fifth the number we did for 1888. This alone speaks louder in praise of the Esterly Machines than volumes of high sounding words. Having increased our facilities, we are prepared to furnish a larger number of Binders and Mowers the coming year than ever before, and with added improvements that have been suggested by experience in the field, we hope to merit the continued favor of the public.

In Binders, as well as every other class of farming machinery, there is in general appearance a striking similarity among those of different manufacturers; so that to a person unskilled in their construction it would be difficult to tell wherein they differ. This fact shows clearly that there are certain underlying principles in the construction of farming machines that are recognized by all manufacturers as being not only important, but absolutely necessary and inseparable from their success. The Esterly Binder has, from its first introduction, taken rank among the best made; and by having carefully watched its operations in the field we have been able to adapt it to the work to be performed, so that for years it has stood without a peer in the great family of binders in its adaptability to all the varied kinds and conditions of grain it has been called upon to handle. In the changes that have been made from year to year we have not lost sight of a growing feeling among agents and farmers that a **departure from the old style** of "wood and iron" machines to something that would be lighter in appearance and more durable, was at least desirable. To meet this demand explains in part why we devised and, during the season of 1886, put out several all-steel machines, the success of which was so complete and triumphant that we decided to confine our manufacture entirely to steel. In doing this, we have not forgotten nor lost sight of the great success that has followed the Esterly during these many years, and we have, as a matter of **ordinary business prudence**, retained the general features in our

NEW STEEL BINDER

that made the Standard Esterly so popular. And now, after having placed several thousand of them in the hands of farmers in nearly every grain growing State and Territory in the Union **during the past three years**, where they have been most thoroughly tested and tried in all kinds and conditions of grain, we feel like congratulating ourselves on the wonderful success with which our machines have withstood the ordeal. Inasmuch as many of the prospective future agents and purchasers may not have seen and compared it with those of other manufacturers, we herewith

80

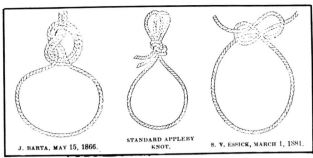

THE VICTORS

In all Contests (including the World's Exposition at New Orleans), where strength, durability, and simplicity of construction were considered. The success of the Plano Machines has been unprecedented in the history of Harvesting Machinery. This is explained by the Machines being

Honest in their Make and Perfect in their Work.

We invite an examination of our Machines for positive knowledge of what we represent. Shall be pleased to promptly answer agents' applications for territory.

ADDRESS, PRINCIPAL OFFICE.

PLANO MFG. CO., COR. W. LAKE AND CANAL STREETS
CHICAGO, ILL.

Top: More knot variants. The knot was the central feature of the twine-type binder (Farm Implement News, 1886); Center: The Plano twine binder, from an advertisement in Farm Implement News, 1886. The Plano Mfg. Company, was established in 1881 in the original Marsh Harvester Works after William Deering moved his operation to Chicago; Opposite Page: Esterly's steel twine binder with folding platform, (Farm Implement News); Right: Advertisement for National Binder Twine in Farm Implement News, February, 1888. The growth of the twine binder business was contingent upon the availability of a reliable-quality twine.

McCormick's Twine Binder

The McCormicks' launched their Appleby-type twine binder in 1881. A year earlier, Cyrus McCormick, Jr., had joined the company. He made the twine binder his first project. The fact that the McCormicks were unable to lead in this development was due in part to the growing disunity between the senior partners, an unhappy state of affairs which led to older brother Leander ultimately relinquishing all interests in the Company.

In 1883, the Company dropped the wire binder altogether and switched entirely to twine-type binders, promptly launching into a price war with Deering.

The story of the later development of the harvester and the binder during the 20 years following the Civil War forms a maze so intricate and perplexing that a clear narrative is virtually impossible. Hutchinson's monumental two-volume work on Cyrus McCormick which, of course, closed with the death of this great man in 1884, is a magnificent attempt to unravel some of the tangle, although often employing others as a foil to better show the hero.

By 1884, the McCormicks' had sold 54,841 machines, including about 15,000 twine binders. Their market share increased to a healthy 20 percent of the harvesting machinery business in the competent hands of the founder's son.

Twine Binders Gain the Ascendency

The speed with which harvesters and binders overtook the reaper is graphically reflected in the tremendous sales growth of harvesting machinery. In 1880, total industry sales of all types of harvesting machinery was 60,000 units. By 1885, this number had multiplied more than fourfold to 250,000 units, even though the acreage sown to wheat had not increased significantly. One hundred thousand "Harvesters", (carrying men to bind by hand) had been put to work by 1879, two-thirds being Marsh types. In spite of the appearance of the twine binders, "Harvesters" and even reapers continued to be produced, although in diminishing quantities.

WHAT WOULD THE FARMERS OF THE PRESENT
THINK OF HAULING THE WEIGHT OF A MAN
AROUND ON THE BINDER?

Above: The McCormicks began building twine binders equipped with the Appleby knotter in 1881. They introduced their light steel binder in 1888. It was lighter by the weight of a man; Below: Adriance reaper-binder with binder attachments, 1886.

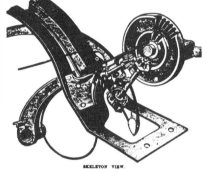

THE McCORMICK
NEW SIMPLE KNOTTER.

SKELETON VIEW.

*McCormick advertisements from the February, 1888 issue of **The Farmers Advance**.*

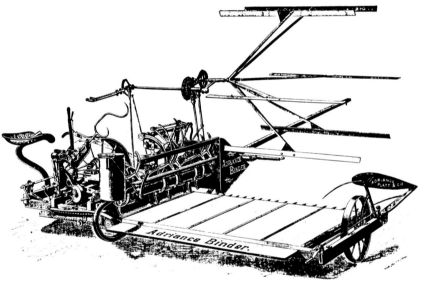

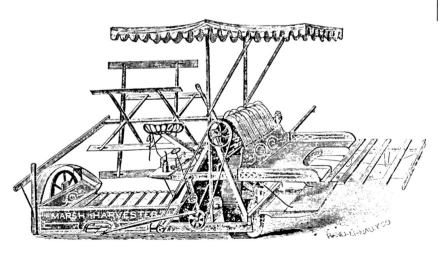

*The 1879 Marsh Harvester (**Farm Implement News**, 1886). The older style of machines continued to sell alongside the self binders.*

Buckeye Banner "Low-down" binder, 1888. This design was said to avoid the need to lift the grain over the drive wheel before tying. Aultman, Miller & Co., Akron, Ohio.

The Toronto Light Binder made by Massey Manufacturing Co., in Toronto, Ontario (Massey catalog 1894).

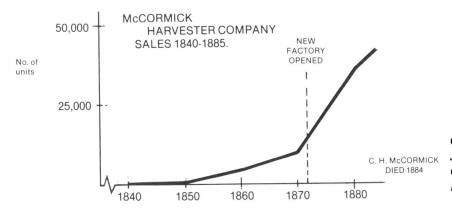

Growth of an industry (1840-1885). Sales of McCormick Harvester Company reapers, mowers and binders.

McCORMICK HARVESTER COMPANY SALES 1840-1885.

No. of units

50,000

25,000

NEW FACTORY OPENED

C. H. McCORMICK DIED 1884

1840   1850   1860   1870   1880

There were attempts to install binders on self-rakers, but binders were not suited to the task. An unusual design was the Hubbard "Gleaner" binder, a machine that picked up the bundles dropped by the self-rakers and mechanically tied the bundles. These enjoyed a limited degree of success in the East where they were simple and cheaper to operate though they did involve a second pass over the field.

By 1888, an estimated 100,000 self-binders and 150,00 reapers and mowers—using 30,000 tons of twine—were being sold annually. The industry was now providing employment for over 30,000 people.

But even as the twine binder foreshadowed the end of the reaper, another development was underway to obsolete the binder—the combine. Still, a great many binders are in use today. The basic Appleby knotter is used on hay balers, still tying round knots using a billhook, but of a sophisticated pattern incorporating modern manufacturing processes. It can also tie hempen or plastic-based twines (Crawford & Old 1972).

A sea of grain binders at work on a "Bonanza" farm in Dakota Territory in the late 19th century.

Artist's drawing of the Michigan combine. The drive wheel visible at the rear is out of proportion because it was known to be about 7 feet high. The original reel was a wooden cylinder into which nails were partly driven. These protruding nails clawed the grain back toward the sickle.

"Mr. Moore, I can invent, but I can't drive the horses."—Hiram Moore to A. Y. Moore, 1837.

10

Hiram Moore and the Michigan Combine

Development of the combined harvester-thresher or "Combine" antedated even the reaper. Samuel Lane of Hallowell, Maine, patented the first combine. Lane's patent on a traveling thresher with harvesting attachment was issued August 8, 1828. Combine patents were issued to Ashmore and Peck in 1835, and Briggs and Carpenter in 1836. Unfortunately, few details of these machines exist—the 1836 Patent Office fire saw to that. There is no record indicating the results of practical trials of any of these machines.

The first workable combine was the Moore and Hascall machine, built in 1834, and patented in 1836, only two years after McCormick's celebrated reaper.

Hiram Moore, son of a New England stonemason, was born in New Hampshire, July 19, 1801, and came to farm in Western Michigan in 1831. His brothers Lovell and John followed him to the newly-opened wilderness. They settled in Kalamazoo County, in a village Hiram called Climax.

Two other men played important roles in this story: John Hascall, a lawyer practicing in New York, became enmeshed in an exposé of Free Masonry in 1820 and found it prudent to leave. Moving west, he settled in Kalamazoo where, finding it difficult to earn a living practicing law, he took up farming.

Andrew Y. Moore, another Michigan pioneer, a neighbor but unrelated to Hiram, brought the first gang plow to Michigan. He later went on to found the Michigan State Agricultural College (now Michigan State University). He recalled John Hascall:

Briggs and G. C. Carpenter's combined harvester-thresher invented at Fort Covington, Kentucky, and patented on February 5, 1836 (Butterworth 1888). The rear pair of the four wagon wheels provided ground wheel drive to the thresher. This 8½ foot machine was demonstrated in Upstate New York in 1836.

"On viewing Prairie Ronde, in Kalamazoo County, of some 20,000 acres, seeing that it was good for wheat and thousands of acres uncultivated, he, Hascall, thought if he had a team perhaps he could hold the plow and put in wheat. He knew he could not harvest it, and there were no men to hire. He spoke of it to the family, and in consequence it caused his wife to dream; so one morning thereafter she stated to her husband that she saw in her dream a large machine going over the prairie drawn by horses and harvesting wheat, and described its motion and appearance.

Mr. Haskall (sic) related the dream to Hiram Moore of Climax Prairie, knowing him to be of inventive turn of mind.

Hiram asked him how he would have it operate, Mr. Haskall replied, holding out his hand with fingers extended, he would run it through the grain, and with the other hand draw over and backward, he would cut it like that. Hiram did not intend to give it much thought, but it troubled his mind for six months, when he concluded he would put his mind seriously upon it and succeeded in the invention, perfect as he thought, and made a model and took it to Washington City, exhibited it at the patent office and obtained a patent. This was in the year 1834. By the harvest of 1835, he, at Flowerfield, Michigan, made a temporary machine, and went into a field for a trial. It cut only two rods and broke something. The thresher was not in; he merely wanted to try the cutting process. In the failure he said, 'I see the shore afar off and it will take a long time to get there, but I will succeed in time'. The next season he intended to perfect a machine for further trial. My brother Abner and I came to Prairie Ronde in October, 1835, he asked me to let three acres stand for trial.'' (Higgins 1930).

Although it was late in the season, the harvester cut and threshed satisfactorily. A cleaner had not yet been incorporated. A. Y. Moore made a comparison of costs of combine harvesting against the traditional hand methods in 1835:

'' . . . at the going rate of prices for men's labor and the hire of horses. Twelve horses were worked on the machine at that time, besides the team hauling to the barn. I took the exact time for cutting the three acres, and made the actual cost of

Left: A. Y. Moore; Right: Hiram Moore in a photo taken about 1854; Bottom: The drawings from the US patent issued June 28, 1836 to Hiram Moore and John Hascall for "Improvement in Harvesting Machines for Mowing, Thrashing, and Winnowing Grain at One Operation."

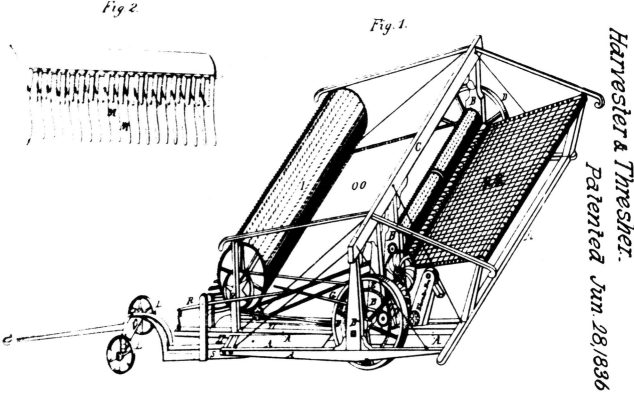

Moore & Hascall.
Harvester & Thresher.
Patented Jun. 28, 1836

82 cents per acre. I inquired of my brothers, who were farmers, as to the cost of harvesting and threshing in the ordinary way of cradling, raking and binding, shocking, stacking, threshing, cleaning, etc., and upon strict calculation it cost in that old process $3. 12½ per acre. The contrast was so great that I took an active part in its future. By the harvest of 1837 it cut only 20 acres, and he found it wanted further perfection, although the threshing and cleaning had been added. Before another harvest he said to me: 'Mr. Moore, I can invent, but I can't drive the horses.' I replied that I would drive the horses and assist him, and I did so each year."

Hiram Moore evidently experimented with five machines in all. Financial support was obtained in the early years from one of Michigan's first Senators, Lucius Lyon. Lyon, a New Englander, came to Michigan as a land speculator, surveyor and promoter. The tone of a letter to Hiram Moore in May of 1839 was apparently intended to stir the inventor to further effort:

"I have very little expectation that it will ever be worked to any advantage anywhere, and I would be very glad to have my money back for my share of the invention. Not that I do not believe that grain may be harvested and threshed by machinery cheaper than it has ever been done by hand, for I do believe it, and furthermore I think the principle of your machine is correct and that it will lead to important results; but a machine to be useful on a farm must be far lighter and more manageable than the one I have been removing. It is too heavy and unwieldy for the average field . . . ".

Top: Moore-Hascall machine believed to be operating in California in the 1850's (Courtesy A. Prauss, Kalamazoo Public Museum, Kalamazoo, Michigan); Below: One of the five Moore-Hascall combines operated in Michigan in the late 1840's. This picture is from one half of an old stereopticon slide. High-hatted Moore is sitting atop the machine and Lucius Lyon is believed to be the man with hat at the rear of machine (F. Hal Higgins Library).

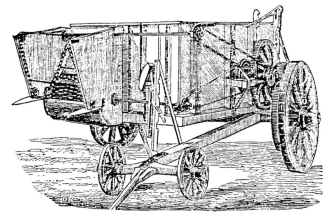

Illinois inventors, Churchill and Danford's combine or "Harvesting Machine." The 7-foot gathering and threshing apparatus is on the far side. Bars on a chain reel bent the grain over the upper edge of the cylinder box where it was instantly threshed by the cylinder. A man shoveled the grain back into the body of the wagon. The left rear wheel provided the drive through gearing (The Union Agriculturalist, August, 1841).

But Hiram Moore was no trifler. He, like Lyon, was a Yankee. He went to Lyon's farm at Prairie Ronde determined to perfect his machine and by November 1839 had Lyon reporting:

" . . . There is no longer any doubt of the success of the Moore and Hascall's harvesting machine. Mr. Moore has had a machine in the field on Prairie Ronde in this county during the past summer which harvested and threshed 63 acres of wheat in very superior style and could have harvested 250 acres with greatest ease, at the rate of 20 acres per day, had it not been for one or two trifling accidents . . . Twenty of the 63 acres were harvested on my farm and every expense attending it does not exceed one dollar per acre. A great

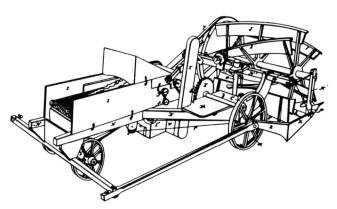

D. A. Church of New York: "Harvester and Thrasher" patented May 4, 1841.

Darling's combine, Adrian, Michigan, 1846 as illustrated in Butterworth (1888). First tee-shaped configuration.

number of farmers witnessed its operation. All are entirely satisfied with its complete success and many, in sowing their wheat this fall, are calculating largely on the benefits to be derived from it next year. I have within the last three or four years, advanced to Mr. Moore between three and four thousand dollars to enable him to bring the machine as near perfection as possible, and am much gratified at the result of his labors."

A year later, Moore transferred to Lyon and a Mr. Robinson the major share (9/16) of all subsequent improvements he might make on his machine, as security on the money that Lyon had invested. In 1841 Hascall assigned his entire interest in the harvester to Lyon for $100 in cash, $200 in personal notes, and a promise of $5,000 from the profits from the sale of the machine. (Chase 1929). Hascall was no longer involved but his name is perpetuated as co-inventor in the 1836 patent.

In the November 17, 1839 letter noted earlier, Lyon asked his friend, Commissioner of Patents Henry Ellsworth, whether certain improvements made by Moore might not be patented without furnishing a model:

"The following are some of the improvements made by him on said machine in the winter of 1836-7, and publicly tested in the neighborhood of Rochester, in the summer of 1837:

1. A new principle or mode, and fixture for throwing the machinery into and out of gear; which may be understandingly exhibited by plates or drafts.
2. A new method and fixture for starting and operating the sickle or cutter.
3. Revolving racks or endless aprons filled with teeth or spikes for bringing the grain to the cutter, and when cut conveying the cut grain from the gathering cylinder to the thresher, when the gathering cylinder is used to aid in gathering and cutting.
4. The revolving wire or network screen for separating the threshed wheat from the straw and carrying off the latter from the machine.
5. A fender or retainer to prevent the loss of such wheat or grain as may be thrown over forward of the machine by the gathering apparatus.

All these improvements may be as well represented by drawings as in any other way, and at the request of Mr. Moore, I write to enquire whether they cannot be patented without furnishing a model of them? I also beg leave to enquire whether a model of the original machine, as it was patented,

cannot be supplied at the expense of the department? If it can the improvements can be added to it with very little additional cost."

Hiram Moore continued to experiment and in 1841, added an angle-edged sickle which would operate the entire harvest without regrinding. His original sickle was said to be "straight, with teeth cut in sections each way."

Interest in the combine began to mount, as A. Y. Moore related:

"In the fall of 1841 Mr. Moore went to Rochester, New York, and procured good mechanics and he completed two machines. I used one for him and Ira Lyons for the harvest of 1842. In the spring of 1843 I bought the Bates farm, near Schoolcraft, and moved there on the last of March. I operated his machine as usual, and in that season had the only complete machine built for myself. It was drawn by 16 horses, hitched two abreast, walking by the side of the grain cutting 10 feet wide, threshing, cleaning and bagging the same, doing 25 acres in a day."

Lucius Lyon continued his support, for in August, 1841 he wrote:

"We have had our harvesting machines in the fields during the past harvest and mine has harvested 150 acres without much delay for alterations . . . but the operation of cutting, threshing and cleaning the grain in the field, all at one time, is so complex and the harvest season, which is the only season for experiments, is so short that it will require some years yet to perfect it so as to make it profitable."

In August, 1841, Lyon asked Arthur Bronson, a partner in past business ventures, to join him in manufacturing the new machines. Bronson declined. Here is part of Lyon's appeal to Bronson, citing the effectiveness of the harvester:

"When the machines are driven with an ordinary degree of care nearly every grain of wheat is saved, while under the old method fully one-fifth was lost. Ira Lyon operated one of the machines and paying all expenses cleared about $300, which is more than 50% on the cost of the machine. In addition to saving one fifth of the crop, he harvested and threshed at $3 an acre, while the usual cost was $5 an acre. The machines will work well on any ground that is free from large stones and stumps and may be operated by any man of ordinary common sense after two days' experience." (Chase 1929).

Then Lyon hit rock bottom financially. Embezzlement

by an Iowa land agent forced Lyon to terminate his extensive agricultural ventures. He remained interested in the machine in which he had invested heavily and continued to act as intermediary for Moore with the Patent Office, advising him in 1844 to patent each of his improvements separately.

The last information on the relationship between Hiram Moore and Lyon is in an 1845 letter which Lyon wrote introducing Moore, who had gone south for his health, to Governor Mocton of Louisiana:

"He has done for the grain growing regions of the north what Whitney and his cotton gin have done for the south . . . He reduced the cost of harvesting and threshing and cleaning grain from more than three dollars to about one dollar per acre, and for this he is destined to be and deserves to be ranked in the first and noblest class of our country's benefactors." (Stuart 1897).

James Fenimore Cooper, speculator, author of classical American novels, and contemporary of Lyon, saw a Moore combine at work in 1847 or 1848. He wrote in "The Oak Openings" :

"The peculiar ingenuity of the American has supplied the want of laborers, in a country where agriculture is carried on by wholesale, especially in the cereals, by an instrument of the most singular and elaborate construction. This machine is drawn by sixteen or eighteen horses, attached to it laterally, so as to work clear of the standing grain, and who move the whole fabric on a moderate but steady walk. A path is first cut with the cradle on one side of the field, when the machine is dragged into the open place. Here it enters the standing grain, cutting off its heads with the utmost accuracy as it moves. Forks beneath prepare the way and a rapid vibratory motion of a great number of two-edged knives, effect the object. The stalks of the grain can be cut as low or as high as one pleases, but it is usually thought best to take only the heads. Afterwards the standing straw is burned, or fed off, upright.

    The impelling power which causes the great fabric to advance, also sets in motion the machinery within it. As soon as the heads of grain are severed from the stalks, they pass into a receptacle where, by a very quick and simple process, the kernels are separated from the husks. Thence all goes into a fanning machine, where the chaff is blown away. The clean

Old timers who worked on Moore's machine got together in 1930 and are seen holding original parts of the combine recovered in Wisconsin in the 1880's (F. Hal Higgins Library).

Hiram Moore wearing his "plug hat" is seen with his force-feed grain drill in Wisconsin about 1870 (F. Hal Higgins Library).

grain falls into a small bin, whence it is raised by a screw elevator to a height that enables it to pass out at an opening to which a bag is attached. Wagons follow the slow march of the machine, and the proper number of men are in attendance. Bag after bag is renewed, until a wagon is loaded when it at once proceeds to the mill, where the grain is soon converted into flour . . . As respects this ingenious machine, it remains only to say that it harvests, cleans and bags from twenty to thirty acres of wheat in the course of a single summer's day! Altogether it is a gigantic invention well adapted to meet the necessities of a gigantic country."

Factors Hindering Adoption of Combine Harvesters

Hiram Moore recuperated in Louisiana. In 1850 Democrat Moore represented Kalamazoo County on the State Legislature. That same year the original 14 years of the patent expired. The Governor of Michigan and The State Legislature petitioned Congress and the Patent Office for renewal. However, Moore's application for an extension became entangled in the controversy between Cyrus McCormick, Obed Hussey and others seeking similar patent extensions. These men, all vigorously involved in the machinery business, had less at stake in the extension than Moore who was not in the business of selling machines. Legislative battles were waged in several State Legislatures and for a brief time even pushed slavery aside on the floor of Congress. But the extension bill died as a result of intense vocal opposition by rural lobbyists who aroused fears that patent extensions would bring higher farm machinery prices. (Hutchinson 1930).

While other men with vision could have developed a combine in the 'forties—Churchill and Danford of Illinois, Church of New York and Jeremiah Darling of Adrian, Michigan (their machines were certainly noteworthy)—none had the financial backing or public support of the Moore-Hascall machine. Darling's machine was, however, the first combine employing the T-shaped symmetrical feeding platform configuration.

These early Midwest combines would work well only if the crop was ripe and dry, the very conditions that would shatter grain as ripe heads were pulled in. Farmers would hesitate to combine earlier when the crop was green but less shatter-prone. Today's shorter-season varieties of cereals were unknown. Finally, the climate was not uniformly dry at harvest time, as was the case in California.

If Hiram Moore had accompanied the machine which A. Y. Moore was to send to California in 1853, he could have shared in the eventual triumph of the combine a decade later. But he was discouraged by his combine's lack of acceptance in Michigan. In 1852, he moved to Brandon, Wisconsin, and farmed 600 acres of fine prairie land. He continued to innovate—building hayrakes, pumping windmills, corn planters and he resolved some grain harvesting problems as well:

- A method of grain cleaning to provide a weed-free crop;
- a grain drill with depth control for even stands of wheat (Patented 1860);
- ventilating screens in the bottom of his granary to prevent heating and spoilage in stored grain.

One of Michigan's lasting tributes to Hiram Moore is the bronze plaque on a boulder at the site of the original Moore farm near Climax, Calhoun County.

As for his combine, contemporaries later recalled for noted agricultural machinery historian, F. Hal Higgins, that Moore was always rebuilding and changing his combine between seasons. He used it season after season, but only for the standing wheat. He harvested only 100 acres a year, but at times managed up to 40 acres a day.

The inventor enjoyed nothing as much as putting on a show for the neighbors using the great machine. In the morning, the first wheat harvested would be rushed over to the mill, ground, brought back to the house and made into biscuits for supper, to be eaten on long tables set out on the lawn.

Hiram Moore died at Brandon, May 5, 1875 at the age of 73. Moore, like many another inventor was ahead of his time. Well over half a century would pass before a combine again harvested grain in Michigan. The Moore-Hascall story was resumed in the West, where it is most fittingly described in Cooper's words as "a gigantic invention . . . to meet the necessities of a gigantic country".

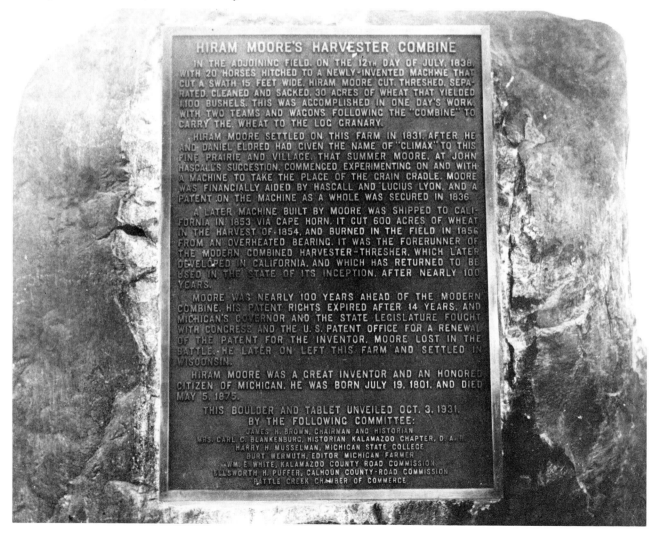

"We claim that one-half of the expense of harvesting would be saved to the Farmer by using the Harvester: in fact, the entire expense of Threshing is saved. Farmers! Come and see if our claims are well founded. John M. Horner, Wm. Y. Horner."
(Portion from notice of the first advertised public demonstration of a combine harvester in history, in California Farmer, August 1868).

11

California's Leviathans

John M. Horner, Bonanza wheat farmer of Mission San Jose, California in 1854. A year earlier, he had invited A. Y. Moore to bring his Michigan combine to California (Higgins 1958).

Gold! Its discovery in California in 1848 was the biggest single factor in settling the American West. In the 12 years following the Sutter strike over 300,000 people flowed into California. When the gold played out, most of the new arrivals, farm hands, returned to the land and began a new bonanza for California—agriculture.

James E. Patterson was one of these men. He came from upstate New York and in 1852 wound up working on the Horner farm at Mission San Jose, Alameda County. At harvest time, young Patterson saw Mexican Indian helpers on their knees cutting grass and grain with sickles. Purchasing a grain cradle, he proceeded to earn $10 a day, or four times the wage earned by the Mexicans.

He made friends with one of Horner's neighbors, Lyman Beard, whom the former New Yorker told about the Pitts threshing machines "back East". They decided to build their own harvester. Given permission by Horner to construct a machine, Patterson worked feverishly to get the combined harvester-thresher ready for the 1853 harvest. The first time the contraption was thrown into gear, however, the 22 draft mules, frightened by the noise behind them, bolted and tore the machine to pieces. Patterson gave up, but his employer did not. Horner invited A. Y. Moore to bring his Michigan combine to California.

Moore's Michigan Combine in California

Moore had his combine built in 1844 after the Moore-Hascall pattern and had used it every season near Climax, Michigan, until he accepted Horner's invitation in 1853. He sold a part interest in the venture to George Leland and in 1854 loaded the combine on a clipper ship bound for San Francisco via Cape Horn. Moore dispatched his son, Oliver, and partner Leland overland to California with six of his best horses. In 1854, they were able to harvest 600 acres of wheat for farmers Horner, Beard and Brifogle near Mission San Jose. Times were hard, however, and none of the farmers could afford to pay their harvest bill that year. The venture had cost Moore and Leland $3,800. Short of cash, they sought work in the goldfields and stored the machine through the 1855 season.

In 1856, a new crew was contracted to harvest with the machine, but the "green" operators failed to lubricate the bearings properly. One of the bearings overheated and started a fire which destroyed not only the machine but a wheat field as well. Thus the year 1856 marked the first combine fire in history. It also sent a disconsolate Oliver Moore back to Michigan. "Thus ended the

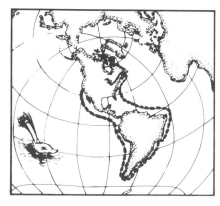

*Above Left: Before the advent of
the transcontinental railroad,
goods were transhipped via Cape Horn. A.Y. Moore and his Michigan combine took this route to California;
Above Right: Andrew Y. Moore (1802-1878) and his sons at home in Michigan. Seated on the left is Oliver
Kidwell Moore, who was dispatched to California to supervise the Moore machine (F. Hal Higgins Library).*

NOTICE!

There will be a Public Exhibition of the

Traveling Harvester,

MONITOR NO. 2,

Upon the farm of RICHARD THRELFALL, in Livermore
Valley, Murray Township, Alameda County, on

Thursday, August 28, 1868,

Commencing at 1 o'clock P. M.

☞ On Thursday, the 3d day of September, commencing
at the same hour, MONIROR No. 1 or 3 will be exhibited to
he public upon the farm of WM. Y. HORNER, near the
Mission San Jose, Alameda County.

We claim that one-half of the expense of harvesting would
be saved to the Farmer by using the Harvester; in fact, the
entire expease of Threshing is saved.

Three men and twelve horses have Cut, Threshed, Cleaned
and Sacked, in good, workmanlike manner, fifteen (15) acres
of grain per day—making five acres per man—a feat, we believe, never performed in America before! One and three-
quarter (1¾) acres to the man, working with the most approved machinery, is about the highest figure yet reached—
one acre per man being nearer the general average.

☞ Farmers! come and see if our claims are well
founded.   JOHN M. HORNER,
v80-5   WM. Y. HORNER.

$500 REWARD

Will be paid by the under-igned for the arrest and conviction of the person or persons who caused the burning
of Monitor Harve-ter No. 2, in Mr. Wilder's field, at
the point of timbers, San Joaquin Valley, Contra Coasta
County, on the night of July 14th, 1869.
                                JOHN M. HORNER,
                                WM. G. HORNER,
                                WILLIAM HORNER.

*Left and Above: Advertisements
placed by the Horners in the
California Farmer.*

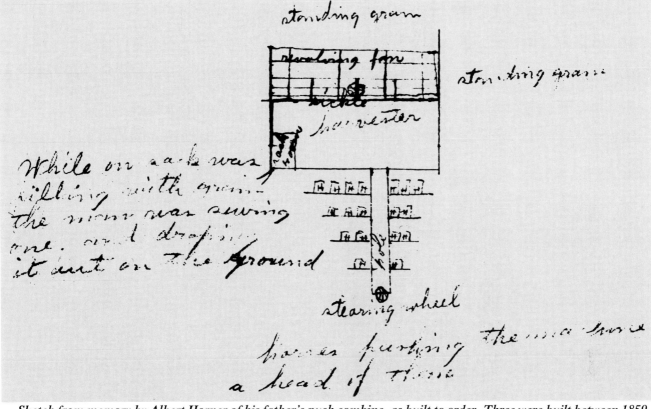

Sketch from memory by Albert Horner of his father's push combine, as built to order. Three were built between 1859-1867. The resemblance to Moore's design is apparent. The machine cut, threshed, cleaned and sacked the grain in one operation at a rate of 16 acres a day using a crew of three (F. Hal Higgins Library).

machinery business with me" recalled his father A. Y. Moore (Higgins 1930).

John Horner had captured the vision and was determined to build his own combine. The pull-type Michigan machine required drivers handling the jerk lines to walk beside each pair of horses on the draft team. Horner now placed the team behind the machine, reducing manpower and lessening the chance of a runaway. Three machines were built between 1859 and 1867, using the same functional principles as Moore's Michigan combine. The first one reputedly cost Horner $12,000 to build. It could harvest 16 acres a day with a three man crew. The Horner combines harvested not only Horner's grain, but were also on contract in Contra Costa and San Joaquin Counties where wheat was under extensive cultivation in the 1860's.

Bonanza Farming

California, with a 16-billion bushel wheat crop in 1869, ranked eighth among states in wheat production. It took just another 11 years to make California number one wheat growing state in the Union. Perhaps it was the philosophy of life that developed during the gold rush that inspired the "bonanza" farm operations of the era. Wheat was grown on a scale never seen before. The young, eager-to-get-rich-quick population had plenty of room to cultivate and wheat fetched $3 per 100 lb bag in the 1870's. California's "bonanza" farms—so named for their enormous size—were found to require management and capital the equal of any

industrial enterprise. Machinery cost represented only a small fraction of total expenses. Big combines were built by and for these big spreads—combines which could harvest as much as 2000 acres a season at rates of 50 acres a day. Big as they were, they were comparatively few in number. In 1886, 10 percent of the Californian wheat crop was combine-harvested. It was the header, also developed in the Midwest, that clipped the rest.

Headers Reach the Pacific

In 1858, McCormick's agents in the harvest fields of Illinois wrote:

"Haines' Headers manufactured at Pekin are cracked up mightily, but there is little danger (competitively) until they are able very largely to increase the manufacture" and, later: "When the fields are reckoned by 100's of acres, they almost uniformly purchase Haines' Harvester." (McCormick Business Letters 1858).

The header eliminated the need for binding. Fewer men were required. With these push machines there was no sidedraft, so wider cutterbars could be used. Cutterbars ranged from 10 to 18 feet in width. Headers typically cut only the top 8 to 12 in. of straw and so more capacity was obtained from threshing machines. In spite of these advantages, the header made little impact in the Midwest because of the weather: the reaping and binding method allowed the harvest to start five or six days earlier, ensuring grain of better quality with less harvesting loss. The header, which could only be used when the crop was ripe, was better suited to the drier

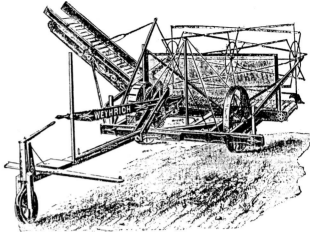

Upper Left: Haines' Illinois Harvester in California, 1872. Harvesting with a header used fewer men than binding, but the unit was more expensive and required more horses than a reaper or binder (Pacific Rural Press 1872); The headers on this page and the Benicia (1883) header on the top of page 95 were made on the Haines pattern. The vertical pole just ahead of the steering tiller was often marked with a scale to enable the header-puncher to gauge the cutting height. By reefing on the adjacent lever, the cutterbar was tilted up to cut higher. Clockwise from top right: Weyrich (1883), Case (1884), McCormick (ca 1890), Randolph (1882).

regions of the Pacific Coast and on the vast expanses of the Great Plains. By 1869, reaper production exceeded 60,000 units, while 2,500 headers were produced for this growing Western market.

The average header, with a 12 foot cutterbar, could handle three "barges" or header boxes hauled alongside and could harvest 15 to 25 acres of wheat per day using a four-horse team. The driver, known as the "header puncher" was located over the rear pivot wheel. He half stood, half sat, with the tiller between his knees and his feet braced on the small platform. A skilled "puncher" had little trouble steering the outfit so that it took a full bite of the waving heads of wheat.

The objective was to take in as little straw as possible. Cutting too low added straw which took up space in the barges and slowed down the threshing crew. The header was invariably driven counter-clockwise around the field. That meant the left hand team was tied more tightly than the right team, allowing the left team to swing the machine around the turns.

The "cockeyed" look of the header boxes or barges was functional. One side was built lower than the other so that the driver of the barge could maneuver his wagon under the moving header spout. The headed wheat was leveled by a forker or "loader" as it cascaded from the elevator spout. A good wagon driver made it easier on his loader by gradually moving his wagon forward of the spout during loading. Two wooden rollers were fitted under the spout for protection in case of careless driving or if a team got out of control. On hilly land, the axles of the so-called "Palouse Wagon" were extended up to 12 feet. Occasionally the wagon had to be chained to the header on very steep areas.

Driving with the knees. Header at work with "barge" or header box alongside. The machine was mounted on three wheels. The front left wheel or "bull" wheel powered the works. The tiller between a header-puncher's knees was used for minor course corrections to ensure a full cut. Note the peculiar shape of the wagon box, designed to accommodate the elevator spout (Deere and Company).

The plan of attack for heading a field was unique to this machine. The header worked from the inside out. Most harvest operations are just the opposite.

The header-puncher got the harvest going by "taking a bead" on the center of the field, then cutting straight through to that point. He would start cutting out in a circle, eliminating corners. Where the wagons followed to the center, a proportion of the wheat would be trampled. To avoid this, some farmers attached a huge canvas bag to the elevator spout to catch the grain until there was sufficient operating area for the barges and teams to accompany the header. A binder might even be used to start the field if the field was so large that a bag could not hold the heads cut during "opening up".

Headed wheat was either stacked loosely from the barges or pitched directly into the threshing machine, the same as bound wheat. If the farmer was going to thresh from stacks of headed wheat, he built the stacks in the hub of the field, thus minimizing the distance the header boxes traveled. (Higgins 1967).

A development peculiar to the Pacific region was the use of a derrick and forks to facilitate the building of large stacks. The Jackson fork for example, was first made in 1872 and 700 were sold during the next eight years. The addition of a net under the load in the wagon box also saved labor by enabling the derrick to unload the entire box at once by pulling up on the net which was fastened to the other side of the box. The practice

Harvesting on the Upper Ranch of Abraham Clark, Berryessa Valley, Napa County, California in 1878. The three headers and six wagons required a 28 man crew, a fairly large outfit in those days. The usual procedure for heading wheat involved cutting from the center to eliminate corners. This also enabled the heads to be hauled directly to the threshing machine located in the middle of the field (F. Hal Higgins Library). On one of Dr. Glenn's ranches, a locally built thresher processed a record 6,750 bushels in a day, but the average was half this amount. On that particular harvest, the thresher was fed by seven headers and 21 barges employing 56 men using 96 horses and mules (Pacific Rural Press 1876).

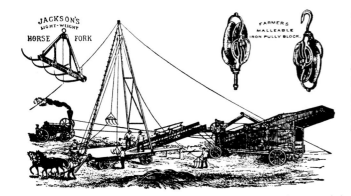

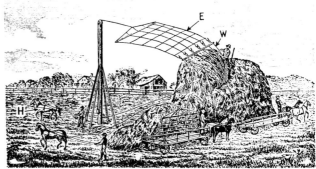

Top and Center: The Jackson Fork and Jackson "Acme" Overshot Stacker for unloading barges and stacking headed wheat, 1885 (Pacific Rural Press 1885); Bottom: Kelley's grain stacker, 1878. Horse H drew a load of up to 1½ tons to the side of the stack. The hooks in the extension net E were attached to the rings in the wagon box net W prior to lifting to the stack (Pacific Rural Press 1887).

was not widely adopted then, but has found a use today in sugarcane country. (Rogin 1930).

Some of the biggest operators headed and threshed simultaneously. Dr. Glenn, the biggest wheat farmer in his day, owned over 100 square miles of Colusi County and for his record-breaking harvest of one million bushels of wheat in 1880, he had 600 men and 800 draft animals at work on his farms. Later, when the price of wheat dropped to $1 per 100 pounds, he switched from headers to combines to save labor and horses and to cut costs.

A New Industry Out West

The combines used and built by the bonanza farmers paved the way for a new industry. In the 30 year period 1858-1888, at least 21 combine manufacturing ventures were started in the Pacific region. In Oregon, the Davis Brothers built their first combine in 1862. When they displayed the machine at the Oregon State Fair in 1867, their exhibit "took the crowds away from the bearded lady". It was a push-pull job, with four horses ahead and six horses astern, and boasted a capacity of 1,500 bushels a day with a three-man crew.

The earliest combine builders in California included such names as Marvin and Thurston, Young, William Patterson, the Housers, Matteson and Williamson, Myers, Shippee, Best, and the Holts. In 1880, Daniel Houser of Stockton started the first factory exclusively devoted to the mass production of combines. A Houser horse-drawn outfit cost $2,000 in 1882. L. U. Shippee, Stockton's banker and a livestock farmer, put together an enterprise that was to eventually make Stockton the first "combine capital of the world". Shippee's first 90 machines were built in 1883, but in 1884 a number of suits were brought against him, charging poor per-

formance. The first suit was settled in favor of plaintiff, John Fox, who was granted a $6000 settlement.

Shippee had purchased the patents of the Davis Bros., whose Oregon Iron Works had been destroyed by fire, and those of another Oregon combine builder, The Warrish "King". To these he added those of several "lone wolf" combine makers, each offering something a little different. The Houser, Benton, Powell, Grattan and Minges machines all wound up being taken over by Shippee's "Stockton Combined Harvester and Agricultural Works".

The Hay Burner Replaces the Hay Eaters

The slump in the price of wheat in the 1880's convinced Pacific grain farmers that their share of the export market could only be maintained by reducing production costs. The biggest costs were harvest labor and horses. Now there was greater interest in combines, which were better constructed and more reliable. Steam power had been used to pull some early ground-driven machines but the first time an auxiliary engine was used to drive a combine was the engine-functioned machine built and exhibited by B. F. Cook at the 1871 California State Fair in Napa. He put together a Haines Header,

*Top Left: One of the first combined harvester-threshers built by Daniel Houser in the first factory devoted exclusively to combine production, Stockton, 1880 (**Farm Implement News** 1888); Top Center: Benton's combine was another machine taken under the Shippee umbrella at Stockton. Note the long crop lifters for lifting lodged straw up to the cutterbar (F. Hal Higgins Library); Top Right: An 1885 Shippee combine built at the "Stockton Combined Harvester and Agricultural Works" (**Farm Implement News** 1928); Right: The Stockton Combined Harvester and Agricultural Works was established by L.U. Shippee in 1883. He introduced the idea of standardized parts for combines (F. Hal Higgins Library).*

Left: The 1885 Minges' "Combined Header and Thresher" model was another of those taken over by Shippee (F. Hal Higgins Library); Below: An 1884 Shippee combine in the San Joaquin Valley. Shippee had bought out the interests of Oregon combine builders and built 90 of these 14 foot cut machines in 1883, but not all brought satisfaction to their owners (F. Hal Higgins Library).

Above Left: First engine-functioned combine. The horses were only used to haul the machine. The auxiliary steam engine powered the outfit. Butterworth's illustration is believed to show the 1871 B.F. Cook machine; Above: Combine Harvester built by the Benicia Agricultural Works could harvest 50 acres a day in 1889 when pulled by a Buffalo Pitts steam engine. Note the double spark arrestor cap over the smoke stack and the horse-drawn water tank being pulled alongside the outfit at Stockton (F. Hal Higgins Library).

a Pitts' separator and used a steam engine to drive the thresher. Horses were used only to pull the machine and fewer were required since the draft was much lower.

Another combine "first" was the use of oil for fuel in a steam plant on Pritchard's engine-operated combine built in California by Judson Manufacturing Co. in 1885. A 12 horse team was used to pull this oil burning outfit.

Meanwhile up in the Sacramento Valley, a farmer, George Stockton Berry, appeared with a steam combine that established more "firsts" than any other combine builder before or since.

- First self-propelled combine.

- First straw-burning steam combine, "fuelled off the land".

- First tractor to work both forward and backwards—when he finished the harvest he used the tractor for plowing.

- Largest header at the time. (40 feet in 1888).

- First machine to harvest over 100 acres in a day and to use lighting to work after dark.

Photo of the world's first self-propelled combine, 1886. Capacity of the first 22-foot cut machine, which was used on G.S. Berry's farm, was 50 acres of grain cleaned and sacked per day (F. Hal Higgins Library).

*Berry's steam combine (1888) with 40 foot cut fueled by straw from the separator (**Farm Implement News 1888**).*

Berry's Self-Propelled Combine

Ex-Missourian Berry of Lindsay, California, toiled five years (1881-1886) on his 4,000 acre farm to develop his epochal machine. He had experimented with steam for motive power on other farm jobs before he conceived the idea of a self-propelled combine. He purchased the parts in San Francisco and the engine in Oakland. The main propelling unit was built around a reversed Mitchell-Fischer 26 HP steam engine. A separate 6 HP Westinghouse steam engine powered the thresher, separator and header. Both engines took steam from the same boiler.

This arrangement, a "live power take-off", enabled the threshing machinery to operate steadily and independently of the tractor's forward motion. The main drive wheels, 5½ feet in diameter and 2 feet broad, were located under the boiler to take most of the weight. The steersman sat in front of the engine and controlled the machine by an endless chain gear connected to the rear "rudder wheels".

The 22-feet wide Benicia header frame was supported by an outrigger wheel. The cutterbar could be raised to cut at 2 feet or lowered to ground level. The Young-built separator was attached to the left hand side of the tractor chassis and had its own support wheel. Thus, there were six ground wheels in all. The threshed straw from the separator fell into a chute and was delivered into a trailed platform, or rack, adjacent to the boiler where it was stoked into the boiler grate for fuel. The use of straw as fuel considerably reduced Berry's operating expenses. Wood and coal were scarce and expensive in the district. An exhaust fan assisted the draft up the stack, built tall to minimize the risk of fire from sparks.

A supply of straw was deposited on the rack at night ready for the 30-minute task of getting up steam the following morning. As the rack filled during the day, the excess straw was diverted and dumped on the ground. Straw was also utilized as fuel when the tractor was used for tillage. The combine components were removed and the tractor, now running forward, could easily pull a 20-bottom plow.

In 1888, Berry rebuilt the machine. He increased the cutting width to 40 feet and added headlamps for night work. He wrote:

"The machine I built this season for my own use . . . is as near perfection as a machine for that purpose can be made. The ground wheels are four feet face and six feet diameter. My header cuts a swath forty feet wide; it is made in two sections, and so arranged that it is handled with ease on very rough "hog wallow" lands, at the same time making a clean cut wherever it goes. The separator is of sufficient capacity to handle all the grain that can be got to it. I have averaged this season about ninety-two acres per day. I cut in two days 230 acres. It does not take any more men than I used last year to handle it, and it does about twice the work."

Local records indicate that four of Berry's combines were subsequently built by Benicia Agricultural Works at a cost of $7,000 each and were used by big grain growers until 1902. Berry's expansive outlook led him into politics. He was elected to the California Senate to represent Inyo and Tulare Counties in the 1890's.

The Best and Holt Era

Daniel Best, an Oregon patternmaker, became interested in combines after he was hired to help build one. He decided to move to California where he could be a part of the action. He started the Best Agricultural Works in San Leandro in 1884 and by 1886 had 15 combines at work. He continued to build the grain cleaners that he had sold successfully in Oregon and in 1888 designed and built steam engines to power his

*Best began manufacturing steam traction engines of 30, 40 and 50 horsepower in 1890 (**Pacific Rural Press 1890**).*

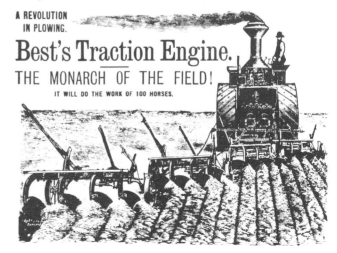

This graphic illustration by Ben Sharpsteen of Calistoga, California, shows four Holt outfits at work. The horseless carriage in the foreground is a 1900 Locomobile. The brash young driver is trying to reignite the burner. His carelessly discarded matches have started a small fire in the stubble (F. Hal Higgins Library).

machines. When he built his first steam powered combine in 1889 (the selling price was $4,500), he stayed a week at the owner's farm to see that the new owner was satisfied with its performance. The selling price was $4,500.

The Holt Brothers had an established wagon wheel business in Stockton. In 1885, they started experimenting with combine harvesters. Early combines used the "tight gear" method of drive, which transferred power from the master gear on the bull wheel to the threshing cylinder and separator by constant-meshing pinions and countershafts. The results were less than satisfactory. Gear wear was rapid and costly to repair and the machine was apt to be demolished during a runaway. The first Holt combines of 1885 introduced hook-link chain and segmented V-belt drives instead of gears.

In 1889, the Holts bought out the Stockton Shippee interests. The first combine exported from the Holt factory was shipped to Melbourne, Australia in 1894, although it was by no means the first combine "Down Under". The Holts soon became the major manufacturer in the combine business. The Holt and Best companies

"Holt Bros. Link and V-belt Combined Harvester", 1885. First combine to use hook-link chain from the main wheel drive sprocket and V-belt drives elsewhere. There is a beautifully preserved machine on display in the Smithsonian Institution in Washington, DC (Caterpillar Tractor Company).

competed head-on for combine and tractor business, until their merger in 1925.

By 1890, two-thirds of California's 2.5 million acre wheat crop was being harvested by combine. A number of crack combine crews could cut up to 6,000 acres a season. Their typical contract prices for harvesting were: *$1.50 per acre—1887; $1.15 per acre—1892; $1.50 per acre—1893; $1.75 per acre—1900.*

By comparison, harvesting costs using horse-drawn headers and a stationary thresher were $3 per acre in 1890. The days of the header were practically over.

"Surely the smooth easy pace of the steam traction indicates even to those who are not mechanically minded that life and repairs are all in favor of the steamer."—1936 steam engine advertising by Foster & Co. of Lincoln, England.

Early steam engines were hauled from job to job by draft animals. This 1892 photo was taken in the Path Head district, north of McGregor, Manitoba (courtesy of W. H. Little, Brantford, Ontario).

12

The Golden Era of Steam and Big Threshers

Steam Engine! For some, nothing can evoke a flood of memories quite like those two words. Although steam power arrived with the threshing machine, the steam engine never really became a farm machine in the broadest sense. Small farmers could never afford one. As a result, threshing by steam was more often a custom operation, the large capital investment only being justifiable for contractors who could keep the equipment in use for the greatest number of hours in a season.

The earliest "portables", typically about 8 to 12 horse-power, were hauled from job to job by horses. When set up and put to work they furnished power to the thresher by belt and pulley or endless rope and sheave. tract steam threshing around 1830 in England and about 1850 in the US. No less a person than Abraham Lincoln urged the use of steam on the farm in 1859.

Self-Propelled Steam Engines

The first commercial self-propelled threshing engine was marketed by Merritt and Kellogg of Battle Creek, Michigan, in 1873. Self-propelled steamers gained some popularity in the closing period of the 19th Century. (Turner & Irwin 1974). These threshing engines were equipped with a simple ground drive, just strong enough to move their own weight but only at very low speed. By the turn of the century though, steam engines were

being designed to power not only the thresher, but to haul the outfit and plow between seasons as well.

Steam Hazards

There were problems with steam. Consider transport. Some states required a threshing or traction engine to be preceded by a man with a flag and, furthermore, insisted that a horse-team precede the outfit whenever it was used on a public highway.

Boiler explosions, frequently caused by low water or by exposure of the top plates of the boiler on slopes, were often tragic and had a baleful effect on steam ventures. "Vertical" engines were claimed to be less hazardous in the latter regard. Fire was an ever-present hazard. Sparks from the smoke stack could be controlled by efficient spark arrestors—if they were fitted. Nevertheless it was considered prudent to keep the engine a great distance from the thresher, to minimize the chances of a spark landing in the straw. The long belt gave better wrap or grip on the pulleys and also provided maneuvering space for the horse teams to pull in their wagon loads alongside the feeding table of the threshing machine.

The steam outfits were large, heavy and cumbersome. Early steamers with their bulky boiler, heavy drive mechanism and massive frame could weigh as much as a ton per horsepower. The Case Model 110 (110 horsepower) was a comparative lightweight, it weighed

Above: On this portable engine, the team driver's seat was affixed to the stack. It folded down for transport (USDA); Upper Right: Waterous "Fire Proof" Champion 12 horsepower portable vertical engine at work near Edmonton, Alberta. Advantages claimed for the vertical engine were faster warm-up and less risk of exposure of the crown sheet. This machine had a water-type spark arrester, hence the name "Fire Proof". Waterous Engine Works, St. Paul, Minnesota (Ontario Ministry of Agriculture and Food); Right Center: 1887 Case "Portable" threshing engine. Note the seat on the front of the boiler used by the driver for transporting the engine (From the German edition of J. I. Case catalog); Lower Right: Rumely straw burning traction engine, 1900. M. Rumely Company, LaPorte, Indiana.

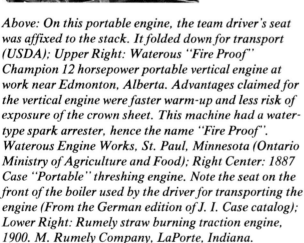

a mere 20 tons and cost a mere 10 cents a pound in 1909.

Bridges were always approached with trepidation. Many a brave driver and his engineer went down with the engine when they broke through a wooden bridge designed only for the horse and buggy. Scarcely a month passed without a report in the pages of *The Thresherman's Review* of an engine and separator plummeting through a bridge. Here is a typical case:

"Bert Sadler and Frank Kenworthy met with a serious accident about dusk Monday night while running a steam threshing engine across a county bridge one mile east of Grant, Iowa. The bridge gave way and men and engine went down. Both men were pinned beneath the debris, and were rescued by the other members of the crew and neighbors, medical

assistance was summoned and it was found that each sustained a broken leg and had been badly scalded by escaping steam." (*The Thresherman's Review,* October, 1903).

Threshing engines required a lot of water—700 gallons or so a day—and good water too. "If you won't drink it, don't put it in your boiler", was the engineer's rule. Finding 20 or more barrels of water a day for the steamer was a task on its own.

Upper Left: The driver of the horse team is seated atop the boiler of this 1901 George White portable engine, manufactured in Ontario (Ontario Ministry of Agriculture and Food); Above: Frank Dority's steam threshing rig at Shelton, Nebraska, 1910 (Nebraska SHSM).

The development around 1880 of a form of firebox to burn crop straw solved two problems—the fuel supply and how to dispose of the straw. As long as the farm used the straw for bedding and feed in the old way, about as fast as it accumulated, there was no problem. But with the arrival of the custom threshers the whole season's crop was threshed at one time and produced veritable mountains of straw.

Glorious Days of Steam

Steam engines enjoyed their heyday in the period between 1885 and 1914. They were the first heat engines to be built, and accordingly became surrounded by an aura of glamor which still persists in the form of "live steam shows", threshing days, special steam train excursions and so on. Here is how one writer recalled the impact of the steam outfits arrival in his community:

" . . . the most exciting event of the year was probably the arrival of the threshing rig for the . . . hectic action that filled granary bins to overflowing and left behind a fresh stack of clean straw. Threshing rigs often moved at night to save precious daytime hours for actual work. There was no other sound quite like the chuckle of the steam engine coming down a moonlit road towing a big dusty separator, plus a tender filled with straw, plus a tank wagon carrying water for the engine. There was always a breathless moment when the heavy rig was maneuvered through the front gate, trailing wisps of straw spilled on the road as the fireman stoked the firebox . . . " (Johnson 1976).

The Threshing Ring

Johnson continues: "The neighborhood threshing ring was one of the high points in pioneer rural America, taking its place with barn raisings and corn husking bees. A threshing ring of even moderate size required something like 20 men and 30 horses to keep it working at full capacity. It was absolutely necessary for five or six neighboring farm families to join hands. This was the crew needed for the shock threshing operation in which bundles (sheaves) were picked up in the field, where they had been drying in shocks, and hauled directly to the separator. To keep a rig going without interruption called for eight or ten bundle teams with racks, depending on whether the ring used field pitchers to hasten the loading or let the driver of the team do his own pitching. It was also necessary to have two teams on wagons equipped with grain boxes to handle the sacks of new grain being transported to the granary or barn." (Johnson 1976).

A Nebraska bridge failure spills engine and separator (The Thresherman's Review, December, 1903).

The man in charge of the steam threshing ring was the engineer. He was often the owner and was held in special esteem by the community as he piloted his ponderous outfit through town during the wheat harvest.

The engineer and crew took quiet pride in their ability to set up the thresher and have the rig working within the shortest possible time. Later, in the 1920's, crews joined in public competitions, which possibly started at Winnipeg, for setting up their rigs in the least time. Points were accorded for accuracy in lining up the belt. Alignment had to be just right, or the belt would jump the pulley. The steam plant might be as far as 60 feet from the thresher. Where the engineer had the choice he would deliberately twist the belt to help it run truer and for better proof against the wind. If there was a strong crosswind the engine might be up to two feet out of line to keep the 150 foot belt on. The machine was usually set up with the stacker pointed downwind so the chaff wouldn't blow back on the crew. If the wind changed, the set was changed accordingly. (Jennings 1968).

On models with wind stackers the discharge could be raised, lowered, shifted side to side or controlled for length of straw. Aiming the blower was usually a job eagerly accepted by the water boy. But the blower was the main dust-producer on the rig and a lot of thought

105

Top: The belt looks dangerous, but few ever "tangled" with it. Where the engineer had the choice, he twisted the belt deliberately to make it run truer and provide more security against the wind. If there was a strong cross wind, the engine might have to be as much as two feet out of line to keep the 150 foot belt on. The great distance between thresher and steamer kept sparks away from the straw stacks and helped increase the grip of the belt on the pulleys; Lower Left: "Belting up" the threshing rig. The steam plant was frequently as far as 60 feet from the separator. Alignment had to be just right or the belt would jump the pulley. A rope drive was used on this 1899 Ontario rig (Ontario MAF); Lower Right: Sometimes the fireman could be coaxed to "let her smoke" for a photograph. Actually, a good fireman ran clear smoke. Kearney, Nebraska, 1910 (Nebraska SHSM).

went into placing the separator so that the dust would blow away from the workers. The sky was scanned and the wind checked to ensure the threshing machine was "spotted just right".

The Fireman

The fireman's day often began at 3 a.m. He cleaned the flues, laid and kindled the fire, and generally encouraged the boiler until a head of steam built up, an operation that usually took two hours or more. Straw burners required continuous feeding, with "little time to wipe your nose". (Brumfield 1968). "Everyone liked to watch the steam engine turn the separator and hear the whistle. The blacker the fireman looked the more popular he was, and important." (Robertson 1974).

The Family Emergency

Threshing was a job for the men and boys. And it

was a family affair too! The host farmer's wife took pride in preparing and serving hearty meals. Threshing ring dinners and suppers (supplemented by morning and afternoon lunch under the shade trees near the rig) could be the talk of the community. Farmers, their boys, and hired men, liked the threshing circuit for both the food and the camaraderie. For the farm wife and her help this period represented an annual event—if not an ordeal—that could take months to prepare for and recover from.

"There was considerable friendly rivalry in the matter of feeding the threshers and there were dark stories told of certain places where they got no raisins in their rice pudding and nothing but skim milk to eat it with and where the pies were made of dried apples . . . The woman who had nothing ready for the threshers was almost as low on the social scale as 'The woman who had not a yard of flannel in the house when the baby came.' "—Nellie McClung, "Clearing in the West" (cited in Robertson 1974).

106

The pictures drawn here of the threshing ring illustrate farmer activities on the small- or medium-sized family farm in Mid-America. On the Western plains and in Canada the crews were largely recruited by the threshing rig owners from transients who followed the harvest. The farmer had to pay the full bill for labor and machines, but was relieved to some extent of the need to enlist the help of his neighbors. In many cases, the mobile rig included not only the engine and separator, the fuel and water tenders, but a cookhouse on wheels as well, complete with cook and assistant. There were compromises between the large rig with its own crew and the smaller rig using mostly local labor. Stack threshing, in which the sheaves had previously been brought together and stored in weatherproof stacks, required a smaller crew.

Labor-Saving Attachments

"The early threshing machines, as I saw them, were utterly devoid of labor-saving attachments . . . A lad stood on each side of the front of the cylinder to cut bands and shove the cut sheaf to the center where the feeder swept it into the cylinder. The straw came out the rear transferred by a carrier and dropped on the ground. Here a man and team bucked it away, taking some around to the engine to fire the boiler. The separator had a low bagger at waist level, the grain being run into sacks and hoisted into wagons. The bagger kept track with a peg arrangement" (Robertson 1974).

Experimentation with self-feeders started about the same time as the "vibrator" or walker-type of machine was introduced in the US, in the early 1860's. The self-feeder would sever the band and feed the sheaf to the separator. By the 1890's, feeders were equipped with governors to regulate the intake of straw into the cylinder. The bundle-pitcher placed the sheaves on the "sheet" of the self-feeder heads-first. He had to

" . . . keep the machine 'up in the collar' without slugging, and had to keep from falling into the clawing, chawing feeder, because meat and bones gum up the cylinder." (Jennings 1968).

By the turn of the century most contractors' threshing rigs or "separators" were equipped with self-feeders, straw blowers, and bag weighing and counting apparatus.

Threshing Ratings:

The mechanical self-feeder, which tended to markedly increase threshing capacity, led to overloading of the separator section. The next step was to make the separator section wider than the threshing cylinder. Threshers were thus rated by reference to the "width of the cylinder opening" and "width of rear opening" respectively, for example, 24 x 36. A 24 x 36 in. machine of the 1930's could typically thresh 60 to 100 bushels of wheat per hour or 100 to 175 bushels of oats per hour.

Mechanical self-feeding conveyor on Glasgow MacPherson "Climax Apron Separator", Beamsville, Ontario, 1880. Note the straw elevator at rear of machine (Ontario MAF).

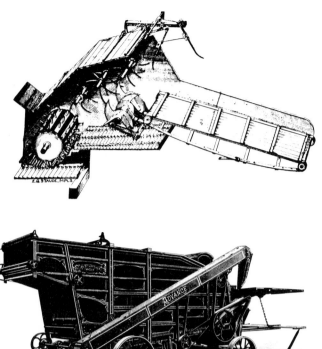

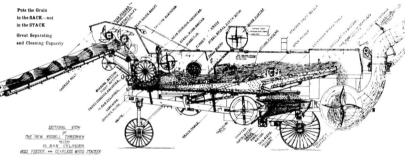

Upper Left: Eloquent Parson's advertisement makes a compelling case for the self-feeder. Parsons Hawkeye Mfg. Co., Newton, Iowa (American Thresherman, 1900); Upper Right: Heine self-feeder advertisement. The whirling knives of this device severed the bands before the loose bundles were teased into the cylinder (Thresherman's Review, June, 1903); Above: 1897 Advance Thresher Separator from the elevator side. Advance Thresher Co., Battle Creek, Michigan; Left: 1918 Russell thresher cross-section. Note dust removing fan over the walkers. Russell & Co., Massillon, Ohio; Lower Left: The Avery "Yellow Fellow" thresher was equipped with a rotary separator device to increase capacity in 1903. Avery Manufacturing Co., Peoria, Illinois.

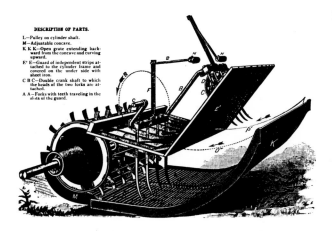

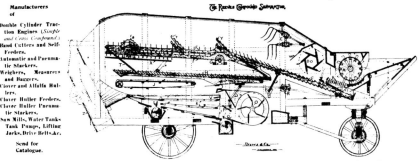

Upper Left: Four-Bar kinematic linkages were employed in the straw agitator over the transition grate of the Advance Thresher in 1897. Advance Thresher Co., Battle Creek, Michigan; Left Center: Reeves "Compound Separator" used oscillating rakes to agitate the straw for additional separator action in 1897. Reeves & Co., Columbus, Indiana; Lower Left: The Closz "No-Choke" and Closz adjustable sieves and chaffers, still used today, have been around since 1890 when Charles Closz built the first "No-Choke" for a Russell thresher. (Thresherman's Review, 1897 and 1903); Above: This Russell & Co. separator advertisement illustrated the "bustle" design typical of the later generation of threshers. The attachment of mechanical feeders increased threshing capacity, tending to overload the separation section. The separator was then made wider than the threshing cylinder.

This was about as much work as 100 men working with flails.

Mechanical Stackers and Ten Dollar Bills— To Prove A Point

The straw, coarse material and chaff were expelled out the rear of the machine, where it tended to pile up rapidly. This led to the development of mechanical stackers to remove material other than grain away from the back of the separator—the dirtiest job of all when performed by a crew with pitchforks. Around 1870, stackers of the elevator-conveyor type were attached. In 1879, the Buchanan straw blower or "wind-stacker" appeared. Old-timers opposed it on the grounds that it would pull the grain out of the pipe or shoe, causing excessive losses. In reply, Buchanan's men arranged a convincing demonstration for their detractors. They turned off the separator fan and then laid ten dollar

bills on the shoe to prove that the wind stacker could not draw grain—or the bills— from the shoe.

In 1891, A. McKain bought out Buchanan's and other important patents of the time and built up a remarkable monopoly. For two years, he gradually built a demand for the wind stacker and then began exacting enormous royalties from manufacturers obliged to adopt his principles when the design caught on. (Horine 1938). The wind stacker "took the men off the straw stack" and was even claimed to "relieve the thresherman from care and trouble, preventing him from using bad language and thus imperiling his soul".

Steam's Swansong

At the height of production, around 1900, 5,000 steam tractors were built annually. The leading manufacturer of steam engines was J.I. Case, whose production at their peak was almost three times that of the nearest

32 x 54 Case "Agitator" thresher equipped with hand feed and common stacker threshing oats on the farm of John Weber, South Milwaukee, Wisconsin, August 24, 1932 (J. I. Case). Threshers were given ratings such as "32 x 54" which referred to a 32 in. cylinder and a 54 in. separator.

In 1902, the recleaner was added to the "Agitator" to produce a better quality of work when threshing very weedy grain. Notice the self-feeder (J. I. Case).

competitor, the Huber Manufacturing Co. of Marion, Ohio. Before production of steam engines was discontinued in 1924, Case had built a total of 35,737 engines over a period of 55 years. The first all-steel thresher design (1904) was a Case—the decision to go to steel being hastened by a factory fire that destroyed 110 wooden threshers. It was Case too that produced the last stationary thresher in North America—in 1953. By that time, total Case thresher sales had exceeded 100,000 units.

The development of the internal combustion engine and the emergence of the more compact and efficient gasoline or oil tractor displaced the steam engine. For a lengthy time, perhaps a generation, considerable hostility existed between gasoline and steam power operators. Some steam engine manufacturers even refused to load their wares on the same freight car if it was carrying a gas tractor.

In the early days of the internal combustion engine the most rapid sales inroads into the steam market were made in California. The big ranches of that State, the leading wheat-raising Sate in the Union by 1890, faced a perennial labor shortage.

Innovative Californians such as G. S. Berry, Daniel Best and Benjamin Holt resolved the labor problem by creating the largest agricultural machinery ever built.

Operations on the great bonanza farms in such areas as the swampy delta country in the San Joaquin and Sacramento Valleys were on such a scale that cost of machinery was secondary. They built equipment of an unprecedented size: on one hand a steam traction engine having drive wheels each 18 feet wide; another, a harvester with a cutting width of 50 feet! And it was also these Californians who were first to put gasoline engines to work on grain harvesters, and build self-propelled and hillside combines.

Harvest Origin of the First Commercial Crawler Tractor

Holt's 45 foot wide "Big Betsy" tractor with its gigantic wheels was recognized as a poor solution to the problem of flotation. Although there was a greater width of ground over which to distribute the weight, the supporting area of the wheels was still inadequate. Holt turned to the "lag bed" as a more practicable means of providing traction and flotation for pulling gangs of plows, harvesters or both. The concept of a track-laying machine was already a hundred years old, but had never become commercially viable. Holt's first practicable machine was tested on Thanksgiving Day, November 24, 1904, near Stockton. Here was a 40 Horsepower machine that could pull better than its 60 Horsepower wheeled counterpart. Demand for tracked machines

Right: Caterpillar "Sixty" tractor pulling a 24 foot Holt combine of 1915 vintage. The "Sixty" is also pulling a 24 foot side disc harrows somewhere in Kansas, 1928 (Caterpillar Tractor Co.).

Left: Holt led in the development of the first commercially successful crawler tractors that were needed to help pull large implements and the enormous combines found on California bonanza farms. Note that the front wheel was retained for steering. A farmer claimed the early crawlers had two speeds: "Slow and damn slow" (Deere & Co.); Below: A sectional view of the Caterpillar Model 36 combine, 1929. Bulk grain handling was an innovation on the Caterpillar rigs (Caterpillar Tractor Co.).

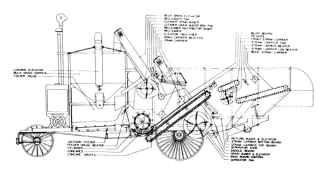

Above: This Holt steamer had a barrel-shaped roller in advance of the steering wheel to smooth the ground and improve flotation (Deere & Co.).

Below: A self-propelled Holt crawler combine with 16 foot cut and 6 foot extension. Note bulk grain unloading, ca. 1913 (F. Hal Higgins Library).

Left: A Holt combine was used in one of the first combine windrow-pickup jobs in 1900. This 1928 Model 36 is being used with windrow-pickup for shock threshing of oats near Manteca, California (Caterpillar Tractor Co.); Right: Caterpillar Thirty-Five Diesel crawler tractor and Caterpillar Model 36 combine harvesting at 4.75 acres/hour in Monterey, California in 1930 (Caterpillar Tractor Co.).

eventually called for an expansion of the Stockton plant, and in 1909, the addition of a new plant in Peoria, Illinois.

In 1906, gasoline began to replace steam and Holt's crawlers were proving versatile and less costly to operate. Holt's crawler became the inspiration and guide for the military tank developed by the British between 1915 and 1916. Holt was first to use a gasoline engine for auxiliary power on a combine in 1904.

Caterpillar Tractor Company

In 1919, the first all-steel Holt harvester was constructed. It was put into production in 1921. In 1925, the Holt and Best organizations merged, to form the Caterpillar Tractor Company. Manufacturing was centered in Peoria for strategic and geographic advantages. Soon after the merger, the Holt facilities in Stockton became a wholly-owned subsidiary, the Western Harvester Company, where production of the 40 year-old line of Holt combined harvesters continued under the old name. Harvesters were sold under the Caterpillar trademark when production was shifted to a new plant in Peoria in 1930. (Caterpillar 1954). The total production of Holts, Bests, Houser-Haines and Northwest models had totaled 14,111 from 1887 to 1929, of which nearly 11,000 units were Holts and 1,351 were Bests. (Higgins 1967). Caterpillar eventually dropped out of the combine business to concentrate on heavy construction equipment. Their combine line was sold to Deere and Company in 1935.

1893 Holt steam traction engine and 50 foot cut combine, complete with American flag. This machine, Holt No. 574, had a 26 foot main header and two 12 foot headers hingeably attached to the main header. It cut 150 acres each day.

"My invention cost me some money, some anxiety and condemned my little ones to all the miseries of poverty and banishment to the bush. If I had been a successful cricketer, good bowler, or a rifle shooter, without pluck, a Blondin or an acrobat, I and mine would have escaped these ills."—Robert Bowyer Smith, South Australian inventor and implement maker, 1876. (Cited in Wheelhouse 1966).

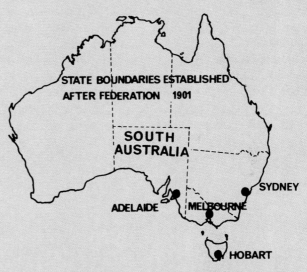

AUSTRALIA 15ᶜ
Pioneer Food

13
Meanwhile, Down Under

Captain James Cook claimed the Australian continent for King and country in 1770. In the beginning, the British Government used the Great South Land as a penal settlement and sent convicts to Australia for 80 years. Some miscreants were even deported from the British Colonies in North America.

The first white residents, convicts and soldiers, depended upon England for practically all their needs. Their lifeline was a slender thread of sailing ships that took as long as six months to reach Australian shores. The first crop, English wheat, was planted near the future site of Sydney in 1786 by James Ruse, a convict. In 1823, gold was discovered. At first a trickle, but with the discovery came adventurers . . . and settlers. The pace of life quickened. In 1836, the colony of South Australia was established with the nation's first "free" settlement at Adelaide. The crop season of 1842 was an especially critical one for the new settlers in the driest colony on the world's driest continent. One Adelaide citizen, on reviewing the grain situation observed:

STATE BOUNDARIES ESTABLISHED
AFTER FEDERATION 1901

SOUTH AUSTRALIA

ADELAIDE   MELBOURNE   SYDNEY

HOBART

South Australia, the driest colony on the world's driest continent. Grain ripened rapidly and uniformly in the hot summers. Shatter loss or shedding was severe when the crop was hand harvested with the sickle or cradle. Hand reapers were scarce and could demand 15 shillings a day plus allowance for wine, beer and rations.

"In reaping this by sickle a very large portion of wheat was shed and lost. Many who witnessed it in 1841 thought the loss so great that wheat would not pay on the plain if reaped by hand. The matter was so important that it was the one topic of conversation in Adelaide. Search was made for information as to harvesting machines that might meet the difficulty and in Loudon's "Encyclopaedia of Agriculture" was found . . . a drawing of a machine described by Pliny . . . " (Cited in Sutton 1937).

So acute was the problem that Public Offices were closed and South Australia's Governor Grey arranged

Placer gold was discovered in the Australian colonies; first in Victoria in 1823. The discovery of gold eventually changed the whole fabric of society in the colonies.

for 150 soldiers to be released from their usual duties and brought across from New South Wales to assist in reaping the 1842 wheat harvest. Still there were not enough hands. Once the crop ripened, delays meant heavy grain losses—perhaps as much as half the crop would be lost with the sickle. Governor Grey himself pitched in to help, and afterwards he said of his conscripted hands, "not often have soldiers been more nobly occupied".

Loudon's Encyclopaedia (1829) contained a description of Patrick Bell's Scottish reaper, two of which had been imported into Van Dieman's Land (Tasmania) in 1833. (Blackwood 1834). News about American reaping machines was also trickling in, stirring some innovative settlers to build their own harvesting machines.

A group of Adelaide pioneers calling themselves "The Corn Exchange Committee" decided to encourage inventors by offering a £40 prize for the most promising design or model of a harvesting machine suited to the colony's needs.

Birth of the "Stripper"

Seventeen candidates were present or provided exhibits at the meetings called by the Committee in September, 1843. With one exception, all were intended to cut or reap. (Sutton 1937). All, that is, except John Wrathall Bull's model. He brought a model of the first Australian head-stripping harvester. According to the *Adelaide Observer* of September 23, 1843, the model of Bull's machine, considered by one to be "a joke":

"consisted of a long tooth comb, fixed to the back of a close-
    bodied cart; the teeth being operated upon by four revolving

Balhannah farmer, John Wrathall Bull, inventor of the Australian "stripper," a machine that used a comb and beaters to detach and thresh standing grain heads without cutting the stems. This invention was first disclosed in September, 1843 (Sutton 1937).

horizontal beaters, with square edges, which would have the effect of taking off the ears of corn, and depositing them in the body of the cart, a wide bag being placed under the after part of the comb to catch any dropped grain."

Bull's account of the inspiration of his harvester concept is an inimitable classic:

"I will now give an account of my own experiences in the harvest of 1842-43, and in the conveyance of the crop to market. Prices had fallen considerably and buyers were scarce. My crop was in condition for hand-reaping before the end of December, but I could not procure reapers before the 24th, as men had been earning large wages on the plains. Harvesting hands had been so scarce that the soldiers had been allowed to lay down their arms and take up sickles and many soft-handed gentlemen had also turned out to give their doubtful but well-intended assistance in the emergency. On the 24th December, 1842, I was able to induce five men to accompany me and I conveyed them to the farm. I did not allow them to work on Christmas Day, but they had Christmas fare. I engaged to give them 15s. and one bottle of rum an acre, with rations, for hand-reaping. The crop was dead ripe, the heads drooping with the weight of the plump grain. On the 25th a fiery hot wind was blowing and continued on the following day when I expected the reapers to start work, but they were missing. I found them at the nearest grog shop. After some trouble I got them away to start work on the following morning. Before a sickle was put into the crop, the loss in shed wheat was over one bushel to the acre, and a further loss necessarily followed in harvesting.

Immediately on my return I took one of the men, the most sober of the lot, to see the over-ripeness of the crop, and by what transpired it will be seen how providentially, out of the difficulties of my situation, the idea flashed upon me as to the possibility of thrashing a standing crop of wheat, and which idea, on being worked out, has since wrought such a beneficial result for the Colony at large.

On taking this man into the crop, and pointing out to him its over-ripeness, and how careful they would have to be in performing their work in handling the standing crop and in binding, calling his attention to the shed grain on the ground (I was standing a short distance within the crop), and to show how tender the heads were, with the full grain staring us in the face out of the gaping chaff, I passed my left hand with my fingers spread, under and just below the ears, allowing the straws to pass between my fingers, the ears being close to the palm of my hand. I then struck the heads with a sweep of the edge of my right hand, and held out my open hand for the man to see the clean threshed wheat in the hollow of it, most of the chaff having been carried away. (I must here mention that before this occurred I had for many weeks been pondering over plans for applying machinery to a standing crop, and had passed many sleepless hours in bed, and had been remonstrated with by my good wife, who said I should lose my senses.) Before I moved from my position in the standing corn, I stood in a sort of amazement and looked along and across the fine even crop of wheat. The ideas I had in vain sought for now suddenly occurred to me, and I felt an almost overwhelming thankfulness. I did not move, but sent the man for a reap hook and caused him to cut me a small sheaf of wheat which I took into the barn. There, holding a bunch of it in a perpendicular position, I struck the ears with a circular sweeping blow upwards, using a flat and narrow piece of wood, and found the threshed grain to fly upwards and across the floor; and thus I satisfied myself that the grain would bodily fly at a tangent up an inclined plane when struck by beaters and that a drum, as in a threshing machine, would not be required to complete the threshing, and so felt I had gained the correct idea for a field thresher, and that a segment below the beaters would be apt to cause the wheat to be carried round and so be lost. All this occurred in 1842.

I afterwards lost no time in exhibiting a rough drawing to many of my neighbours . . . but I got no encouragement, but from my oft recurrence to the subject was sometimes told I had lost my senses."—J. W. Bull, "Early Recollections and Experiences of Colonial Life", 1884.

Neither Bull's nor any of the other designs were considered worthy of the £40. There was, however, an astute observer at the meetings, John Ridley, flour miller and landed gentleman, of Adelaide. Ridley had operated a flour-milling business in England and when he migrated to Adelaide he brought part of his plant, including an overhead beam steam engine. This had been one of the first six "Grasshopper" types of James Watt's own construction. At first there was barely enough grain to keep Ridley's plant at capacity, so he geared his engine to a saw mill and found a real demand for work from new settlers establishing farms. He also tried farming himself and, by 1842, was sharecropping 300 acres. Ridley clearly saw the need for harvesting equipment, and in 1843, "availed himself of all the best points" of the machines shown to the Committee that September. Ridley's first attempt to build a grain harvester ended in failure. It was in the form of reaper intended to *cut* the crop. He next adopted and tried Bull's comb-and-beaters principle. Following a successful field trial on November 12, 1843, he took immediate steps to patent the principle in England.

Ridley had a considerable advantage—money—over Bull in the development of the stripper. Bull had only been able to demonstrate a model. Ridley proceeded directly to a full-sized machine which demonstrated the soundness of the stripper principle under South Australian crop conditions. The name "stripper", while particularly appropriate, was not original with the machine. At first the locals called it the "locomotive thrasher".

The press lavished praise on Ridley's stripper, claiming that a new era of prosperity had dawned for the colony. No longer would the area of crop planted be constrained by the amount of labor likely to be available at harvest time. So enthusiastic did the rhetoric become that Bull was all but forgotten and Ridley himself apparently began to believe he was the inventor, although he never followed through with the patent proceedings. As for Bull, he eventually did succeed in having two machines built by J. Marshall, a blacksmith, and he used them on his own farm. Marshall received orders from other farmers and he in turn became a prominent manufacturer. Ridley himself sold many strippers at £150 each. Several other firms started into the Australian stripper business and "within a few years" 30,000 machines had been built. (Mellor 1956).

A competitive reaper and stripper field trial was arranged in December, 1855. Four Ridley-type machines were exhibited. J. H. Adamson of Mt. Barker was there with a machine drawn by four gleaming black horses. Joseph Mellor entered his light-draft machine, a pull-type

John Ridley, 1806-1887. He was responsible for producing the first successful head-stripping cereal harvester in November, 1843 (Ridley 1904).

Ridley's stripper of 1843. Invented by J.W. Bull and brought to workable form by John Ridley (Sutton 1937).

requiring only two horses. Marshall entered two machines, one pushed by a team of eight bullocks, the other by a team of six. The agricultural society awarded the prize of a silver and gold medal—the first ever given for a stripping match—to Mellor, for "a very creditable piece of workmanship". Mellor went on to win other Australian prizes for his machinery. He was also successful at the Great London Exhibition of 1862. Here Mellor was awarded a silver medal for a stripper with a cutting knife. Ridley, who by this time had returned with his family to England, also displayed a stripper design with rear cutterbar for mowing the stripped straw at ground level.

Upper Left: A hinged door enabled the threshed mixture to be unloaded by raking it from the rear of the stripper box. The box held about 20 bushels of grain (Science Museum, Victoria); Left: Closeup of the comb teeth, beater and studded concave of the Australian stripper. The beaters detached the heads from the standing crop, further threshing the crop as it passed over the concave and conveying it into the box by impact and aerodynamically (Courtesy of Norman Grove-Jones); Above: The stripper design with rear straw cutter which Ridley displayed at the Great Exhibition, London, England, 1862, was labeled "The Australian Reaping Machine." The purpose of the S-shaped "flyer" (t') was to carry the straw over the stubble knife (q) (**The Engineer**, *November 1, 1861*).

Ridley's departure from the Colony in 1853 was accompanied with considerable ceremony. He was acclaimed a Public Benefactor and formally recognized by Governor Grey and the Legislature. The *South Australian Register* reported in 1853:

"During the present and past year's harvests when almost all our labourers had left for the Victorian goldfields, machines built on Ridley's principle were the means of preserving our grain crops and enabling us to export during the year £190,000 worth of wheat and flour."

In South Australia the wheat crop sown was increased from about one thousand acres in 1840 to 168,000 acres by 1856—the increase due in large measure to the development of the stripper.

Back in England, Ridley maintained his interest in farm machinery. Twelve of his strippers were built by a Leeds firm, but the Australian machine proved unsuited to England's climate. A British writer recorded that Ridley's machine was "tried in Northumberland, with an application for cutting the straw as well, with questionable results; in fact, our uncertain and damp climate will be sufficient to prevent its extended use in this country." (Wilson 1864). All 12 of Ridley's Leeds-built strippers were shipped to South America in 1868.

As for Bull, the controversy that arose over who had really invented the stripper eventually resulted in the South Australian Government awarding him a £250 premium in 1882— 40 years after he discovered the comb-and-beaters principle for harvesting grain.

Some companies continued to produce strippers until after WW2 and even today plant breeders still make use of strippers for plot research work.

Reapers and Binders in the Antipodes

The first American reaper—a McCormick—was imported into South Australia in 1851. The earliest Marsh Harvester arrived in New South Wales in 1879. These machines and later twine binders, although challengers, were never as economical as the simpler and more effective strippers in the low rainfall wheat regions of the 19th century Australia. Later, in 1887, Massey binders were sent from Canada to test the market, with favorable results—annual sales exceeded the one thousand mark by 1894.

Grain from Chaff

The Australian stripper did not complete the job. When the stripper-box was raked empty, after every few rounds, there was still the onerous job of winnowing the grain from the chaff, straw pieces and the few unthreshed heads. This was a job for the winnower-cleaners.

J. S. Bagshaw was the first Australian winnower manufacturer. Beginning in Adelaide in 1838, he produced thousands of hand-cranked winnowers over the years. Frank May of Gawler followed him with a combined winnower and cleaner. A treadle enabled the operator to drive the machine with leg power by shifting

Each of these five-foot strippers could harvest about 7 to 8 acres a day depending on the weather and the horses. Note the one-horse power winnower. Scene at Gawler Plains, South Australia, 1876.

weight from one foot to the other. The May Brothers provided the further refinement of an optional bagging elevator. For decades after the invention of the stripper, the majority of farmers continued to use hand-cranked winnowers. Two men were needed on one winnower to keep up with the dump-loads from three five-foot strippers. It was a hot, dusty, and unpleasant task.

In the 1860's Thomas Illman of Balaclava, South Australia developed a thriving business manufacturing horsepower driven winnowers that were reputed to produce the cleanest sample of wheat in the Colony. The particular method of attaching the horse-tread power to the box-winnower was uniquely Australian.

The Australian strippers were simple in design, low in cost and required only one man, but there was still the bottleneck of cleaning the grain from the chaff.

Australia's First "Combine"

The idea of combining the stripping and winnowing operations in one field machine had occurred to Ridley. In 1857, Mellor patented a combined harvester designed to strip, thresh and winnow the grain, but neither he nor Ridley pursued the project to the commercial stage, although Mellor's design was Australia's first "combine" patent. Mellor's company, founded in 1842, developed many colonial-built implements and manufactured strippers for half a century. After Mellor's death in 1880, his sons moved the factory from South Australia to Braybrook Junction, Victoria. Later models of the "Mellor Bros." stripper built at Braybrook were so light to pull that they could be drawn down a hard road by a man pedaling a bicycle—hence the label "Mellor's Patent Bike Strippers."

There were some charitable things said about Adamson's axial flow stripper-harvester. Apparently there was one unit in California as early as 1872 which, according to Adamson, "cleaned better and wasted less than their best machinery." Several of Adamson's machines were built under license in 1873 and sold in the Livermore Valley. It worked this way: after gathering and threshing by the stripper comb the crop was funneled into the wider end of the slowly-revolving perforated cone, through which an air stream passed. The cleaned grain passing through the axial separator/cleaner cone was finally conveyed up to a bagging spout. Revived in 1888/89 by L. Smith of Oakdale, California, about 25 of the machines were sold in the hilly grain areas of the district. In spite of this, the editors of *Pacific Rural Press* loyally commented that it was regrettable that the "San Joaquin combined harvesters were not shown (in Australia), for their work thus far has been far more satisfying than anything yet reported from Australia".

An International Competition

In October, 1877, the South Australian Government voted a sum of £4,000 as an incentive for the

"best machine combining within itself the various operations of reaping and cleaning, fit for bagging in the field the various cereal crops of South Australia . . . the trials will take place in December, 1879."

The competition, advertised internationally, attracted 27 applicants. Only 14 actually competed in the field trials at Gawler: 11 from South Australia, two from Victoria and one American, S. L. Gaines, who brought his combine from Oregon. The Government's repre-

Wheat stripping on a grand scale. There were 37 strippers at work on this 10,000 acre property of pastoralist C.B. Fisher in 1872. Equally impressive was the tremendous amount of effort needed on the handles of the winnowers to dress the grain delivered in batches by the strippers. One winnower was required for cleaning the batches from three strippers. The men were paid one penny per bushel for putting the wheat through once and twopence for cleaning it twice (Morphett 1945).

sentatives were not suitably impressed with any of them to award the major prize. Instead, they parted with only £250 and that sum was shared by four Australians. One reason for their reluctance to declare any machine a success may have been the failure in the design of the combined stripper-winnowers to overcome the problem of tilting of the sieves (or "riddles", as they are still known in Australia) when the gathering comb was raised or lowered.

Yet Adamson's machine, which shared this minor prize, did overcome the problem. He employed a rotating axial separating cone, suspended so that it retained its position relative to the ground. The American, Gaines, retired from the field soon after starting up. He claimed that he had lost a pulley on the voyage over and that the smaller one he had substituted did not serve the purpose. Returning to the US, he eventually succeeded in California.

The £4,000 prize was never awarded. In subsequent Government trials held in Victoria in 1884 and 1885, the judges again declined to provide anything beyond token encouragement.

Some Californian Interest in the Competition

The Australian stripper was known to Californians as soon as it became popular in Australia. Contemporary articles in the *Pacific Rural Press* showed an active interest by farm writers in Australian machines. A few strippers had been used in California before 1880, and they were said to meet a need by small farmers for a low cost machine. There seems to have been an irrepressible

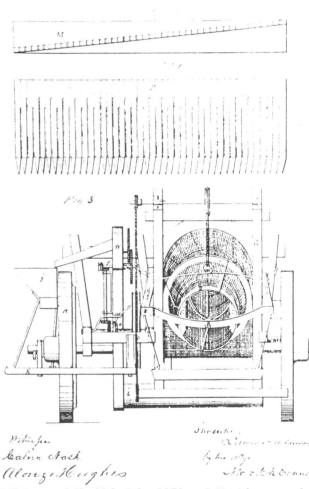

J. H. ADAMSON
Combined Harvesters and Thrashers.

US Patent 140, 396, July, 1873 to J.H. Adamson of South Australia for "Combined Harvester and Thrashers." His design was possibly the earliest axial flow separator to have been successfully used in a field machine. In this case the crop was gathered by the Australian stripping comb and conveyed into the conical separator/cleaner. Means were provided to keep the cone in its correct relative position as the comb was raised or lowered.

First Australian "combine." Joseph H. Mellor's Victorian patent No. V58, October 24, 1857. The first Australian patent on a combined machine to strip, thresh, winnow and bag the clean grain.

conflict between the Californian and Australian points of view however. The Australians were said to:

"scout our combine harvesters because the hitching of two or three dozen animals to one machine seems to them preposterous. We return the compliment by believing that their little, dinkey strippers require too many men to operate . . . If one man could drive all the mules in the State it would be the acme from our point of view."

The Sunshine Harvester—Hugh Victor McKay

At the ripe old age of 16, Hugh Victor McKay was already a veteran harvest hand. He had driven his father's stripper and had spent hours on the crank-handle of a winnower cleaning grain on the family farm at Drummartin, Victoria. How he detested that winnowing job! One day in 1882, as he was driving the stripper, the solution dawned on him. He announced to his family that he'd build a machine that would, in one operation, strip the standing heads, thresh the grain and clean it for market. Friends scoffed. After all, hadn't there been an international competition and no one had been able to win the award?

Undismayed, the then 17 year old and his older brother John built a crude smithy's shelter and set to work. Their father, Nathaniel McKay, let the boys use scraps from an old rear-delivery reaper, a binder, a winnower, and the comb from a stripper. The father built pulleys from red gum logs, split with gunpowder.

In February, 1884, the prototype, with its diminutive 2 ft. 9 in. comb, was prepared for operation in two acres of wheat that had been set aside. It worked well. H. V. McKay had solved the problem of keeping the riddle box level while the gathering comb was raised or lowered. But the youthful McKay was not in a position to manufacture his "stripper-harvester". Only after a number of frustrating encounters with existing stripper manufacturers, including the Mellors, did he reach an agreement with plow builders McCalman & Garde in Melbourne to build five of his machines for the 1885 harvest.

The first machine was sold in 1885. In 1886, a small public company, the McKay Harvester Co. was formed with an office at Ballarat to handle manufacturing contracts and sales. Sixty machines were sold in 1888. Construction of the stripper-harvesters was farmed out to various contractors on a royalty basis.

In 1892 an economic downturn forced McKay to start all over again. He redesigned the machine and began building them himself.

The theme of an address given by a visiting American evangelist, Dr. Talmage, provided the brand name—the "Sunshine". The first 12 Sunshine harvesters were built in 1895. From then on production expanded rapidly at Ballarat. By 1905, his crude facilities produced a record 1,926 harvesters. In 1902, McKay faced yet another crisis—prolonged drought. After a succession of crop failures, farmers could ill-afford new harvesting equipment. Undaunted, McKay sent his brother with

Above: Hugh Victor McKay, harvester builder and entrepreneur at age 20. He was 17 when he developed his stripper-harvester; Below: Facsimile of an invoice for the stripper and winnower purchased for the McKay farm in 1877 and subsequently used by McKay.

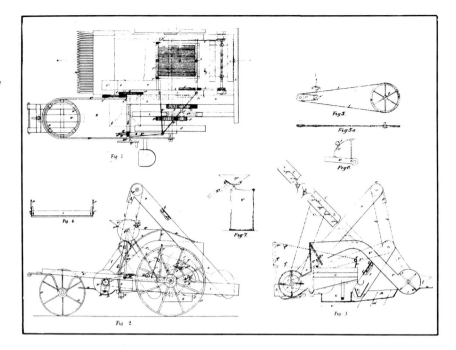

*One of the five original McKay
stripper harvesters was sold to
Mr. William Anderson of M ̧ama,
Victoria. He used it for m ̧ny
years and, 43 years later, it was
still capable of good work
when it was returned to the
Sunshine works.*

50 stripper-harvesters to Argentina. The export venture was highly successful; the backlog was sold. Thus began an export trade that would send over 10,000 Sunshine harvesters overseas until stopped by the first World War.

The McKay's solution to the 1902 drought created another problem. The rapid expansion in export production put a heavy strain on the Ballarat facilities. Transportation became an important production consideration. Raw materials were shipped 60 miles from Melbourne; finished machines were sent back down the tracks for shipment out of the docks of Melbourne.

In 1906, the works were moved to Braybrook Junction, into none other than the now-defunct Mellor's Braybrook Implement Co., one of the manufacturers who had years earlier rejected McKay.

Over the years, McKay had worked at one time or another on most of the jobs in his plant, earning the respect and loyalty of his employees. As a result, over half of his 466 factory workers moved to Braybrook with him. Growth at the new H. V. McKay Harvester Works was prodigious. By 1910, 1300 were on the payroll. Under McKay's leadership implements of considerable importance to the Australian broadacre wheatland economy were built, including the world's first one-way disc cultivator,—the "Sunflower"; the "Sundercut"—a stump jump plow; the "Suntyne" seed drill, and other machines and implements under the

"Sun" label. Even the name of the new settlement and its rail station were changed from "Braybrook Junction" to "Sunshine", the name this Melbourne suburb and home of Massey-Ferguson (Australia) Ltd. has today. In 1909, McKay developed one of the world's first gasoline-powered self-propelled combine harvesters, an experimental model having a 24 foot cut. It was judged too costly to produce, however, and never reached production.

McKay was now the largest farm equipment manufacturer in the Southern Hemisphere. As wheat production expanded by leaps of half a million acres a year into the higher rainfall areas of Australia, there were regions where crops were weedy and prone to lodging. Under those conditions the stripper front was largely ineffective, and the binder was used. McKay had secured agreements before 1900 with the firms of Johnson and D. M. Osborne in New York for twine binders and, in 1901, for binders and mowers from the Milwaukee Harvester Company. In 1911 he came out with his own reaper and binder.

In 1912, a stripper-harvester cost £90 and together with 10 horses (two shifts of five) a total of £400-490. The machine harvested the crop and bagged clean grain. On the other hand, an 8-foot stripper cost £70 and needed only six horses (two shifts of three) for a total outlay of £310. If a farmer already owned a win-

The ''Sunshine'' Harvester built by H.V. McKay Harvester Company at the Ballarat works in 1901. The machine was equipped with a tailings return and rethreshing system.

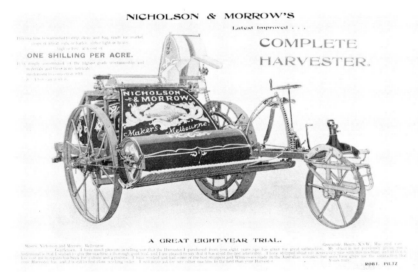

Nicholson & Morrow stripper advertisement, 1901. James Morrow's design was among the earliest stripper-harvesters built in Australia, but was not as well promoted as McKay's.

nower—most of them did in those days—and used family labor, he could thus save £180 in capital expenses. But the stripper-harvester could save 10 pence a bushel if the cost of winnowing after the passage of the stripper was taken into account. Local manufacturers sold both types of machines. The cost per bushel of using the binder and stationary thresher on the other hand, was almost ten times higher. The Australian farmer's isolation and small profit margin made him "one of the shrewdest and most demanding customers in the world", according to a contemporary Canadian farm machinery salesman.

Massey-Harris' Australian Market

Overseas manufacturers soon discovered that the market "Down Under" was a lucrative one. For Massey-Harris, Australia became the Company's second largest export market—second only to Europe. In 1894, 1000 Massey-Harris binders and 2000 other implements were sold in Australia—a prodigious growth, considering that the first machines had been imported seven years earlier.

In 1897, the Canadian firm purchased a McKay stripper-harvester for scrutiny, then experimented with four of their own in 1900. They sold 350 in Australia in 1901 and, that same year, test-marketed the machine in the Argentine. By 1904, Massey-Harris had stripper-harvesters at work in three Australian States and in South America. International Harvester Company soon followed suit, building stripper-harvesters in North America solely for export in the Southern Hemisphere. Their first, the Model No. 1 Stripper-Harvester, was sold in Australia in 1910.

The Reaper-Thresher and Tariff Protection

McKay was not amused by Massey-Harris' activities. He referred to

"The Canadian and American pirates . . . they took the result of our brains in the stripper harvester". (Kendall 1972).

121

The stripper continued to be used for many years after the stripper-harvester was launched. Harvesting at Flinty Ranges, Mulwala, New South Wales in 1903. Operators of 14 strippers and three box winnowers with integral horse tread-power pause for the photographer. A total of 80 men and 24 strippers worked on Sloane & Sons 50,000 acre "Savernake" and "Mulwala" stations (courtesy Ian and John Sloane).

This action did nothing to diminish total sales of Massey-Harris equipment. Threatened in their Australian stripper-harvester market, Massey-Harris concentrated on increasing market penetration in South America. They did not give up on Australia—far from it!

Massey had two competent local men, Matt H. East and J. S. Charlton, working in Australia at the turn of the century altering the Canadian built stripper-harvesters to meet local requirements. These men met during the 1901/2 season at Mallala and together came up with another approach to combine development. They proposed the use of a knife under the long-toothed comb and a draper platform to feed into a regular thresher instead of the stripper beater. Anticipating the tarriff restrictions, they spent considerable time in Canada working on their machine so that they could test it in Australia. The result was Massey's first combine, the "No. 1 Reaper-Thresher", first shown in 1909. (Thyer 1952). In 1910, 83 of the 8-foot cut No. 1's were sold. The name "Reaper-Thresher" was coined in deference to the tariff act, but it is worth noting that in Australia the word "combine" had already been coined for combination cultivator-drills. Sales of the Reaper-Thresher increased to 1260 per year just before WW1, by which time the Australian market was absorbing 40% of Massey's total equipment sales.

Matt East's career with Massey-Harris spanned a full 27 years, during which time he made 17 trips to Canada and 12 around the world.

Australian manufacturers paid for tariff protection. The "price" was "fair and reasonable" wages. The unions claimed McKay's wages of 36 shillings for a six-day week both unfair and unreasonable. The arbitrator of the Commonwealth Court ruled in favor of the unions in a landmark settlement known as the "Harvester Judgement", which determined in 1907 "that seven shillings a day was the basic wage for an unskilled worker living as a human being in a civilized community". It defined for the first time the concept of the basic wage in Australia.

The battle lines were drawn between the rivals competing for the Australian market when Massey-Harris advertised their Reaper-Thresher as "the perfection to which the McKay stripper-harvester had been but a

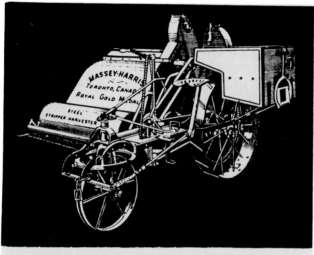

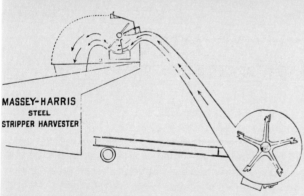

Above: The Massey-Harris all steel stripper-harvester. Built in Toronto in 5- and 6-foot sizes exclusively for the export market. Note roller over stripping comb used for tall crops; Below: Schematic view shows trajectory of crop being delivered through the peg drum or "damp weather" thresher. The beater drum typically revolved at 600 rpm.

Prodded by Australian manufacturers, particularly McKay, the Australian government passed stiff tariff protection measures. The tariffs, which were levied in 1906, placed prohibitive duties on competitive imported machines such as stripper-harvesters and stump-jumping plows.

The reaper-thresher made its debut in South Australia in 1909. This picture was taken during an official demonstration on the property of Thomas Irish, near Kadina. The demonstration was attended by the Premier of South Australia, Sir Richard Butler (Massey-Ferguson, Australia).

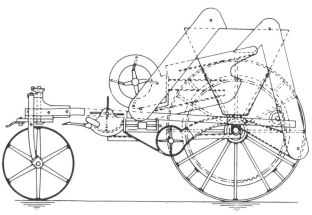

Australian patent No. 4394, October 19, 1905 to M.W. Charlton and D.E. Chapman was one of the foundation patents of the reaper-thresher produced by Massey-Harris for the Australian market. The patent disclosed a head which folded up for transport and a reciprocating cutterbar under the long-tooth comb. Drapers were a later addition to the design.

stepping stone". This proved to be somewhat optimistic. A further development was about to appear out of New South Wales—the "Header Harvester".

McKay's Self-Propelled Stripper-Harvesters

In 1909, H. V. McKay built the first of a series of self-propelled stripper-harvesters. Although he experimented with SP Harvesters for over a decade and even had one evaluated as far away as Spain, none was ever put into serial production. This work was contemporaneous with the Holt SP gasoline-powered combine, the first commercially successful SP combine.

Headlie Taylor of Henty

Headlie Shipard Taylor was born in 1883 and raised on the family farm at Henty, New South Wales. He showed mechanical abilities early in life, and as a young

Upper Right: McKay built several self-propelled stripper-harvesters, but they were never produced in volume. The first appeared in 1909, the same year that Holt released their gasoline-powered SP in California; Right: The Massey-Harris "No. 2 Reaper-Thresher" was introduced in Australia in 1912 with 8-and 9-foot cutting widths. Note the use of straw walkers located transverse to the frame.

Massey-Harris No. 2 Reaper-Thresher. 8-ft. cut

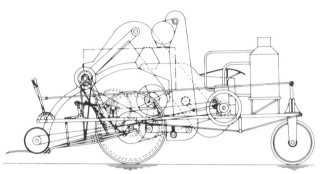

Clockwise from Upper Left: McKay's early SP Harvesters had tiller steering; An SP stripper-harvester with 24-foot stripper comb was evaluated by McKay around WWI. This experimental machine shows the method of conveying the crop from the individual stripper head extensions on the McKay machines; Australian patent No. 13,492 to H.V. McKay and H.S. Taylor, July 23, 1923 disclose an SP stripper-harvester.

man he rose to the challenge of improving the stripper-harvester to handle lodged and tangled crops;

"It was this failing that caused me to wonder if a machine could be devised which could handle the tangled mass of down crop without such loss," Taylor said. "In January, 1911, I was ready to start building the machine of my dreams. I had to enlarge the little farm workshop . . . I realised that progress would be slow and two or three years might pass before success or failure could be proved. But I was determined to give my plans a practical test—and there was no turning back. I had confidence in my ability to do the job itself but I was worried about finance. Would I be able to hold out long enough to accomplish my desire? I could only hope for the best."

Between 1911 and 1914, Taylor exhausted all his capital building two machines. The second succeeded in harvesting 200 acres, some of it down and tangled. He had taken out patents in October 1913, and borrowing funds to build a third machine he exhibited at the Henty show in 1914. In field trials that December, he consistently recovered several more bushels an acre than a stripper-harvester working alongside.

A key feature of Taylor's "header" design was the use of a knife under the comb and augers on the platform. There was no reel or draper canvas. The front auger swept the heads away from the comb and knife to the second auger which conveyed the crop across to the feeder-elevator, ready for transport to the threshing cylinder. The inspiration to use the augers had come to him in a flash as he contemplated two posthole digging augers leaning up against a shed.

The favorable publicity which his machine received at Henty attracted financial backing and so Taylor entered into a contract with Robinson & Co. to build headers at £1500 each. Three headers were built in 1915 by the Melbourne contractor, the firm later going into the business of producing their own headers under the "Federal" logo. The three Taylor machines were sold to farmers Nottle, Shipard and Kendall of the Henty district, New South Wales. Following this success, Taylor then wrote to McKay, the leading manufacturer in the business, and arranged for him to see the header working in a tall and rank crop on Mr. Kendall's farm. Said Taylor in 1916:

"It was with pardonable trepidation that I prepared to demonstrate the capabilities of my header in Mr. McKay's presence. I drove into the crop and McKay accompanied me, sometimes walking behind to observe its action and work. We went round the paddock and the shrewd McKay closely examined the machine and its work from every angle. He said the header was very simple, light in draught and incorporated many novel features. 'But,' McKay asked, 'what advantages do you claim over the Stripper Harvester?' I replied that I could handle a heavy lodged crop and get practically the whole of the grain from it. McKay then said he was prepared to negotiate for the patent rights. This was a tremendous thrill for me and we parted on the understanding that I was to visit Sunshine a few days hence to open negotiations."

McKay recognized in Taylor a man of similar cast. Hired in April, 1916, Taylor was installed at Sunshine, not only in charge of producing his machine, but responsible for all other machinery development as well. Taylor's new six-horse "Sunshine Headers", built at the McKay works, performed well during 1916-1917 harvest.

The demand for the Taylor machines grew and one thousand Sunshine Headers were scheduled for the 1920-21 harvest. The factory worked day and night to meet the demands of an anticipated bumper crop. When the time came, the machines were proven beyond dispute, for late storms throughout Eastern Australia flattened the fence-tall crops. The Headers were equipped with "Headlie" crop lifters (long wooden fingers that lifted the crop up to the comb), and the machines astonished farmers by gathering up to 30 bushels an acre from wheat crops they thought ruined. Never before had such tangled crops been harvested without sustaining severe loss. Australian farmers rescued millions of bushels of grain and the Sunshine Header won a reputation as "the greatest harvesting machine of all time".

Given "the freedom of the works", Taylor piled success upon success—the first once-over harvest of field peas; auxiliary gasoline engine-driven headers; Sunshine rice headers that were used in the first Australian-grown rice, and his crowning achievement; the Sun Auto Header.

The Sun Auto Header

The Sunshine SP Header or "Sun Auto Header" was a joint Taylor-McKay development. They took out their first patent in July, 1923 and others the following year. The Sun Auto Header was the first commercial combine with a T-shaped configuration of platform and separator. The great virtues of this arrangement that no crop was run down and there was a symmetrical feeding to the cylinder. The first production models, released in 1924, had a 12-ft cut and were propelled by a Fordson engine. Ground drive was through the single left hand side driving wheel. The Auto-Header continued with

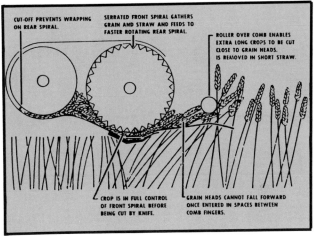

Above: Headlie Taylor with his 1914 header. The label ''header'' has stuck to Australian combines ever since Taylor's day. The term ''combine'' is reserved in Australia for combination cultivator-drills; Below: Cross-section shows the relationship and functions of the twin-auger platform designed by H.S. Taylor.

One of the 1915 headers made in Melbourne under contract for Taylor and purchased by A.J. Kendall of Henty. It was used by Kendall for 13 years to harvest 135,000 bushels of cereal grains. That's Taylor in this 1954 picture.

Taylor's comb-front and twin auger combination. By mounting two narrow fans on either end of the threshing cylinder shaft and placing ducting under the separator, the need for a separate winnowing fan for the riddles was eliminated.

Massey Acquires Australian Plant

Massey-Harris sales in Australia gained considerable momentum after WW1, a momentum that in 1930 was again checked by tariff legislation. Confronted with

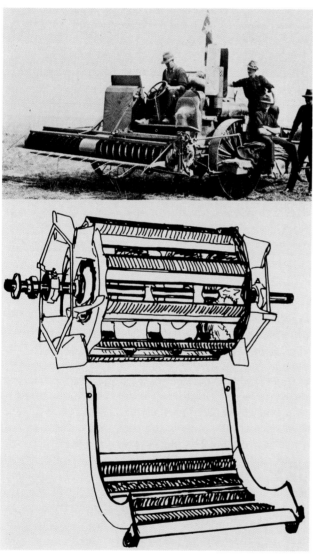

Clockwise from Top Left: The Sun Auto Header, introduced in 1924. The first models had a 12-foot cut and were propelled by a 4 cylinder Fordson engine. Ground drive was through a single wide driving wheel on the left-hand side; Australian patent issued to Taylor and McKay in 1923 was the foundation of the Sunshine Auto Header, the first commercial SP combine with a T-configuration; The Sun Auto Header was manufactured at Sunshine from 1925 to 1945; Semi-sectional elevation of the Sunshine pull-type header Model HST shows the twin fan arrangement on the cylinder shaft in place of a separate fan and housing.

prohibitive trade restrictions, Massey-Harris sold all its Australian assets to the H. V. McKay Company—in exchange for a minority interest in the re-formed H. V. McKay-Massey Harris Proprietary Limited. (Massey-Harris sold the assets but not the hyphen). Under the terms of the 25-year contract, the Australian company became the distributor for Massey-Harris products. After the Depression, there was rapid expansion at the Sunshine Harvester Works and by WW2 the plant covered almost two million square feet, all on one floor. A larger range of pull-type stripper-harvesters was introduced, along with headers, including the AL series, started in 1928 and produced continuously until 1954.

At the end of WW2, the company discontinued production of the Auto Header after building 1356 units. Only pull-type machines were built thereafter. In 1952, McKay-Massey Harris began importing Massey-Harris SP combines from Canada, again offering Australian farmers a machine of the type they had themselves pioneered earlier.

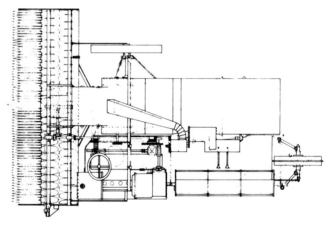

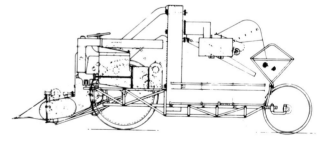

Above: Stripper-Harvesters were produced along with headers at the Sunshine Harvester Works in Australia continuously until 1954. The pull-type pto driven AL Sunshine harvester is shown. Note the "choke cutter" over the comb – a device for clearing weed blockages; Below: Sunshine pull-type No. 6 Header, 1949. Only pull-type machines were produced at the Sunshine Harvester Works after WWII until discontinued in 1955. The pto-driven machine is coupled to a Sunshine Massey-Harris tractor for one-man harvesting.

End of the McKay and Sunshine Brand Names

The 1930 agreement between McKay and Massey-Harris contained a clause that caught Massey-Harris by surprise in 1955. McKay had apparently acquired the right to use the Massey name on its machines in perpetuity! Thus the Massey-Harris-Ferguson merger of 1953 became extremely awkward in Australia. The Ferguson Company was already well entrenched. Ferguson distributors were outselling McKay-Massey Harris tractors 5 to 1. Massey-Ferguson found itself represented by two organizations! The problem was resolved, but not without rancor, after Massey-Ferguson acquired the Sunshine Harvester Works. Massey-Ferguson unceremoniously dumped the old Sunshine logo, installed new management, and effected drastic changes in the distribution network.

In 1957, production of self-propelled headers resumed at Sunshine with the release of the Australian-designed MF 585 SP header. This combine was offered with the option of either the Australian "closed" stripper comb harvester front or the American style "open" front. The advantages of the Australian "front" are: considerably higher forward speed capability; lower gathering losses; and reduced loading on the separator since less straw is ingested, due to the combing action of the stripper-type front. A drawback is that weedy and certain long-strawed crops cannot be effectively harvested, hence the optional open front. (Brown and Vasey 1967).

Practically all makes of combines offered in Australia—imported or local—are available with the closed front option. Massey-Ferguson's combines have retained the parallel lift arrangement for their closed front designs.

In 1969, Massey-Ferguson (Australia) Pty. Ltd. produced the 542 combine series, with 42 in. cylinder width. These combines have several unusual features: adjust-able open grid-grate under the cylinder beater; transverse flow fan (first in the industry); rear fuel tank and towing ball built into the separator hood. The 542 pull-type combine has a high lift capability on the comb, enabling it to clear gate posts as high as 48 in.

Connor-Shea's Auto Header

During the change from McKay to Massey-Ferguson, two engineers, Tom Connor and Les Shea, left the company to branch out on their own. They founded Connor-Shea Limited at Sunshine in 1952 to build a range of cultivating, seeding and harvesting equipment, including a tractor-mounted harvester. Theirs was a new approach. The tractor was reversed and the header mounted over the top, with the controls and operator's station strategically located as on a regular combine.

Although the Connor-Shea combine was discontinued in 1976 and production never exceeded 50 units per year, it was a significant contribution to the art. It had the following features: widest gathering front on a post-war combine; closed fronts from 15 to 27 feet wide were available; the gathering front and elevator were raised parallel, up to a maximum transport clearance height of 53 in. on one version; it was possibly the fastest combine in its day—being able to utilize the tractor transmission, with road speeds up to 20 mph; it was designed with ease of cleanout in view and could be cleaned out faster (4 hours) than any other field combine—an important feature in Australia where over-wintering insects in combine residues may re-infest the grain harvested at the beginning of the next season.

Horwood-Bagshaw/Shearers of Mannum

The 1975 merger of these two firms brought together two of the oldest farm machinery enterprises in Australia.

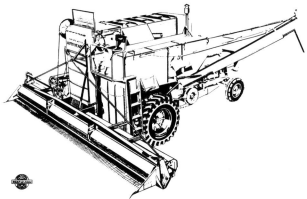

Above: MF 585 SP was the first post-war self-propelled combine produced in Australia; Below: The largest front offered on the Connor-Shea Auto Header was 27 feet – possibly a record width for a post-war combine. The more popular size was the 23-foot front.

Massey-Ferguson (Australia) Pty. Ltd. produces the SP 542 Series 2 Header in Australia. This 42 in. cylinder-width SP combine has several unusual features, such as parallel lift front, adjustable grid grate, transverse-flow fan, integral rear fuel tank and tow bar built into the separator hood.

John Stokes Bagshaw, founder of the Horwood-Bagshaw line in Adelaide in 1838, was Australia's first manufacturer of winnowing machines. It was J. S. Bagshaw who in 1843 was contracted to fabricate key assemblies for Ridley's first stripper. His company expanded into tillage and other equipment, but did not produce combines until 1924, when the first H series header-harvesters were sold. Stripper-harvesters were also produced at the Mile End works in South Australia from 1932-1936. The company's first SP combine was the model OH, produced between 1964 and 1969. In 1975, the company assumed the management of Shearers of Mannum, and began selling the Shearer line of harvesters, at the same time becoming Australian agents for the Italian Laverda SP combines.

The Shearer company was founded on the banks of the Murray River at Mannum, in South Australia in 1877 by brothers John and David Shearer. Beginning with tillage tools, the Shearers' product lines grew rapidly, and in 1883, expanded into strippers, wrought

steel plow shares and even a steam automobile, the first in the world with a differential. In 1910, the partnership dissolved. David Shearer remained at Mannum making harvesters, harrows and plow shares, while John Shearer established his business at Kilkenny making tillage implements only.

In 1912, with the expiration of the original McKay Stripper-Harvester patents, Shearers of Mannum began making ground-driven stripper-harvesters in 6-, 7- and 8-ft sizes. The first Shearer header, a 9-ft pull-type model, was produced in 1927. This was the first of a series of well-received pull-type combines that have been produced continuously at the Mannum works. The company developed their own SP machine, the XP88 in 1967, and produced this machine until the acquisition of Shearers of Mannum by Horwood-Bagshaw in 1975. Only the pull-type model "1070 Centenary" header was produced at Mannum after the takeover.

International Harvester in Australia

The two multinational corporations, Massey-Ferguson and International Harvester, were building combines for Australia before they started building combines for North America. The first IH model was the No. 1 Stripper-Harvester, produced in the US in 1910 exclusively for markets in the Southern Hemisphere. It was not until 1914 that IH began manufacturing combines for the North American market.

In 1924, the Melbourne firm, Gaston Brothers began producing pull-type header harvesters. Eight years later, IH signed an agreement with Gaston to manufacture Gaston's pull-type machines under the IH logo. Following WW2, IH began manufacture of their own

Top Left: Shearer Stripper, 1914; Top Right: Shearer pull-type combine, 1930; Bottom: Shearer 1070 pto header; Left: David Shearer XP88 SP header.

A new dimension in header design.

Packed with big capacity features

Shearer of Mannum
1070 PTO header

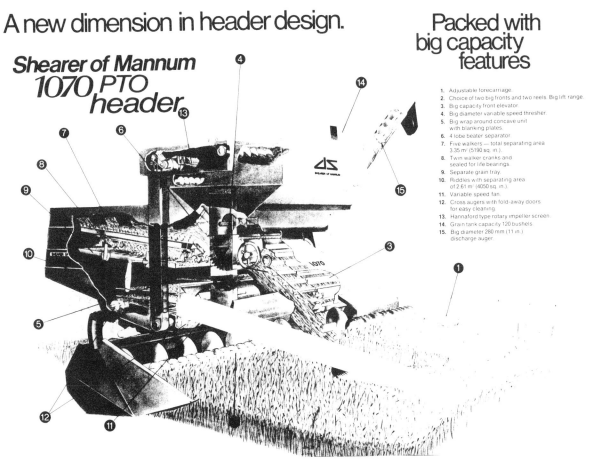

1. Adjustable forecarriage.
2. Choice of two big fronts and two reels. Big lift range.
3. Big capacity front elevator.
4. Big diameter variable speed thresher.
5. Big wrap around concave unit with blanking plates.
6. 4 lobe beater separator.
7. Five walkers — total separating area 3.35 m² (5190 sq. in.).
8. Twin walker cranks and sealed for life bearings.
9. Separate grain tray.
10. Riddles with separating area of 2.61 m² (4050 sq. in.).
11. Variable speed fan.
12. Cross augers with fold-away doors for easy cleaning.
13. Hannaford type rotary impeller screen.
14. Grain tank capacity 120 bushels.
15. Big diameter 280 mm (11 in.) discharge auger.

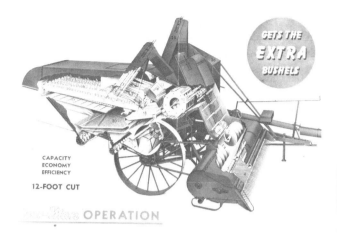

GETS THE
EXTRA
BUSHELS

CAPACITY
ECONOMY
EFFICIENCY

12-FOOT CUT

OPERATION

Left: The GL200 pull-type combine was the first Australian-built IH combine. The 12-foot comb front machine was produced between 1947 and 1956; Below: The A8-2 was the first IH Australian-built SP combine. It was offered with an 8-foot open front or 10-foot closed front option.

combines at Geelong, Victoria, with the GL200 pull-type, first produced in 1947. The first IH SP combine built in Australia was the model A8-2 in 1959, which came with an 8-foot open front or 10 foot comb (closed) front. In 1978 production centered around the 711 SP and 710 pull-type, 38-in. recessed-cylinder machines with six-walker separation.

Summary—Australia's Unique Contribution

Around 1840 a Government official stated categorically that Australia would never keep itself in food! Nowadays, Australia is one of four major grain-exporting nations. Two bushels of wheat out of every three produced are shipped overseas. This success has been due to the Australian farmer's perseverence with low fertility soils under an often perverse climate, the rapid adoption of advanced agronomic practices, and major innovations in farm machinery. Australian farmers have been pioneers in the business of broadacre farming under low-rainfall dryland conditions. (Thompson 1976).

By its very nature, dryland cereal production involves large acreages which need to be covered at high speeds. Harvesting at 7 mph or faster is not uncommon. That grain loss levels can still be kept low is in large measure due to the unique stripping-comb gathering head design used in Australia. Developed in stages from the stripper, to the stripper-harvester to the header, the closed front design ensures that the grain-to-straw ratio is considerably higher than found on American machines.

Thus, with less straw, far higher work rates are possible for a given horsepower and size of machine. At the same time, the combing action restrains the heads, prevents grain shatter and saves heads from dropping onto the ground at the front.

Practically all Australian grain is harvested and handled in bulk and has been since around WW1. Much of it is passed through secondary cleaning systems on the machine in the field, ensuring a cleaner sample in the bin.

The Australian market is not a large one—sales average about 3000 combines a year—yet it is lucrative enough that most of the free world's manufacturers attempt to distribute machines Down Under. The result is that, per capita, Australians have a higher density of combines, or headers as they still call them, than in the US.

Tariff restrictions still apply—although only on combines of 48-in. cylinder-width or less. This has ensured that practically all the smaller SP headers and most of the pull-types—a steady 40 percent of the market—are made in Australia.

14
Massey-Harris/ Massey-Ferguson

The Massey implement business began on a farm. Daniel Massey, the founder, was born in Upper Canada, and as a teenager plowed new land near Coburg in the future Province of Ontario. In 1830, while visiting relatives in New York, he purchased and brought home Canada's first threshing machine. He established a small smithy in 1840 for repairing implements and building plows, scufflers, maple sugar kettles and repair parts for threshing machines (Massey-Harris 1947).

In 1847, Massey purchased a factory and equipment at Newcastle, at a site on the new Grand Trunk Railroad, a mile from Lake Ontario's northern shore. He began manufacturing tools and farm utensils, eventually adding harrows, cultivators and fanning mills. His son, Hart A. Massey, was named superintendent of the works shortly after, and, in 1855, became sole proprietor. Hart Massey remained head of the Massey enterprise until 1896.

In 1852, the Masseys acquired the Canadian rights to produce the Ketchum Mower, the first Canadian-built mowing machine. Manny's Combined Hand-Rake Reaper and Mower followed in 1855. By 1857, while McCormick was producing thousands of reapers annually, the little Newcastle plant recorded total sales of 166 implements, including a few sales beyond Canada's boundaries. In 1861, Massey introduced the famous Wood's Self-Rake Reaper, first of this class of machine in the Dominion.

An unprecedented demand for Canadian wheat in the 1860's resulted in great fleets of ships shuttling across the Atlantic by way of the recently-completed St. Lawrence Canal system. They carried wheat and lumber eastward and returned with settlers for Canada's

Above: Daniel Massey, 1798-1856. Pioneer Ontario farmer and founder of the Massey works. He purchased a "Bull" thresher, the first thresher to be brought into Canada, in 1830; Below: The original "Newcastle Foundry and Machine Manufactury" established by Massey at Newcastle, Ontario in 1847. The name was changed to H.A. Massey and Company when son Hart took over as sole proprietor in 1855.

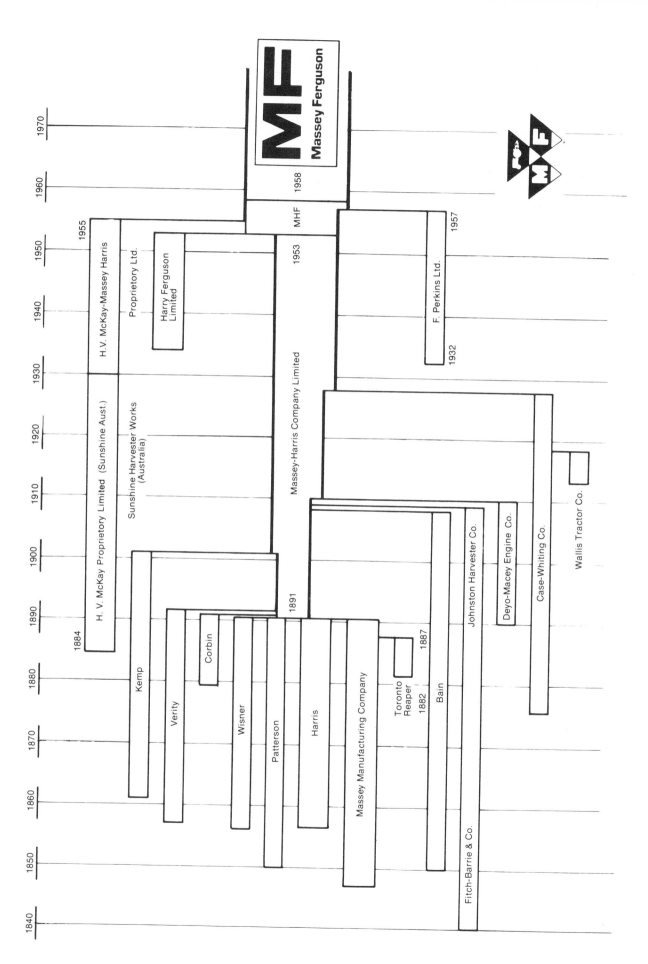

Major lines of development of Massey-Harris, Massey-Ferguson Companies.

Harris Works, Beamsville, 1857

Above Left: Alanson Harris, 1816-1894, founder of A. Harris and Son implement works at Beamsville, Ontario; Below Left: The flop-over or revolving hay rake was one of the first implements to be built by A. Harris and Son; Above: Harris' Beamsville, Ontario works founded in 1857.

expanding farm lands. The Newcastle plant worked overtime to meet demands for farm implements.

The Harris Branch of the Massey-Harris Company

The development of the Harris company paralleled Massey's and that of other Canadian enterprises; they were predominantly local in character and diversified in output (Phillips 1954). Alanson Harris, the founder, was the son of Elder John Harris, an itinerant American preacher who had moved into Brant County to farm and to preach. He loathed farm drudgery and resolved to develop machines to lighten the burden. An early invention was the first flop-over hay rake which he and his son built at their mill near Brantford. The beginnings of the Harris implement business were even less pretentious than Massey's. Production of farm tools began at Beamsville in 1857 when Alanson Harris purchased a factory alongside the tracks of the Canadian Great Western Railway which connected Buffalo and Detroit.

Before long, Harris was joined by his son, John, and the business was named A. Harris and Son. In 1862, the Company acquired the rights from D. M. Osborne and Co., New York, to make the Kirby Hand Rake Reaper & Mower. They eventually developed a wide line of equipment. In 1857, another American contributed to Canada's youthful farm equipment industry. J. O. Wisner left New York State to settle in Brantford where he established the firm of J. O. Wisner & Son, specializing in seeding and tillage

machinery. That same year, W. H. Verity, a new arrival from the British Isles, began casting his famous iron plows at Francestown, Ontario.

In 1861, the first Canadian census (which contained figures on implement production) showed 46 firms with American licensing arrangements. These companies, producing for a protected market, enjoyed lower costs than their American competitors who were paying as much as 20 percent in tariff duties. For the American companies holding key patents, royalties provided a source of revenue not subject to Canadian tariff duties.

The 1871 Canadian Census recorded 36,874 reapers and mowers in use on Ontario farms—indicative of the tremendous rate of expansion since Massey produced the first machine just 19 years earlier.

First Overseas Markets for Canadian Equipment

Hart Massey decided to specialize almost exclusively in harvesting machinery during the 1860's. Massey's harvesting machinery was chosen at an Industrial Exhibition in Toronto to represent Canadian manu-

Hart Massey obtained the rights to make Walter A. Wood's combined reaper and mowers in 1861.

133

Family officers of the Massey Company, 1853-1903 grouped for a photo taken in the early 1880's. Left to right: Chester D. Massey, president 1901-1903; Hart A. Massey, father of Chester and Charles, president of H.A. Massey & Co. 1853-1870 and Massey Mfg. Co. 1870-1896; Charles A. Massey, vice president 1871-1884.

THE TORONTO MOWER No. 2 IN THE FIELD.

Above: The "Toronto" No. 2 mower in the field. The quiet running Whitely wobble gear drive made this mower a popular machine in Canada by 1879; Below: Eight 7-foot cut Brantford all-steel Harris binders at work on the 14,000 acre Edgeley farm, Qu'Appelle Valley, Manitoba.

facturing at the 1866 Paris International Exposition. The company was awarded several medals at Paris and received a German purchase order for Massey-Woods equipment. This sale marked the beginning of Canada's world-wide export trade in manufactured goods.

Unprecedented sales growth resulted in a building boom at the Newcastle works which eventually boasted 10 buildings and a 60-horsepower steam engine. In 1878, the company decided to move to Toronto, the financial and distribution hub of the Province. The transfer of the renamed Massey Manufacturing Company to the six acre site known as "The Old Exhibition Grounds" took place in 1879.

The Toronto Reaper and Mower Company, an enterprise formed in 1877 with American and Canadian capital, was perhaps the first American branch operation in Canada. This rival was located close to the Massey works on the "outskirts" of Toronto. Each day, as he passed it on the way to work, vice-president Charles A. Massey felt increasingly apprehensive, especially after just two years of operation the burgeoning company boasted that they would sell two-thirds of all the mowers made in the Dominion. Their "Toronto" line was an improved version of the Whitely line of wobble gear "Champion" harvesting machines.

Expansion of the Massey Company

The "Toronto" Company, however, was in financial trouble by 1881. They had overstepped the mark in their attempt to launch a light binder and Massey Mfg. Co., was then able to purchase all the assets of this precocious rival. With the acquisition, Massey doubled its output. One hundred Massey "Toronto" twine binders were successfully test marketed in 1882. The following item appeared in the 1887 Massey catalog:

"We have this year for the first time completed a systematic and efficient organization in Great Britain and the Continent for the wider distribution of our goods. In England, Scotland, Ireland, France, Germany, Belgium, Russia, Asia Minor, South Africa, South America, West Indies, Australia our machines are at work, and during the past season have given remarkable satisfaction".

Opposite: 1893 Massey-Harris advertising boasted of Canadian production reaching 49,737 machines and implements which were distributed world wide.

In 1889, the Massey Toronto light binder defeated the world's foremost makes of harvesting machinery to win first prize in the International Field Trials held in connection with the Paris International Exposition at Noisiel, France. Victory was achieved because the Massey entry was lighter in measured draft and was operated by a sole driver-serviceman, in contrast with the retinues of mechanics and larger horse teams furnished by other competitors.

The Harris Twine Binder

In the spring of 1881, John Harris went to Texas to follow the progress of the grain harvest northward so he could evaluate American twine binders. He returned to Brantford in September convinced he could build a superior self-tying binder. Twenty-five "Brantford" steel-frame binders built in 1882 were the result. Based on successful tests, 1,000 were advertised for the 1883 season.

Massey-Harris binders of 1904 were of the "open-end" type and could readily handle varying straw lengths. They were manufactured in 5-, 6- and 7-foot cutting widths.

The opening of the Canadian West, high national tariffs and a widening market for the Canadian firm demonstrated signs of approaching maturity and growing independence from American designs. Although the smaller of the two leading Canadian organizations, the Harris Company was no less enterprising, as demonstrated by its aggressive marketing campaign to sell open-end binders in Europe. Some 5,000 binders were in use for the 1883 Canadian harvest—2,000 produced by Massey and Harris, with the remainder coming from 25 to 30 other manufacturers or sales agents.

Massey and Harris Unite

By 1890, the harvesting machines produced by the two leading Canadian companies were practically identical, and both had highly competitive products. Their production, sales, distribution and service organizations were duplicated across Canada and in a dozen foreign countries. The growing success of the Harris Company abroad spurred the Masseys into initiating merger talks.

After two months of negotiations, the Massey-Harris Company Limited was formed in May 1891, with a capital of $5,000,000. Hart A. Massey was named president. Joining the new firm were executives and family members of both companies. Consolidation brought immediate advantages—the best features of the Toronto and Brantford harvesting lines were combined and distribution costs lowered by eliminating duplicated lines and agencies in Canada and abroad. Within a year these savings were passed on to farmers in the form of lower prices. The merger made Massey-Harris the largest manufacturer in Canada, but added no new product lines.

Two Canadian competitors, alarmed by the threat to their position, took similar measures. In Ontario, The Patterson and Brothers Company, of Woodstock, and J. O. Wisner, Son and Company, of Brantford, were merged just three months after Massey-Harris. Though smaller in size, the new Patterson, Wisner Company had a formidable advantage—it offered a full line of machines. The Pattersons were producing the Johnston self-binder, along with tillage and harvesting implements, while Wisner brought seeding and cultivating machinery to the enlarged company.

Massey-Harris had ample reason for wanting this diversified company under its control. Before the year was out, there appeared another announcement, the absorption of Patterson, Wisner Co., into Massey-Harris. Other acquisitions followed, as the Canadian company sought to round out its product line with a year-round variety of machines.

In 1893, Massey-Harris entered the American market. The shipment of a train-load of exhibits to the World's Columbian Exposition at Chicago was hailed as an exciting and conspicuous event. *Farm Implement News* issued an 1893 World's Fair special edition in which the following comment appeared:

"We must frankly acknowledge that one Canadian concern, the Massey-Harris Company, with head office at Toronto, and factories at Toronto, Brantford and Woodstock, have the largest, the fullest and the finest exhibition in the whole building."

In 1910, Massey-Harris began US operations by acquiring control of the Johnston Harvester Company of Batavia, New York. This long-established firm manufactured grain headers and had dealt with Massey as far back as the 1870's.

In 1892, Massey-Harris produced a total of 41,474 machines and implements. By the eve of WW1, production had risen to 105,858 units, of which 59 percent were exported. The intervening decades had seen the company play its part in the agricultural conquest of the Canadian West and the establishment of a world-wide export business.

Massey's First "Combine"

The Australian-designed stripper harvester was the progenitor of a long line of combined harvesters pro-

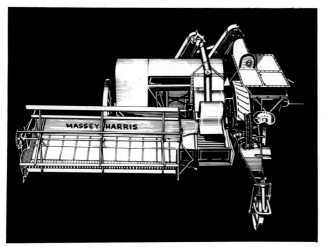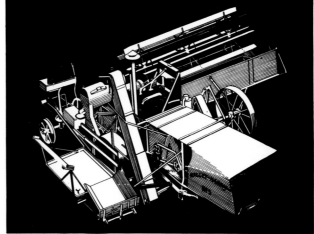

Left: 1912 Massey-Harris No. 3 Reaper-Thresher designed for the Argentine market was similar to the Australian No. 2, but had a large bat reel and knife to replace the comb-type front of the No. 2; Right: The No. 5 auxiliary engine-functioned Reaper-Thresher, 1922.

duced under the Massey name. Manufactured in Toronto for the market Down Under, the stripper was first sold in Australia in 1901. Soon after, markets for strippers were found in South America, and then North Africa. The stripper-harvester was never seriously adopted in North America, although it was tried in several wheat growing regions. The self-binder continued to maintain its universal appeal.

Massey-Harris Enters the Combine Business

In 1910, Massey-Harris offered its first combine for sale, but not in North America. The No. 1 Reaper-Thresher was designed by Australian Massey-Harris representatives for the Australian market as a means of bypassing prohibitive tariffs on imported stripper-harvesters. Local industry considered the Canadians "pirates".

Two men could harvest and bag 12 to 25 acres a day with the 8½-ft cut MH No.1 Reaper-Thresher. This machine incorporated the comb-stripping principle, but used a reel and cutterbar type of gathering mechanism. The improved No. 2 Reaper-Thresher was released in 1912, again for the Australian market, and in that same year the Model No. 3 with large bat-reel was designed specifically for Argentinian crop conditions.

The onset of WW1 created a labor shortage and an unprecedented demand for farm machinery. The Company's export sales far exceed domestic volume. But with the end of the war came economic chaos. The Company found profits inextricably tied to the price of wheat—and wheat prices were disastrously low in North America. Fortunately, export sales held their strength. For example, in 1921-23 Massey-Harris binder exports from the Canadian works totaled 24,867 units, or the equal of one-third of all the American competitors added together.

No. 5 Combine Becomes the Massey-Harris Line Leader

Meanwhile, the No. 5 Reaper-Thresher had been developed, the first auxiliary engine-driven harvester in the Massey-Harris line. The No. 5 could be either horse- or tractor-drawn. The engine provided the power for the machine in place of the old ground-wheel-drive system. The No. 5 was released first through the Argentine Branch which opened in 1922. The sales response was immediate—a refreshing response in view of depressed conditions at home. In 1923, 1500 machines were sold in the Argentine alone.

Curiously, although Reaper-Threshers had been manufactured in Canada since 1910, there were few sales of these machines in Canada's own wheat growing regions. That the local development should enjoy greater prestige abroad than at home was a phenomenon all too familiar to Canadians in many fields of creative endeavor (Brown 1967).

In 1922, the Company loaned a No. 5 to the Swift Current Experimental Farm in Saskatchewan. Tests by Dominion agricultural engineers demonstrated the practicability of the Reaper-Thresher by proving that harvesting costs were less than one-half those of the binder-thresher method. The No. 5 could harvest 25 to 40 acres in a ten-hour day with a two or three man crew.

The Dominion station purchased the machine. Encouraged, the Company proceeded to loan machines to selected farmers in Alberta and Saskatchewan. The door had been opened. Sales of the Reaper-Thresher exceeded 100 in the Prairie Provinces by 1927. With the acceptance of these first combines, the curtain began to descend on the annual Summer steam threshing festivals that had become a familiar scene on the Prairies.

1924: Soybeans, and the Beginnings of Modern Power-Farming

The Garwood Brothers, farmers of Stonington, Illinois saw a future for soybeans as an oilseed rather than as a forage crop, but harvesting was a problem. Elmer J. Baker, "The Reflector", of *Farm Implement News* fame was instrumental in having Massey-Harris deliver a Canadian combine to Illinois for the 1924 soybean harvest. Domestic combine manufacturers had shown no interest. Working with the slightly modified No. 5 Reaper-Thresher, the Garwood Brothers made a commercial success of combine-harvesting their soybean crop. Yes, Illinois! Strategically a very important territory for the Canadian machine. This was claimed as the first use of a combine East of the Mississippi. Successes in Indiana followed, then the Louisiana rice crop fell before the cutterbar of the Combined Reaper-Thresher. "The Reflector" wrote in the November 20, 1924 issue of *Farm Implement News:*

> "The adaptation of the combined harvester to soybeans may open up a market of profitable proportions... Heretofore there has been no machinery that harvested soybeans for seed to the satisfaction of the growers... With the harvester-thresher it has been shown possible to cut and thresh the beans in one operation with minimum shattering and at low cost. The price received for soybean seed is sufficient to justify the large grower to purchase a machine as expensive even as a combine."

The year 1924 was a pivotal one for farm mechanization. The post-war depression was over. Improvements in the combine and the real beginnings of tractor power farming can be traced to this period:

- the lightweight tractor began to replace the horse;
- the tractor power take-off and power lift were developed;
- the one way disc seeder gave the wheat farmer a high-capacity soil preparation and seeding combination to match the major advance in harvesting capacity offered by the combine;
- the number of trucks used to transport produce to market was increasing dramatically.

The closely-knit family character of the Massey-Harris Company was abruptly changed in 1925 when M-H president Vincent Massey announced his resigna-tion to pursue a political career. In 1927, Massey-Harris became a publicly held corporation. That same year the Wallis Tractor Co., was acquired to round out the Company's product line. And in 1927, the No. 9 Reaper-Thresher, with bulk grain delivery and 12- or 15-ft header, became the line-leader. By 1928, Massey-Harris was turning out 35 pull-type combines a day... one every 14 minutes.

Australian Reaction in the Depression

The winds of prosperity that blew to gale force in 1929 were exhausted by 1930. The period of chaos that followed made the depression that had followed WW1 seem trivial. The price of wheat fell to the lowest cost per bushel ever. In Canada, rockbottom was 38-3/8 cents per bushel at the Lakehead in December, 1931. Fortunately, a six million dollar tractor order from the Soviet Union kept Massey-Harris going, but during the depths of the depression, plant operations plummeted to as low as 10 percent of capacity. To add to the gloom, an important segment of the Company's export business was seriously impeded in April of 1930, by the Australian Government's decision to prohibit the importation of certain classes of machinery and to levy a heavy import tariff on the remainder. To meet this situation, Massey-Harris merged their Australian sales organization with H. V. McKay Proprietary Limited of Melbourne. As the basis of the merger, Massey-Harris sold its assets and goodwill in Australia to H. V. McKay and bought stock in the Company amounting to 26 percent of equity. Massey-Harris retained their identity in the Australian market by a change in name. It was now H. V. McKay-Massey Harris Proprietary Limited. McKay was given the rights in perpetuity to manufacture M-H machinery in Australia, a deal that was to plague Massey-Ferguson 25 years later.

Thomas Carroll, Massey's Man Down South

The importance of the 1924 Australian Taylor-McKay Sun Auto-header self-propelled combine was not lost on another Australian farm boy, Thomas Carroll, who became Massey's harvester expert in South America. Carroll later recalled how his early farm experiences led him into the farm machinery business. In particular, there was an old German migrant thresherman who:

> "each year made the rounds of the farms with his big lumbering steam engine and thresher to thresh the oats in our district. Usually by the time he came our way he decided he needed a break and would hit the bottle. Far be it for me to have wished him the worst for such indulgence because on such occasions it gave me the chance to run his steam engine and threshing machine; sometimes for several days at a time. For a boy of ten years this was a memorable experience and not without its influence... I fortunately dedicated myself in later years to farm machinery and not the bottle." (Carroll 1961).

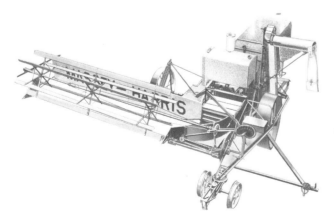

The No. 9 Reaper-Thresher, 1927. First Massey-Harris combine with bulk grain delivery and a separate re-thresher.

Carroll gained machinery production and field experience, first with the Buckeye Company, then with J. J. Mitchell & Co., in Australia. In 1911, he was hired by Massey-Harris' Argentinian distributor to help introduce the No. 1 horse-drawn Reaper-Thresher. The No. 1 was not entirely suitable and Carroll's recommendations led to the development of the modified No. 3 for the Argentine. In 1917, he joined the Toronto staff as a product design leader. The No. 5 Reaper-Thresher was one of the first machines designed under his direction. During his career, he helped introduce such mechanical improvements to the combine as welded sections, roller chains, oil-bath gears, anti-friction ball bearings and the quick-detachable table for road transport. Carroll was a tireless traveler. He returned to the Argentine in the 1930's where he saw engine powered pull-type machines converted locally to SP combines similar to the Sun Auto-Header, but using a second engine coupled to the drive wheels. Then in 1938, a Sun Auto-Header harvested 3300 bushels of Australian wheat in one day, a record that stood unbeaten for the next 33 years.

Carroll realized in 1936 that his company must produce a self-propelled machine for Argentina or local manufacturers would meet the demand, seriously jeopardizing Massey's largest export market at that time.

In 1935, James S. Duncan, fresh from reorganizing the Massey's Argentine business, was appointed general manager in Canada. Duncan's attitude, experience and energy were just the ingredients needed to pull Massey-Harris out of the Great Depression. He had two personal goals:

(1) to develop the US market, which he perceived had always been, and would be in the foreseeable future, the world's largest consumer of agricultural implements;
(2) to develop a Massey-Harris SP combine.

When he was made Company director in 1936, he proceeded to implement both goals.

One immediate result of the new leadership was the development of the "Clipper" combine at Racine, Wisconsin, under the engineering direction of E. C. Everett. The Clipper was direct competition for the Allis-Chalmers' ALL-CROP, a "scoop-type" or straight-through machine built for smaller acreages. The popular Clipper combine put the ailing Johnston plant at Batavia back on its feet. Clippers were manufactured in that plant from 1938 to 1952, an unusually long time for a basic model.

Development of the SP Models 20 and 21

Tom Carroll was appointed chief engineer in charge of developing Massey's first SP combine, the Model 20 SP. Designed without constraint on cost or weight, the No. 20 was tested late in 1937 in the Argentine, just eight months after Carroll had been given the

The Massey-Harris No. 15 Reaper-Thresher, 1937. An 8-foot machine, 3200 lbs. lighter than the previous model. It was tractor pto powered and mounted on rubber tires.

go-ahead. Limited production followed during the next two years. The 16-ft cut model No. 20 was according to Carroll, "somewhat heavy and expensive. We realized that we did not have the complete answer yet to the volume market". With the No. 20, the name Reaper-Thresher was dropped forever by Massey-Harris. The term "combine" had come to stay.

Carroll began developing a lighter, lower cost 12-ft cut model with "a capacity and price to suit the average farmer in most countries of the world". The outcome, on the eve of WW2, was the famous No. 21 SP combine, first tested in 1940. In spite of war commitments, the company rushed the design into mass production. The first units came off the line in 1941, just as the emergency-measures directive came to freeze all design

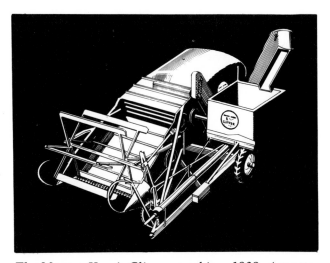

The Massey-Harris Clipper combine, 1938. A tractor pto-driven scoop-type with 6-foot table and straight through 60 in. threshing and separation sections. Designed primarily for the post-depression American market.

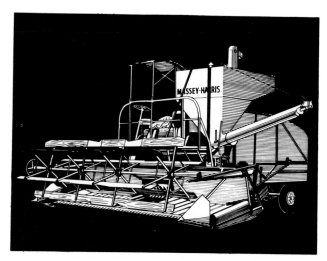

The No. 20 self-propelled combine. Developed by Thomas Carroll from ideas nurtured in the Argentine; 16-foot cutterbar, 65 HP Chrysler engine. After eight machines were evaluated in Argentina, 925 units were built in 1938.

work on farm machinery. The first move to push ahead with the No. 21 gave Massey-Harris a greater lift than could ever have been imagined at the time, and that boost came first from the prized US market.

The Massey-Harris Harvest Brigade

Once more a world war brought farm labor shortages and an unprecedented demand for food. Fortunately, a highly mechanized agriculture existed in North America to meet the challenge. As the demand increased, so did the acreage and intensity of cultivation. The 1942 cereal crop was the largest in the history of the Continent. For the first time, the American corn crop exceeded three billion bushels, the wheat crop approached a billion bushels and the Canadian wheat crop topped all previous records with over one half billion bushels. Grain storage facilities were filled to overflowing. Yet, despite the record of abundance, the insatiable demand for food continued and rationing was eventually introduced. Farmers were pushed to produce even more, at a time when farm machinery and parts were in short supply and manpower was drained by the armed forces.

In 1944, a goal of one billion bushels of wheat, with the demand for other crop plantings in proportion, was set by the US food production board. Implement makers, acutely aware of the shortcomings in parts and machines, urged that they be permitted larger allocations of scarce raw materials. But farm machines were built of steel, and steel was needed for ships and guns and tanks and other war materiel. The best that the Combined War Production Boards could promise was some lessening of restrictions. Implement men brooded over the situation; none more so than Joe

Tucker, vice president and sales manager of Massey-Harris, USA. Tucker found a solution. Through dogged persistence he obtained permission to build 500 combines beyond the quota as a unique aid to the war effort. Tucker convinced the Board that SP one-man combines could harvest more bushels of grain for a given investment in steel than any other machine or combination of machines in existence. The combines were to be sold to custom-operators who would contract to harvest a minimum of 2000 acres each, under Massey-Harris supervision. Operating on the scale projected, the machines of the SP "Harvest Brigade" could be expected to harvest over one million acres and bring in a minimum of 15 million bushels of grain. They would also release over 1000 tractors for other work, save a half million gallons of fuel and a half million bushels of grain that would otherwise be lost if tractor-drawn combines were used to open up fields.

Early in March, 1944, memories of earlier "delivery days" at Massey-Harris were recalled by the sight of 30 flat cars being hauled out of Toronto on rail with the first installment of the 500 red No. 21's, brightly emblazoned "MASSEY-HARRIS SELF-PROPELLED HARVEST BRIGADE."

The Massey-Harris Harvest Brigade claimed an SP victory as ripening crops were harvested by 500 machines operated with military precision to sweep over one million acres in 1944.

The long and well organized campaign came to a halt 1500 miles north of its starting point in September 1944. Massey-Harris was able to report to Washington that they had made good their promise. Altogether the 500 machines had harvested 1,019,500 acres, added over 25 million bushels of wheat and other grains to the granaries, had saved a third of a million man-hours and a half million gallons of fuel.

The top operator, Wilf Phelps of Chandler, Arizona, had cut 3,438 acres. In all, 500 American farmers, many seeing a combine for the first time, had their crops cut by the Harvest Brigade.

Right: Harvest Brigade promotional material included a map showing the area traversed by the ''Brigade'' of custom operators who, during World War II, paved the way for the general acceptance of the self-propelled combine in the US heartland; Below: Massey-Harris No. 21 self-propelled combine first tested in 1940. This was the machine used by the Harvest Brigade.

In 1944, *Fortune* magazine reported:

"The self-propelled carried out an elegant light-armoured blitz of the U.S. wheat belt... For Massey-Harris he (Joe Tucker) had driven his peaceful column of commerce through U.S. heart land in a campaign that has few parallels in farm tool history."

When the company completed its major arms contracts for the Canadian and British Governments during 1944, it received significant recognition by the Canadian Government and was relieved of further war work so that it might concentrate on the farm machinery needs of North America. The Canadian implement concern was able to convert to peacetime manufacturing almost a year before Japan's surrender. At about the same time, the US-Canada free trade agreement removed the custom barrier, enabling Massey-Harris to continue the integration of their North American operations. Thus, in late 1945, Massey-Harris was in a unique position, thanks to early reconversion to peacetime production, to take advantage of their enhanced reputation in the American market and of the enormous postwar demand for farm machinery in the more than 70 countries where their implements were sold (Massey-Ferguson 1964). New products quickly added to the line included a self-propelled

version of the "Clipper" combine, and new lines of tractors and implements.

North with the Wheat Cutters

In 1946, American sales of Massey-Harris machinery exceeded Canadian sales for the first time. The SP combine continued to gain in popularity, extending its revolutionary influence on American agriculture. Just how extensive was fully realized in 1947 when the numbers and mobility of combines gave rise to custom-combining on an unheard of scale. Custom cutter crews with clusters of machines capable of harvesting 500 acres a day and 50,000 acres a season rolled across American wheatfields.

In one small Kansas town in 1947, no less than 2,449 SP combines, mostly Masseys, were counted passing through in just five days!

For Massey-Harris SP combine was the key that opened the doors to the US market. In 1938, with only pull-type combines to offer, Massey-Harris sales represented a mere 3 percent of the US Combine market. By 1948, thanks to the SP machines, the company had captured 52.9 percent of total US combine sales (Neufeld 1969). SP sales increased overseas too, but these were all supplied from the North American plants—until 1949.

Massey-Harris self-propelled ''Clipper 50'' combine was introduced in 1946 with 7-foot cutterbar. Designed in Racine, Wisconsin under the supervision of E.A. Adams, who chose the name for this series of machines produced at Massey-Harris' Batavia plant.

Massey-Harris Produces Combines in Europe

In 1946, Massey-Harris decided to manufacture combines in Britain. The first European-built M-H combine, the SP model 726, rolled off the Kilmarnock, Scotland, production line in 1949. Four years later, the Company's division in Marquette, France, began producing the Model 890, and in 1954 the Eschwege works in Germany released their SP Model 630. German combine sales enjoyed a remarkable growth and exceeded 5000 units a year to make the German division the most profitable in the company with gross sales exceeding even those of North America in 1957.

the driver to stop, got out, and produced a silver half-crown. They agreed to a toss over the matter. "Heads he called, but tails it was". Massey-Harris won and the "million dollar coin" was subsequently mounted and set in silver on a gift box given to Ferguson as a token of his sportsmanship and esteem.

Massey-Harris-Ferguson was now a force to be reckoned with in the world tractor market and, indeed, had the second highest sales volume of any farm machinery company in the world, exceeding even Deere and Company.

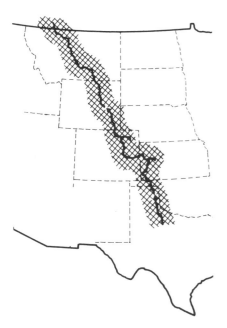

Left: North with the wheat cutters. Map shows typical custom cutters' Northward trek following the ripening crop. Considerable advance planning is needed to keep the equipment harvesting as much of the time as possible. The ripening wave moves north about 15 miles a day in the season; Above: Large scale harvesting operations. Massey-Harris No. 21A combines in Turkey in the late 1940's.

The Ferguson Imprint

Harry Ferguson's oral agreement with Henry Ford concerning the Ferguson system for the tractor three-point linkage was breached by Ford's successors in 1946. Ferguson promptly negotiated with the Standard Motor Co., at Coventry, England, to produce the now-legendary TE 20 tractor. Ferguson tractors and implements were the rising stars of the post-war period and Ferguson began looking for a small pull-type combine to sell in conjunction with his tractor. He was also contemplating plans for a "wrap around" combine to be partially mounted on the tractor. The first high-level contact between Massey-Harris and Harry Ferguson Limited occurred in late 1947. By 1953, negotiations were coming to a head. There were mutual gains for both sides; the tractor system and engineering talent for Massey-Harris, and the harvesters, organization, and world wide market for Ferguson. But there was considerable haggling over the book value of the Ferguson operation: was it worth 16 or 17 million dollars?

As Ferguson and M-H executives made their way by car to a machinery demonstration on Tuesday morning, August 11, 1953, Harry Ferguson ordered

Above: The Massey-Harris SP Model 27 with 32 in. cylinder replaced the epochal Model 21A in 1949. The Model 21 was replaced by the Model 26 that same year; Below: The SP Model 726 was Massey-Harris' first European-built combine. The plant was at Kilmarnock, Scotland. The 780 combine shown was developed from the 726.

Massey-Harris Model 90 SP combine, 1953, with 37 in. cylinder, 12- 14- or 16-foot cut, 60 bushel bin.

Massey-Harris chief engineer Thomas Carroll standing beside one of the Harvest Brigade No. 21 combines during the war effort.

Harry Ferguson proved a restless partner, and within a year left the organization. Paradoxically, the Massey organization experienced its worst period since the Depression following this merger. Undue dependence on outside suppliers, a market slump, and the insistance on continuing two lines of distribution (the red M-H line and the grey Ferguson line), were the key factors in the decline.

Management upheavals, with repercussions throughout the company, took their toll in 1956. In March 1958, with the name shortened to Massey-Ferguson Limited, the company consolidated their distribution network into a single product line. Sales doubled between 1956 and 1965. The company began realizing the advantages of geographically dispersed markets. In 1963 for example, 11 percent of sales were from the Canadian group, 31 percent from the US and 38 percent from Europe. Massey-Ferguson could claim 18 percent of the world combine market in 1965; second only to Claas of Germany, while Deere was third, followed by International Harvester, Clayson, Allis-Chalmers, Bolinder Munktell, Case and Braud (Neufeld 1969). Massey-Ferguson's tractor line eventually displaced combines as product sales leader and by 1966 combines accounted for only 17 percent of total company sales.

Tom Carroll Recognized by ASAE

Carroll retired from Massey-Ferguson in 1961. For his contribution to the advancement of agriculture, and for his efforts in development of self propelled combines, he was awarded the C. H. McCormick gold medal by ASAE in 1958. He was the first engineer residing outside the US to receive the coveted honor.

His engineering leadership at Massey did not stop at the combine. He had been instrumental in the development of their first hay baler and also worked on plows, cultivators and other implements.

An inveterate traveler, he held a commercial pilot's license until just before his death in 1968 at the age of 80. The "Harvest Brigade" concept in which he had so actively participated, lives on today in Massey-Ferguson's

continent-wide system of accelerated spare parts delivery and mobile service vans.

Turret Unloading System for Combines

Massey-Ferguson took the initiative in the introduction of the first "closed" or turret-type bin unloading system in 1962. This design, released on the Model 300 SP combine, enabled the operator to effortlessly extend the unloader spout for truckside delivery by operating a hydraulic lever next to his seat. Grain tank leakage that had occurred on earlier hinged stub-unloaders was eliminated. The spout was also designed to be retracted out of harm's way.

The operator can not only extend or retract the unloader spout "on the move", but he can also "spot" the position of the grain discharge into the grain truck alongside. Some operators even use the extended spout as a signal to get the attention of the truck driver to come out into the field when needed.

Massey-Ferguson's Combine Test Facility

Massey-Ferguson may have developed the most elaborate combine test and development laboratory in the world. Located on a farm near Toronto, this facility includes a torture test track, product reliability testing rigs, indoor laboratory with crop storage, conditioning and feeding facilities, and full instrumentation for the analysis of all phases of combine research (Cooper 1971).

Left: The 300 SP with 30 in. cylinder has been in continuous production since 1963. The turret-type or closed bin unloading system was introduced to the industry on the MF Model 300 combine; Below: MF Model 540, 37 in. cylinder, showing the standard cyclone forced-vortex radiator air cleaning system – standard on all MF combines for 1978.

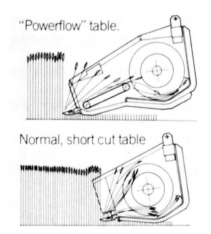

"Powerflow" table.

Normal, short cut table

Left: MF Quick-Attach system enabled heads to be changed rapidly; Above: MF claims their Power Flow System is effective in different crop conditions. The header auger diameter has been increased to 24 in.

The laboratory has played an important role in the development of a number of MF combine exclusives:

- The MF cascade shaker shoe, released in 1968, utilized an initial aerodynamic separation stage as the gram-chaff mixture "cascaded" from the grain pan filters through a proportion of the fan airflow before reaching the triple deck shoe
- An improved MF rethresher design discharged the tailings back onto the grain pan and reduced crop loading into the thresher and walkers

- Dynamic balancer for the cutterbar. This Lanchester-type dynamic balancer introduced in 1974 for the reciprocating cutterbar can accommodate different sized headers and dampen vibrations emanating from this source
- High Inertia Threshing Cylinder. This was an optional cylinder for heavy threshing conditions, permitting superior threshing in crops such as rice
- Cyclone-type filter for radiator cooling air, used a forced vortex to eliminate the dust and chaff that would otherwise clog the radiator. Released on all 1978 combine models.

The MF Universal Power-Flow Table

It has long been recognized that feeding crops heads first to the threshing cylinder improves threshing/separating performance and reduces grain cracking. In 1977, MF's Kilmarnock factory began production of a new header design—the Power-Flow Table. This design extends the distance between the auger and cutterbar to provide space for the cut crop to lay down heads first. Positive feeding over this critical space is assured by an endless fabric-reinforced rubber conveyor, The results claimed are:

- increased work rate because of higher forward speed capability and reduced risk of blockages that allow the conveyor to feed the heads into the auger faster than the unassisted reel;
- improved threshing and separating efficiency because of heads-first presentation and more regular crop feeding;
- improved performance under adverse crop conditions;
- ability to more efficiently handle lodged and tangled crop;
- reduced gathering loss, particularly in "difficult" crops such as rice, rape, and rye;

- reduced grain damage;
- reduced operator fatigue, because of a better view of the table (MacNaught 1977).

The Power-Flow Table is standard on the European MF 525 and 625 series in all header widths and is available to fit other 1978 MF combine models.

Consolidation of Combine Production

Set against a worldwide decline in total combine sales, MF has consolidated combine production into four main centers: North American combine design, test and production in Ontario, Canada; the South American plant in Brazil; European production at Marquette, France; and the Australian design and production facility at Sunshine, Victoria.

In 1959, MF acquired F. Perkins Limited and has since seen Perkins strengthen their position as the world's largest diesel engine manufacturer. A Perkins diesel plant in Ohio, wholly-owned by MF since 1975, supplied engines for North American equipment and in 1978 all MF combines were diesel-powered.

MASSEY-HARRIS/MASSEY-FERGUSON NORTH AMERICAN COMBINE PRODUCTION YEARS

Model Identification	Production years	SP or PT, GD, EF or PTO*	Cylinder or body width (in.)	Header sizes (ft.)
Under Massey-Harris Name				
# 1 Reaper-Thresher	1910	PT, GD		8.5
# 2 Reaper-Thresher	1912	PT, GD		8.5
# 3 Reaper-Thresher	1912	PT, GD		8.5
# 2 Stripper-Harvester	1901-	PT, GD	8′ stripper beater front	8 stripper comb
# 4 Reaper-Thresher		PT, GD	24	9
# 5 Reaper-Thresher (Bagger)	1922-	PT, EF		12 (First EF in the line of bagger attachment)
# 6 Reaper-Thresher (Bagger)	1925-	PT, EF Buda 4, 20 HP	33	10
# 7 Reaper-Thresher (Bulk bin)		PT, EF		12
# 9 A Bulk grain bin	1927	PT, EF	33	12
# 9B Bulk grain bin	1928	PT, EF	33	12-15
#11 Quick-Detach platform		PT, EF Hercules 26 HP	24	12
#14		PT, EF Hercules 45 HP	28	12, 16
#15 First MH with rubber tires	1937-1950	PT, EF Hercules 31 HP or PTO	24.5	6, 8
#17	1937-1950	PT, EF Hercules 31 HP or PTO	24.5	10, 12
#20 First MH self-propelled	1938-1940	SP (first) Chrysler 65 HP	37	16

*PT = Pull-Type   SP = Self Propelled   GD = Ground Driven   EF = Auxiliary Engine Functioned   PTO = Power Take-Off Driven

MASSEY-HARRIS/MASSEY-FERGUSON NORTH AMERICAN
COMBINE PRODUCTION YEARS

Model Identification	Production years	SP or PT, GD, EF or PTO*	Cylinder or body width (in.)	Header sizes (ft.)
#50 Clipper	1938-1958	PT, PTO or EF 21 HP Wisconsin	60	6, 7
#21 SP	1941-1949	SP Chrysler T125	32	12, 14 Last canvas
Auger type headers from hereon; canvas drapers dropped				
#21 A (Auger instead of canvas table)	1943-1949	SP Chrysler 6T112	32	10, 12, 14 Auger
Clipper SP	1946	SP 30 HP	60	7
#26 Clipper SP	1948-1952	SP Chrysler	28	8.5, 10, 12, 14
#27 SP	1949-1952	SP Chrysler Ind. 8A	32	12
#27L Rice type (Live axle)	1951	SP	32	12, 14, 16 Auger 12, 14 Canvas
#222	1947	SP Continental F162-4B1	24	8, 10
#60	1953	PT, PTO or EF	28	7, 8
#60 Spl	1952-1956	Chry. Ind. 30A	32	
#90	1953-1956	SP Chr Ind 8A	37	142, 166, 190
#90 Spl	1953-1956	SP Chr Ind 8A-251	37	12, 14, 16
#60	1953-1959	SP Chr Ind 30	28	10, 12
#70	1953 only	SP Chr Ind 5A	24.6	
#82	1957-1963	SP Chr Ind 30	32	10, 12, 14
#72 (PT)	1959	PT, PTO or EF cont. F140	21.9	8, 10
#92	1957-1959	SP Chr Ind 251	37	12, 14, 16
#35 (PT)	1960	PT, PTO	24	7, 8
#92 Hillside Special		SP	37	
Under Massey-Ferguson Name				
#35	1958-1963	SP Cont F140, 31.6 HP	24	7
#72	1959-1963	SP Chr Ind 30 56.5 HP	28	10, 12
Super 92	1960-1963	SP Chr Ind 265	37	12, 14, 16
#410	1964-1972	SP Chev 292 Gas Perkins A4.300 dsl	37	12, 14, 16 18
#405 (PT)	1966-1972	PT, PTO	37	12
#205	1966-1970	SP Chr Ind 170	26	10, 13
#300	1963-	SP Chr Ind 225 gas or Perk AD 4.203 dsl	30	13
#510	1964-1977	SP Chev 327 gas or Perk AD 354 dsl	45	12, 14, 16, 18, 20
#750	1973-	SP Chev 350 gas or Perk AD 372 or 6.354 dsl	50	13, 14, 15, 16, 18, 20, 24
#760	1971-	SP Perk AD 6.354 dsl or AV 8.540	60	13, 14, 15, 16, 18, 20, 24
#751	1977-	PT, PTO	50	12' grain head 12' pickup
#540	1977-	SP 4 318 dsl	37	12, 18
#550	1977-	SP 354 6 cyl dsl	45	12, 20

*PT = Pull-Type   SP = Self Propelled   GD = Ground Driven   EF = Auxiliary Engine Functioned   PTO = Power Take-Off Driven

International Harvester Company logo used until 1945 when the present IH symbol was adopted.

15

International Harvester Company

International Harvester Company was formed in 1902 by the merger of five leading harvester manufacturers: The McCormick Harvesting Machinery Co. of Chicago; The Deering Harvester Co. of Chicago; the Plano Manufacturing Co. of West Pullman, Illinois; Warder, Bushnell and Glessner Co. of Springfield, Ohio; and the Milwaukee Harvester Co. of Milwaukee, Wisconsin.

As early as 1890, a group of leading binder producers, including McCormick and Deering, had reached a tentative agreement to consolidate into the American Harvester Co. The company was incorporated in Illinois, but never was established as negotiations broke down over valuations. At least two other conferences took place between McCormick and Deering before the 1902 merger. It was the House of Morgan (J. P. Morgan & Co.) that finally brought these old rivals together and successfully concluded negotiations. The Bureau of Corporations estimated the assets of the new International Harvester Company to be worth $100 million at the time of the merger. Control of the policies of the Company during the first decade of its existence was placed in the hands of a three-man trust, composed of C. H. McCormick, Jr., as the new president, Charles Deering as chairman of the board of directors, and George W. Perkins, principal partner of the House of Morgan, representing the smaller constituents.

Although they already controlled 90 percent of the domestic grain binder market and 80 percent of the mowers in the US, International Harvester Company proceeded to acquire three more well-known firms in the first year of its existence: D. M. Osborne & Co., of Auburn, New York; Aultman Miller Co., of Akron,

McCormick Harvesting Machinery Company advertisement, 1898. The closing decade of the 19th century was possibly the most aggressively competitive period in the farm machinery business.

Ohio; and Minneapolis Harvester Co., of St. Paul, Minnesota. The latter two acquisitions were kept secret for nearly two years for account liquidation purposes—an action which was to prove embarrassing to the Company some years later.

In addition to these eight companies which made up the "Harvester" merger, several others were seriously considered. Among these were Massey-Harris Co., of Toronto, Canada; the Walter A. Wood Mowing and Reaping Machinery Co., of Hoosick Falls, New York; and the Acme Harvester Co., of Peoria, Illinois. At one time Charles Deering, son of founder William Deering, actually advocated that the *entire harvester industry* be absorbed by IHC, but this was opposed by the McCormick principals.

Shortly before the merger, Deering had acquired iron ore leases. In addition, Deering controlled steel works, twine works and forest reserves. Thus the merger gave IHC direct access to most raw materials

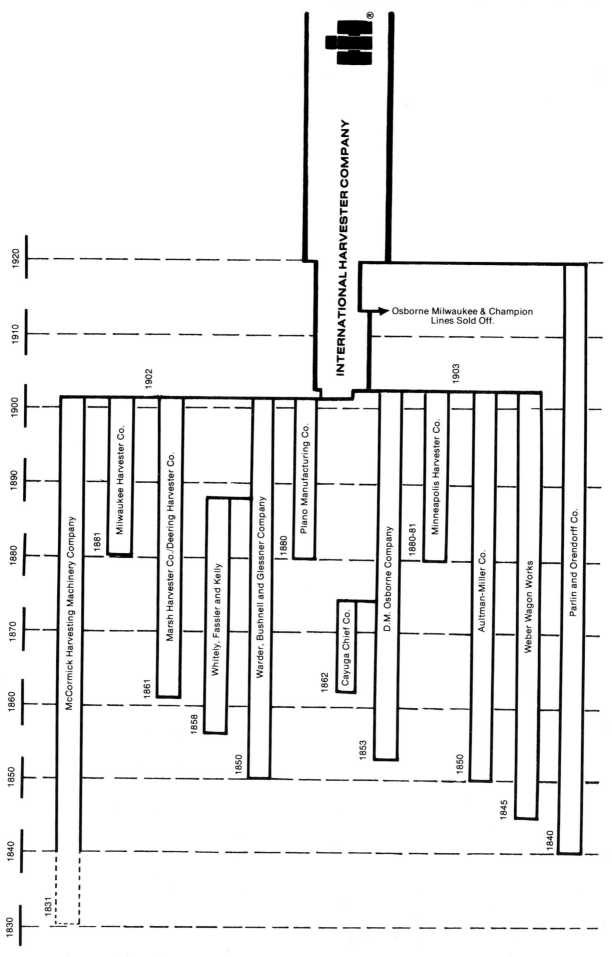

Chart of Development Of The International Harvester Company.

Left: IHC's Deering "New Ideal" tractor binder, about 1918; Below: The McCormick-Deering tractor binder had a 10 foot cutting width in 1925. The redesigned grain binder was the first to appear under the new trade name following consolidation of dealership networks.

and possession of the most extensive plant facilities in North America at the time.

By 1890, the US had begun to run out of frontier. Wheat acreage leveled off and wheat production remained practically static for the first 15 years of the 20th century. As a result, IHC farm equipment sales declined. Prior to the merger, the constituent companies had been selling no less than 152,000 binders and 217,000 mowers a year. By 1910, IHC sales had dropped to 91,000 binders and 170,000 mowers. Lower sales volume meant higher overhead costs per unit and provided strong incentive for the company to reorganize existing facilities. There was also a subtle change in the whole framework of the farm equipment industry. With the elimination of the harvesting bottleneck brought about by progress in mechanization, the need for better soil-working tools and other implements became more apparent and so IHC efforts were redirected.

Plant space that had been idled by the lack of strong demand for harvesting equipment was converted to the production of other lines as IHC sought to diversify the nine original equipment lines. The Company's production and sales activities were spread over a greater part of the year and into more markets by moving into tillage implements, gasoline engines, trucks, and eventually tractors. With the purchase of the P & O (Parlin and Orendorff) plow business, IHC was able to offer 54 different classes of farm equipment by 1919.

Old Trade Names Give Way to the "McCormick-Deering" Name

For years the company had found it expedient to retain the trade names of its subsidiaries. Although McCormick and Deering products were the leading lines in 1912, IHC also marketed products under the "Champion", "Buckeye", "Milwaukee" and "Osborne" logos.

McCormick and Deering had established sales agents in practically every town and village in the Union during the competitive 1890's. After the merger, IHC retained both these identities and the widely scattered dealerships of the other brands. In fact, it was possible for the company to be represented by as many as five dealerships in a given town. Each dealer had to be provided with complete sets of implements distinct from each other, not only in name but often in design as well. The Company justified this action in the interests of maintaining the largest share of the market, but in 1918 this view was modified by the Supreme Court as a result of action under the 1912 Sherman Anti-Trust Act. A 1918 consent decree left IHC with no more than one dealer in any city or town in the US. The Company was also ordered to sell off its Osborne, Milwaukee and Champion harvesting lines. The number of IHC dealers dropped dramatically.

In 1923, farmers were introduced to a re-designed grain binder under the now famous McCormick-Deering trade name as the Company moved to consolidate

Upper and Lower Left: McCormick-Deering "Plain Type Header" could be fitted with this binder attachment according to the 1927 IHC catalog. Lower illustration shows how the horse team was controlled on a sharp 90 degree turn; Below: Norman E. Bunting's US patent No. 1,222,730, "Combined Harvester and Thresher" shows the workings of International's first combine. Filed May 1, 1913, patented April 17, 1917.

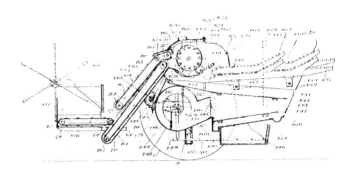

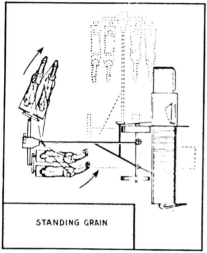

The No. 1 pull-type Harvester-Thresher was first released by Deering in 1914.

its dealership network under the new logo. A host of other McCormick-Deering implements followed. IHC's 15 year battle with the courts over anti-trust issues finally ended in 1927. By this time competition from other full-line companies, notably Deere and Co., had whittled the IHC market share down to 64 percent from 85 percent of the total farm machinery business in 1902. (Phillips 1956).

The First IHC Combine

International Harvester Company entered the combine business in 1914 when the Deering division produced the No. 1 pull-type Harvester-Thresher. Stationary threshers were added four years later and were produced for many years at the IHC Hamilton Works in Ontario, Canada.

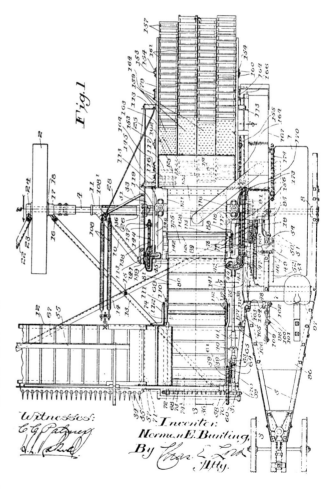

The First IH Self-Propelled Combine

The first self-propelled (SP) combine manufactured by IH was the Model 123, produced from 1942-1948. An interesting feature of this machine was that the steering wheels were at first tandem-mounted in line with the left-hand drive wheel. It was believed at the time that an SP combine should make square turns like a pull-type combine and that such a wheel configuration would allow square turns.

No other new models were released during the war years—production remained steady at approximately 10,000 harvester-threshers and 7000 corn pickers annually. But in 1947, the market for combines and corn pickers exploded. The Model 123 was replaced by the 125 after a production run of 11,000 units.

The Leader in the Corn Picker Market

One-row corn pickers—both horse- and tractor-drawn, were sold in the years through WW1, but for practical purposes the corn picker was not widely adopted until the tractor power take-off had become a commercial reality between 1918 and 1924.

IHC introduced the first two-row corn picker-huskers and the first tractor-mounted pickers (one row) in 1928. They were driven by the tractor power take-off. Two-row mounted units appeared in 1929. In the 1930's, lighter weight, better-designed units appeared at the same time. The improvements coincided with the introduction of hybrid corn. Both factors contributing enormously to the overall acceptance of mechanical corn picking. Sales for tractor-mounted and pull-type pickers soared. IH sold a total of 113,087 two-row mounted (M-series) pickers between 1938-1952.

The arrival of the combine corn head in 1956 ultimately doomed the corn-picker. Although picker sales across the industry climbed to a peak in 1959, they plummeted shortly thereafter.

The IH Model 141 combine

IH's Model 141 combine introduced in 1954 established several firsts for the company:

- The 141 was IH's first combine with a corn head. In 1956, their first two-row (40 in. rows) combine corn head was released for the 141 combine.

- It was IH's first combine with an auger-unloading system for the grain bin.
- The first four-row corn head, the No. 40, was released on the 141 combines in 1959.
- The first fully self-leveling combine was developed around the 141 and designated the 141 H.

Above: Model 123, the first IH self-propelled combine; Below: Model 62 pull-type with a 6-foot cut, rasp-bar type thresher. 42,889 were produced between 1941 and 1951.

IH Model 2 MH D tractor-mounted corn picker on a Farmall 560 tractor, 1964.

8 and 10-ft.
Cutting Widths

The IHC No. 22 pull-type Harvester-Thresher was manufactured between 1935 and 1945. Some 13,276 units were produced, with 8-foot or 10-foot cut.

Above: The No. 42 pull-type was produced between 1940 and 1944. This "scoop-type" machine was equipped with an angle-bar thresher, the first IHC machine to deviate from the spike-tooth thresher configuration; Below: IH's first combine corn head was released on the Model 141 combine in 1956 as a two 40-in. row head.

Origin of Flexible Straw Spreader

Stuart Pool, a veteran combine designer and one time IH engineer, tells of the origin of the flexible straw spreader—a design that is now widely used on combines all over the world:

"Originally straw (from the back of the machine) was dropped onto a steel shelf and swept off by an arm so that it would drop in a windrow. Then the shelf was dropped and a double steel paddle flailed the straw to one side. Either of these designs would plug the combine and bust the walkers if the V belt drive got loose and the paddle stopped, as the straw would build up on the stalled spinner. So enough of that. We had some old tire carcasses and so I made a spinner that hung down when stopped and spun out to whip the straw when going... the flexible spinner became an instant industry standard." (Pool 1976).

Fully Self-Leveling Combine

Since 1928, IH had been a manufacturer of pull-type side leveling combines for use in the Northwestern US and in North Africa. Side-leveling hillside combines opened up fertile land that had previously been considered unsuitable for cropping because of the rolling nature of the terrain. But side-levelers have drawbacks.

Any time the side-leveler varies from following the contour, particularly going up or downhill, grain losses mount. Grain cascades out the back of the combine when the tail is low, or piles up on the grain pan and upsets the cleaner when the combine is traveling directly down a hill. These factors were overcome with limited success on some Harris hillside combines by installing a self-leveling cleaning shoe.

Safety considerations called for fore-and-aft leveling as well as lateral or side leveling. Under an IH contract, Frank Farber of Moscow, Idaho, built several fully self-leveling combines for the company to test and to evaluate.

The development of the fully leveling combine required two elements not found in side levelers:

(1) raising and lowering ability on the rear (steering) axle;

(2) a link connecting the rear axle to the header so that height would remain the same, regardless of fore-and-aft leveling position.

The mechanics of fore-and-aft leveling did not prove difficult, but there was considerable developmental effort involved in the hydraulic control functions to effect automatic control of "four-way" leveling. The key to the solution lay in the use of a 100-pound rigid-rod pendulum and careful attention by IH designers to ancillary valve, spring and dashpot control and damping functions. The first production models of the fully self-leveling Model 141H were released in 1955.

The combine leveled automatically 16° to either side, 18° when climbing, and 6° going downhill, but could be operated on override on hills far steeper than its leveling capacity. It is interesting to note that the new SP combine side-leveled about 12° less than the maximum side leveling of the old pull-types, but operators generally found the SP's leveling ability satisfactory. (Pool 1976).

In 1962, the company redesigned their hillside SP at Moline and introduced the model 151 and 403 fully

In 1955, IH offered the first automatic fully self-leveling hillside combine, the model 141 H with fore, aft and lateral leveling.

152

self-leveling versions. In 1973, IH released an updated fully leveling hillside combine, the larger model 453, with 39.3 in. cylinder width. The four-way leveling system is fully hydro-mechanical (no electrical circuits), as was found on the 141H.

Hydrostatic Transmissions for Combines—an IH First

IH pioneered the use of hydrostatic transmissions on agricultural equipment and actually evaluated the new transmissions on a limited number of their combines beginning in 1964. Hydraulic components became common with the transmissions in the 400 series IH tractors. In 1966, hydrostatic combine ground drives were in full production as an alternative drive for the models 303, 403 and 503 SP combines. This development simplified the combine ground propulsion system by eliminating belts and clutches. Oil under pressure from the engine-driven variable output pump is piped down to the final drive axle hydraulic motor which powers the differential pinion through a three-speed transmission. The hydrostatic transmission provides stepless speed changes on the move within each gear range, as well as providing transmission braking and instant forward/reverse motion without declutching.

Within a few years hydrostatic transmissions had become far more popular than plain mechanical transmissions. In fact, when IH released their 815 and 915 models in 1969, they were only available with hydrostatic ground drive. This policy continues today on all IH SP's.

The IH hydrostatic drive introduced several novel features such as the "foot and inch" valve which allows hydraulic declutching for gentle and accurate maneuvering and greater safety, and the use of hydraulic steel tubing instead of flexible hoses between engine-mounted hydrostatic pump and ground drive motor.

The IH "Even Flow" Cleaning System

In 1969, IH introduced a unique air flow system for the cleaning section of their newly released 815 and 915 combines. The patented "Even Flow" cleaning system developed by East Moline engineers was designed to provide a uniform airflow across the cleaning shoe. As originally conceived, a single fan drew in cool air through the radiator and then forced this air down through a plenum, on through ducting and flow straighteners to the cleaning section. Certain problems with air flow, and, strangely enough, with warm air causing snow to melt on the grain pan and seives led to modifications and the eventual release of the 815 and 915 "Even Flow" system. The air was collected by a twin-inlet furnace-type blower fan high on the machine, out of the trash and dust. The 815 and 915 machines were also unique in having one of the earliest turret unloader systems and fin and wire straw racks.

Combine Monitors

Pool was one of the engineers associated with the earliest development of combine monitors wrote:

"The driver, on a steel seat, close to the left hand combine side could see, hear and 'feel' the combine operation. Any plugging or failure anywhere interrupted the cadence and he reacted promptly.

"That same driver got a new fangled combine with enclosed, air-conditioned cab, (radio) and a specially designed soft spring seat and he hadn't the faintest idea how his combine was doing. So we designed a readout panel that showed with red lights whatever went wrong. But the combine operator watching his header never looked at the readout. So we hooked it onto a warning bell and the driver couldn't stand the noise. He promptly disconnected the bell. Then I had a great idea. I hooked it up so that it shut off their radio whenever anything went wrong. What a holler! Their favorite program was shot to pieces and they missed the markets. None of that! So we put a windshield arm out in front that dropped down in front of the driver (to warn of) any failure. That worked fine. But by the time the drivers were acclimated to the readout—even glancing at it occasionally—and didn't seem to mind a buzzer. So it goes." (Pool 1976).

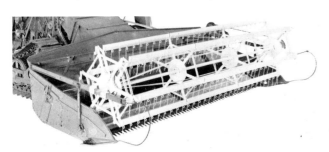

The IH Model 503 Rice Special combine with tracks, diesel power and special IH draper platform for long strawed crops. This was essentially a "California Rice Special" machine. Machine was one of three models to have first hydrostatically-propelled transmission.

IH "Even Flow" patented system for providing uniform air flow across the cleaning section. With conventional combine winnowing fans there is a tendency for reduced air flow in the center of the shoe under heavy crop loading.

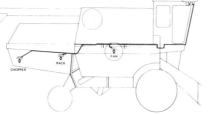

IH was one of the first companies to offer electronic monitoring of machine functions on its combines. IH introduced this system in 1969 on Models 815 and 915.

IH Model 915 combine, 48 in. cylinder shown equipped with No. 854, five-row corn head.

Top: IH Model 815 combine, 39 in. cylinder equipped with soybean header with pickup reel and flexible floating cutterbar; Center: IH Model 715 combine, 39 in. cylinder shown equipped with No. 844 four-row corn head. Production on this machine began in 1971; Bottom: IH Model 453, completely automatic and fully self-leveling hillside combine, is also equipped with hydrostatic drive and diesel engine.

International's Axial-Flow Combines

Mid-September 1977 marked the announcement of the 1440, 1460 and 1480 Axial-Flow series of combines. These machines were a radical departure in combine development for IH and the industry. The 1440 and 1460 models, with 24-in. diameter single longitudinal rotor, were sized to eventually replace the 815 and 915 conventional machines. The 1480 Axial-Flow combine, with 30-in. rotor diameter, 190 HP engine and 208 bushel grain bin, became the highest-capacity machine that IH had ever produced.

The release of this new series of combines came just in time to be featured at the first International Grain and Forage Harvesting Conference and at the Farm Progress Show, both held in Iowa in late September, 1977. IH engineers had, for the past 15 years, been pursuing the development of a high capacity harvester that would eliminate the troublesome straw walkers and contain fewer moving parts. The Axial-Flow machines eliminated walkers, while reducing the number of moving parts from 36 to six in the thresher/separator. The single axial rotor replaces the cylinder, cylinder beater and straw walkers of the conventional design.

How the IH Axial-Flow Combine Works

The crop gathering and feeding systems are conventional. Header tables are interchangeable, within size limits, with previous IH combines. The heart of the machine is the single axial rotor, aligned lengthwise within the body of the machine. Delivery of the crop to the threshing rotor is completed by a three-bladed

IH Model 1460 Axial-Flow combine. The Axial-Flow design marked a major departure in combine design. The number of moving parts was reduced from 36 to six in the thresher/separator areas.

IH Model 1460 with corn head at work.

impeller on the front end of the rotor, beginning the process of rotating the crop. Once the rotation has started, the rasp bars and forwarding vanes in the fixed housing swirl the crop rearward over concaves and grates. The raspbar section of the axial rotor, immediately behind the impeller blades, has helically and axially-mounted sections. The helically-mounted rasps continue the swirling crop flow for several passes over the concaves before the crop reaches the separating zone. It is this multiple-pass crop flow over the grates that distinguishes the behavior of the axial design from that of the conventional thresher.

The separator zone, immediately behind the threshing zone, consists of three-section grates under the axial rotor and rotor bars which impel the grain through the swirling crop and past the grate perforations, directly onto the cleaning shoe. A positive discharge beater across the rear of the separator grates ejects the straw out the rear hood of the combine. The cleaning system has reverted to the conventional style—a departure from the "Even Flow" design of the 815 and 915 models.

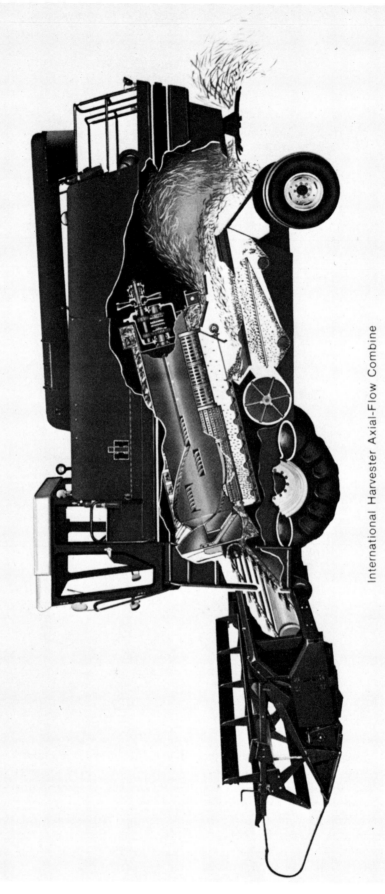

International Harvester Axial-Flow Combine

IH Axial-Flow combine semi-sectional elevation. The 9-foot long, single axial rotor is the heart of the machine. Over one million man-hours of engineering and development were expended in the design.

COMBINE PRODUCTION YEARS — INTERNATIONAL HARVESTER COMPANY NORTH AMERICAN OPERATIONS

Model Identification	Production years	SP or PT, GD, EF or PTO*	Cylinder or body width (in.)	Header sizes (ft.)
# 1 Deering	1914-1915	PT, GD	35⅝	9
# 1 McCormick	1914-1915	PT, GD	35⅝	9
# 2 Deering	1915-1921	PT, GD	35⅝	9, 12, 14
# 2 McCormick	1916-1921	PT, GD	35⅝	9. 12, 14
# 3 Deering	1923-1927	PT, GD	35⅝	9, 12 w/extn., 15 w/extn
# 4 McCormick	1924-1926	PT, GD	35⅝	12, 15 w/extn
# 5 McCormick	1925	PT, GD	35⅝	12
# 6 McCormick	1922-1924	PT, GD	35⅝	9, 12
# 7 McCormick-Deering	1925-1932	PT, GD	24	12, 16 w/extn
(First combine, known as Harvester-Thresher, under the joint McCormick-Deering name)				
# 8 McCormick-Deering	1926-1935	PT, GD	24	10, 12
# 9 McCormick-Deering	1926-1927	PT, GD	24	12, 16 w/extn
#10 McCormick-Deering	1927-1936	PT, GD	24	9, 12, 16, 16 w/extn
#11 McCormick-Deering	1926-1936	PT, GD	24	9. 12, 16, 16 w/extn
#20 McCormick-Deering	1930-1936	PT, GD	22¾	8, 10 w/extn
#21 McCormick-Deering	1932-1937	PT, GD	22¾	5 & 7
#22 Thresher	1926-1957	Stationary	22	—
#22 McCormick-Deering	1935-1945	PT, GD	22¾	8, 10 w/extn
#28 Thresher	1934-1940	Stationary	28	—
#31 Spike tooth	1935-1940	PT, GD	28	12, 15 w/extn
#31 Rub bar	1935-1942	PT, GD	28	12, 15 w/extn
(First combine with rasp-bar cylinder)				
#36 Thresher	1942-1946	Stationary	35½	—
#41 Spike tooth	1935-1940	PT, GD	28	16
#42 Rub bar	1940-1944	PT, GD/EF	35½	4
#51 Hillside	1935-1950	PT, GD/EF	28	14, 16 w/extn, 20
#52	1943-1950	PT, GD/EF	35½	5
#60	1937-1938	PT, GD	28	6
#61	1939-1940	PT, GD	28	6
#62	1941-1951	PT, GD	28	6
#64	1950-1954	PT, PTO/EF	64½	6, 7
#76 McCormick-International	1955-1958	PT, PTO/EF	64	6, 7
#80 McCormick-International	1959-1965	PT, PTO	42	7
#82 International Harvester	1966-1974	PT, PTO	42	7
#91 International Harvester	1959-1962	SP	41¼	8.5, 10
#93 International Harvester	1962-1968	SP	41¼	8.5, 10
#101 International Harvester	1956-1961	SP	28	10, 12, 14
#102 International Harvester	1938-1939	PT, GD	28	8, 10
#103 Rub bar	1935-1939	PT, GD	28	12
#103 Spike tooth	1935-1939	PT, GD	28	12
#104 Spike tooth	1935-1040	PT, GD	28	10
#105	1967-1970	SP	41¼	10.75, 13
#122	1946-1949	PT, EF	31⅛	12
#122-C	1949-1953	PT, EF	31⅛	12
#123-SP	1942-1948	SP	31⅛	12
(First IH self-propelled combine)				
#125-SP	1948-1949	SP	31⅛	12
#125-SPV	1950-1951	SP	31⅛	12
#125-SPVC	1951-1952	SP	31⅛	12
#127-SP	1952-1954	SP	31⅛	10, 12, 14
#140	1954-1962	PT, PTO	31⅛	9, 12 w/extn
#141	1954-1957	SP	31⅛	10, 12, 14
(First IH Combine to use a corn head)				
#141 Hillside	1955-1958	SP	31⅛	16, 18
(First IH SP Hillside combine. First combine with four-way leveling system)				
#150	1963-1968	PT, PTO	31⅛	9.75, 12.75 w/extn

COMBINE PRODUCTION YEARS — INTERNATIONAL HARVESTER COMPANY NORTH AMERICAN OPERATIONS

Model Identification	Production years	SP or PT, GD, EF or PTO*	Cylinder or body width (in.)	Header sizes (ft.)
#151	1957-1961	SP	$37^3/_{16}$	12, 14, 15, 16
#151 Hillside	1959-1961	SP	$37^3/_{16}$	16
#160 Hillside	1951-1953	PT, EF	28	16
#181	1958-1961	SP	$46^3/_8$	12, 14, 16, 18
#203	1963-1966	SP	$41¼$	10, 13
#205	1967-1971	SP	$41¼$	10.5, 13
#303 International Harvester	1962-1967	SP	30	10.5, 13, 14
#315 International Harvester	1967-1970	SP	$41¼$	10, 13, 14
#402 International Harvester	1966-1971	PT, PTO	$39^3/_{16}$	13
#403 International Harvester	1962-1970	SP	$39^3/_{16}$	13, 14, 15.5, 16.5, 18.5
(First combine with hydrostatic drive)				
#403 Hillside	1962-1969	SP	$39^3/_{16}$	16.5, 18.5
Side-leveling only				
#403 Hillside	1962-1972	SP	$39^3/_{16}$	16.5, 18.5
Fully self-leveling, four-way				
#453 Hillside	1973-	SP	$39^3/_{16}$	16.5, 18.5
#503 Rice special	1962-1968	SP	$48^3/_8$	13, 14, 15.5, 16.5, 18.5, 20.5
#615	1971-1975	SP	30	10, 13, 14, 15
#715	1971-	SP	$39^3/_{16}$	10, 13, 13.5, 14, 15, 15.5, 16.5, 17.5, 18.5, 20
#815	1969-1977	SP	$39^3/_{16}$	10, 13, 14, 15, 15.5, 16.5, 17.5, 18.5, 20, 20.5, 22.5, 24
(815 & 915: First combines with monitor systems)				
#914	1970-	PT, PTO	$48^3/_8$	12.5
#915	1969-	SP	$48^3/_8$	10, 13, 14, 15, 15.5, 16.5, 17.5, 18.5, 20, 20.5, 22.5, 24
Axial-Flow Combines				
#1440	1977-	SP	24*	10, 13, 15, 16.5, 17.5, 20, 22.5, 24
#1460	1977-	SP	24*	10, 13, 15, 16.5, 17.5, 20, 22.5, 24
#1480	1978-	SP	30*	10, 13, 15, 16.5, 17.5, 20, 22.5, 24
			*rotor diameter	

*PT = Pull-Type, SP = Self-Propelled, GD = Ground Driven, EF = Auxiliary Engine Functioned PTO = Power Take-Off Driven

IH: One of the "Big Three"

Considerable attention was given to the operator's comfort in the Axial-Flow combine series. All models are equipped with a pressurized, sound-insulated cab. The same engine (436 cu. in. IH diesel) is used, but with different power ratings for each model. The engine is mounted behind the grain bin for greater sound isolation. Swing-down steps at the rear of the machine provide ready access to a large service platform.

IH is today one of the "Big Three" of the farm equipment industry, with total agricultural equipment sales in excess of two billion dollars annually. IH machinery is being sold in over 100 countries. In 1977, combines were being built by IH in plants in the US, Australia and France. North American combine production is concentrated in the East Moline Illinois plant—a factory that covers 156 acres, 51 acres under one roof.

The Agriculture of the US: Its high productivity is a result of development policy, low prices for land and other things needed for farming, stable prices for farm products, and the promotion of innovation.—Earl O. Heady (Scientific American, September, 1976).

16
Deere and Company

Deere and Company, the largest combine manufacturer in North America, was founded in 1837. The Company began with one product—the steel plow, when John Deere, who had been a blacksmith in his native Vermont, moved to Western Illinois. There he observed that a hand paddle was essential equipment to scour the sticky prairie soil from the moldboard plows of the day. He solved a problem by fashioning the first steel plow, using a broken saw blade formed over a template hewn from a log.

By the close of the 19th century, the company he founded had become the largest manufacturer of tillage equipment and check-row planters in the US.

And it had also diversified into a number of other lines—wagons, carriages, and even bicycles bore the Deere name.

The growth of the International Harvester empire and their decision in 1903 to market tillage equipment goaded Deere and Company into becoming a full line farm equipment manufacturer too. As a direct result, in 1911 Deere re-incorporated and expanded by purchasing eight smaller firms (Phillips 1956). 1956).

Deere Moves into the Harvesting Business

In October 1909, the Company's board directed that a suitable location be found for building harvesting

John Deere, 1804-1886. Right: "The Light Running" John Deere grain binder was produced from 1910 to 1945.

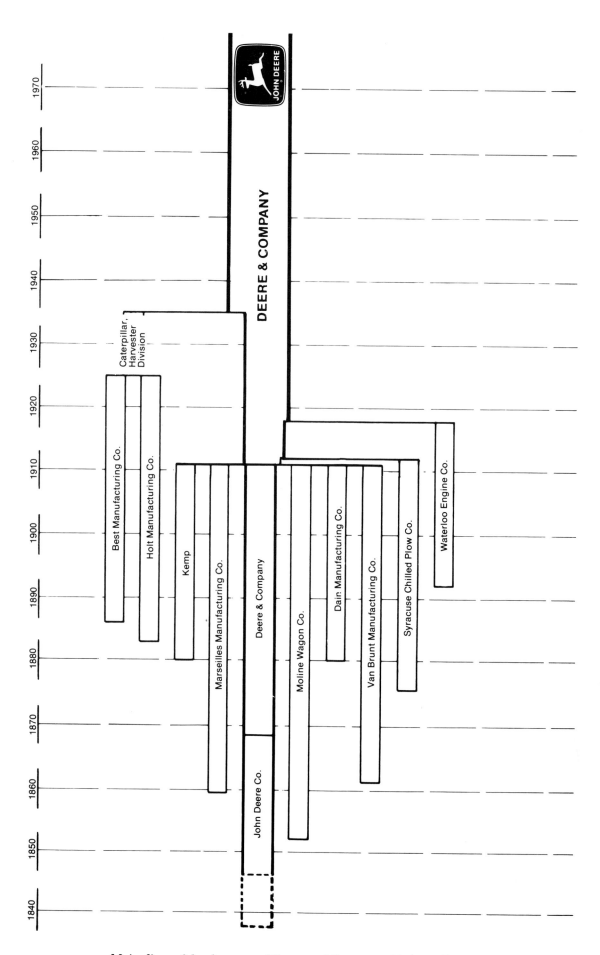

Major lines of development of Deere and Company, Moline, Illinois.

John Deere Tractor Binder, 1929. Available in 6-, 7-, 8- and 10-foot cut, the tractor could be controlled from the binder seat with optional tractor steering and control kit.

equipment and that work on the development of a grain binder begin immediately. Because of the demands of the Canadian market, priority was to be given to locating the manufacturing site in that country.

During the 1909/10 winter, seven prototype twine binders were fabricated at the John Deere Plow Works in Moline. The 1910 testing program was so promising that the decision was made to build 500 binders the following winter. But where?

Coincidental with Deere's first ventures into harvesting equipment, a number of Moline businessmen had purchased the Adams Manufacturing Company of Marseilles, Illinois. They moved the plant and inventory to Moline and renamed the venture the Marseilles Manufacturing Company. Deere rented part of this factory for the 1910/11 winter to build their 500 binders, while the search for a Canadian facility continued. During the 1911 reorganization, the factory became the John Deere Marseilles Works and 2000 binders were built there for the 1912 season. The search for a Canadian harvester factory was called off in August 1913, when ground was broken on an adjacent site in Moline for the present Harvester Works. Manufacturing of binders, rakes and mowers began in the new plant in 1913 and has continued ever since. With the surge in demand for crops brought on by WW1, Harvester Works produced in 1914 no

less than 33,000 machines—all horsedrawn and ground-driven. This included 11,600 grain binders, 72,000 mowers, 6,500 rakes, and 2500 corn binders— a new line for the company.

John Deere Combines

As the third decade of this century began, Deere was under pressure from customers, dealers and branch houses to develop a combine. A $26,000 budget was approved in 1924 for prototype development on the Model No. 1 with 12-ft platform and on the Model No. 2 with 16-ft platform. As luck would have it, the No. 2 reached the market first, in 1927, when 40 units were produced. Model No. 1 followed a year later. Both models had a 24-in. cylinder and were superseded shortly after by the 30-in. Model No. 3 and 24-in. Model No. 5 respectively (Model No. 4 never saw the light of day).

Deere's Caterpillar Acquisition

The emergence of the Caterpillar Tractor Company from West Coast combine manufacturing enterprises has been detailed earlier. Caterpillar management decided in the 1930's to concentrate their energies on tractor and construction equipment lines. In 1936, Deere acquired the Caterpillar combine line and with it a rich heritage of combine experience,

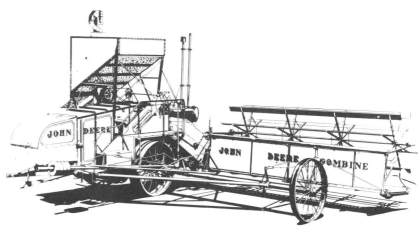

The Model No. 2 was the first John Deere combine, 1927. It was a ground driven, pull-type machine with 24 in. cylinder, 12 or 16 foot canvas platform, 60 bushel grain tank and a separate re-cleaner screen and fan.

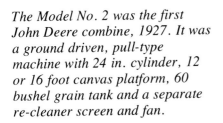

161

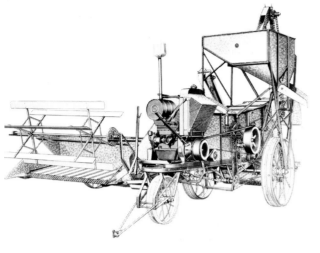

Upper Left: John Deere Model No. 1 combine was produced a year after the No. 2 in 1928; Above: John Deere 5A Combine, 1938; Left: John Deere No. 17 combine replaced the Model 3 in 1932. This popular 12-or 16-foot cut machine was on the market for 16 years. Rubber tires were introduced on this model in 1935.

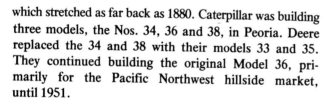

which stretched as far back as 1880. Caterpillar was building three models, the Nos. 34, 36 and 38, in Peoria. Deere replaced the 34 and 38 with their models 33 and 35. They continued building the original Model 36, primarily for the Pacific Northwest hillside market, until 1951.

Left Hand Cut to Follow the Binder

The singular circumstances of the post-depression period created a rising demand for smaller combines. In 1937 and 1938, Models 10, 11 and 12 were field tested. Only five of Deere's smallest-ever combine, the 3½-ft cut Model 10A were built. Each of the models was right-hand cut, as usual. During the late 1930's combine production in North America began to overhaul sales of binders. Company branch representatives in windrow-harvesting areas knew that their dealers would be swamped with traded-in binders. They insisted on left-hand cut combines so that these binders could be re-used by converting them into windrowers. Thus the 11A and 12A combines, introduced in 1939, were left-hand cut machines. The 12A proved to be Deere's most popular PTO combine model. During its 13-year life, 116,027 units were built and shipped.

Above: The Model 7 combine was produced from 1932-1941 for the smaller farmer. It had an 8 foot cut and could be either pto, ground or auxiliary engine powered. Model 7 is shown cutting soybeans with Deere Model A tractor near Assumption, Illinois; Below: The Model 36 Hillside combine was the original Caterpillar Tractor Company Model 36 taken over in 1936 when toolage and parts inventory were transferred from Caterpillar's Peoria plant to Deere at East Moline, Illinois.

Above: Deere's first self-propelled combine, the Model 55, with 30 in. cylinder and 12- or 14-foot cut was released in 1947; Below: The pull-type No. 65 with 30 in. cylinder, 12-foot cut was developed on the Model 55 design and marketed between 1949 and 1966.

Above: The Model 9 came out in 1939 with an auger platform in place of the canvas apron type. The war interrupted production, but the No. 9 was again produced following the war. It was the first of Deere and Company's combines to employ the rasp-bar cylinder; Below: The 11A and 12A combines were 60 in. cylinder straight-through designs with left-hand cut to follow the windrower when windrowing was used. The 12A was Deere's most popular combine model with 116,027 built. The 12A shown was in production for 13 years and was replaced in 1952 by the No. 25.

With the post-war demand for more capacity, the 12A was modified to accommodate a seven-foot apron and released as the No. 25 in 1952. In 1956, Model 25 was, in turn, replaced by the 60 in. No. 30 which had an auger platform to replace the bothersome canvases.

By 1945, the company had come under competitive pressure to develop a self-propelled combine. The SP Model 55 was tested in 1945 and 1946, and released in 1947. Nearly 84,000 units of this popular model were built in 22 production years, a volume likely to stand unchallenged at Deere, since model changes have become more frequent.

The "windrow areas" required the 55's capacity, but were looking for a machine at the cost of a pull-type

combine. The result was the development of the Model 65, a pull-type released in 1949. The Hillside SP Model 55H, with lateral leveling, followed in 1954.

Deere Introduces the Combine Corn Head to the Industry

In 1954, Deere became first to successfully market a combine corn head. Engineers at the Deere and Company Harvester Works had been experimenting with the use of the combine in corn since 1950, at first using modified grain heads to harvest the whole plant. The first corn heads or "corn attachments" as Deere preferred to call them, were tested in 1953. The first 19 commercial units, designated the No. 10, were sold in 1954. In the same year the 26-in. cylinder Model 45 combine with a spot-welded body of more robust construction was released as Deere's new SP "Cornbelt Combine". It was equipped for corn with the No. 10 Corn Attachment. A popular combination, 43,361 Model 45's were sold between 1954 and 1969.

Production of the No. 10, with its unique twin-auger feeder design, was undertaken at Deere's Des Moines Works, a logical choice considering this ex-munitions plant had been purchased by the company immediately

Left: The Model 45 combine was more robustly built to take the punishment of corn combining, heavier grain bin loading, and the often rougher field conditions that prevail later in the season when corn is harvested; Above: The Model 45 and No. 10 Corn Attachment Head, introduced in 1954, were a popular combination. 43,361 No. 45's were built between 1954 and 1969.

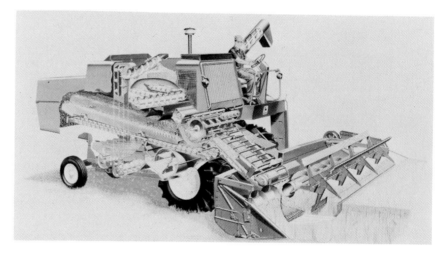

Four models, the 3300, 4400, 6600 and 7700, Deere's "New Generation" SP combines were introduced simultaneously in 1970. The basic design utilized a combination auger grain pan and augers beneath the walkers. The front auger system ("1" in the drawing below) is claimed to deliver positively across the width of the cleaning shoe. From the rear, reversed flighting on the common drive shaft (2) pulls separated grain forward onto the shoe. Cleanout is simplified by a hinged rear auger trough and an inspection door for the front auger bed. The model shown is the 3300.

after WW2 specifically for cornpicker production. Those first 19 units were all recalled and replaced with updated versions when the 1955 season began. In 1956, components were produced to enable the No. 10 series two- and four-row corn attachments to be fitted to the Model 55 combine and, in 1958, to the 95 combine. A smaller corn attachment, the 205, was produced for the 24.6-in. cylinder Model 40 SP combine which was in the line from 1960 to 1966. The 34 and 35 series followed, with further upgradings, and a variety of sizes were made available up through eight rows in both "wide" (38-in. to 42-in.) and "narrow" (20-in. and 30-in.) row widths.

The 40 series corn attachments were introduced with the 1970 "New Generation" combines. But by that time, they were East Moline Harvester Works products, the corn attachments having been reassigned to the combine factory from Des Moines the previous year. The Des Moines Works remained in the corn harvester business, having introduced in 1970 the pull-type Model 300 Corn Husker line for those farmers who still prefer to harvest ear corn. The 40 series corn attachments feature low-profile (25°) gatherers, through-case hex-shaft drives, and hinged gathering points which can be lifted readily for inspection or repair.

The "New Generation" Combines

Plans were begun in 1963, for a new generation of Deere combines to meet anticipated needs of a decade ahead. Project "New Generation", which

Left Top and Bottom: Air drawn through the rotary screen by the radiator fan is forced through the engine compartment and down across the feeder house and right hand drive areas. The vacuum pickup arm sucks chaff and trash off the screen during each revolution of the screen; Top Right: The Model 6602 Hillside with 44 in. cylinder continues Deere's association with the Western hillside combine business. The machine has automatic lateral leveling to keep the separator body level on sideslopes of up to 45 percent; Below: "Posi Torq" ground drive was the standard transmission on the New Generation combines. Deere's hydrostatic drive is optional. Engine power transmitted by belts is converted to hydraulic power by a back-to-back pump motor team. Hydraulic power is converted back to mechanical power at the four speed gear transmission. Mechanical drive is on the left; hydrostatic drive on right.

had a seven-year gestation, was conceived in the shade of a salt cedar tree on the Arizona desert as a group of combine designers and executives watched 1964 models being tested. By July 1963, a new development team had been formed and at the end of the year the New Generation began to take shape in a functional test laboratory. A full scale glass-sided separator embodying the threshing and separating mechanisms was in use, with a full scale cleaning shoe test stand alongside, for the evaluation of design alternatives.

In the summer of 1964, the first field test machine, utilizing a 95 combine as a base, was placed with a custom operator in North Dakota. When he summarized his experiences with the machine at the end of the season by describing it as the "95 with the fast insides", the design team felt they were off and running.

The first prototypes were built in 1965, followed by another group in 1966, and a third in 1967. The prototypes harvested over 2½ million bushels of every combineable crop in the US. Upgrading continued in 1968 and, in 1969, the final pre-production units— four different sizes of self-propelled and one pull type— were built and evaluated in the field.

The PTO Model 6601 was introduced in the Fall of 1969. The four SP models in the New Generation line were released simultaneously in 1970. They had new styling, auger grain pans, on-the-go adjustments, new engine cooling and radiator screen systems, and higher productivity. It was the first time in the history of the industry that any combine manufacturer had introduced an entire line at once—and retired all of that line's predecessors.

165

Keeping Chaff Out of the Engine Compartment

Engine air cooling systems are notoriously difficult to keep free of chaff and trash in the grain harvest environment. The worst problems are clogging of the radiator core causing engine overheating and buildup of flammable material around hot engine components. problems.

Engines on the first New Generation combines were cooled by a reverse airflow system. The rotating radiator screen system had been considered, but then discarded in favor of the reversed fan system which drew air through patented louvers on the top and front of the engine compartment then through perforated screens through the fan and radiator. Debris which penetrated the louvers was conveyed to the far side of the combine by virtue of the shape of the louvers, and discharged through a duct. This system had inherent advantages, such as reducing the heat load on the operator's station, but created its own set of problems, particularly that of contamination of the air in the engine compartment with resultant buildup of deposits such as smut spores and chaff in the engine area.

The solution involved a return to the rotating rad screen concept, but with a significant addition; a patented self-cleaning concept—a pickup arm which vacuums debris off the positively-driven radiator screen, assuring that the screen is cleaned on each rotation. Screened air forced through the engine compartment is then directed down and across the feeder house, helping to keep this area clean while improving operator visibility (West et al., 1974).

John Deere Detachable Platforms and "Quik-Tatch"

Traditionally, the feeder house was an integral part of the header or platform. This was satisfactory in an era when there were relatively few combine models and few header sizes. But combine models and headers have proliferated. By 1961 for example, Deere had four SP combines with headers ranging from 12-ft to 22-ft widths with almost one foot increments. The problem was further complicated by two lengths of feeder houses—the regular short feeder for small grain and corn, and a long feeder for the rice specials, which are usually equipped with larger tires. There were also special platforms for windrow pickups.

Beginning in 1961, means were sought to make any given header fit any of the various models of combines. The solution, when it was worked out, was remarkably simple: the feeder house was removed from the header and made an integral part of the separator. A front closure on the feeder house mated with a standard-sized opening on any of the various headers.

Five crop specialized header types are available for Deere combines. Shown are windrow pickup head, corn head, row crop head, rigid cutterbar grain head and flexible cutterbar head.

Model Identification	Production years	SP or PT, GD, EF or PTO*	Cylinder or body width (in.)	Header sizes (ft.)
# 1 Canvas table	1928-1929	PT, GD	24	8, 10
# 2 First JD combine	1927-1929	PT, GD	24	12, 16
# 3 Repl. #2 (only 50 built)	1928-1932	PT, GD	30	12
# 5 Repl. #1	1929-1934	PT, GD	24	10, 12
# 5A	1934-1941	PT, GD	24	10, 12
# 6	1936-1939	PT, GD/EF	24	6
# 7	1932-1941	PT, GD/EF	24	8
# 9 Auger platform	1939-1946	PT, PTO/EF	30	12
#10 & 10A (10A with 3½' cut, only 5 built)				
#11	1939 only	PT	50	5
#11A	1940-1942	PT	60	5
#12	1939 only	PT	60	6
#12A Various engines	1940-1952	PT, EF	60	6, 7
incl. Model L tractor engine				
#17 Repl. #3	1932-1948	PT	30	12, 16
#25	1952-1955	PT	60	7
#30	1956-1960	PT	60	7
#33	1940-1943	PT	22	10
#35	1937-1940	PT	24	12, 14
#36 Ex-Caterpillar combine	1936-1951	PT	26	16.5, 20
#40	1960-1966	SP	24.6	8, 10
#42	1961-1966	PT	24.6	9
(John Deere Introduced their first SP, the model 55 in 1947)				
#55SP	1947-1969	SP	30	12, 14
#45 First with corn heads	1954-1969	SP	26	8, 10, 12
#55H Hillside	1954-1960	SP	30	14
#65 (Based on 55)	1949-1966	PT, PTO	30	12
#95 &				12 to 20
#95H	1958-1969	SP	40	16, 18
#96	1963-1967	PT, PTO	40	12
#105	1961-1969	SP	50	13 to 22
#106	1967-1968	PT, PTO	50	14
#111 (Peanut special)	1965-1966	SP	47.5	(P-U)
#3300	1970-1976	SP, 70 HP dsl	28.75	13, 15
Current Combine Model Lineup				
#4400	1970-	SP, 100 HP dsl	38	13 to 20
#6600	1970-	SP 120 HP dsl	44	13 to 22
#6600 SideHill (to 18% slope)	1975-	SP 128 HP dsl	44	13 to 22
#6601	1969-	PT, PTO	44	13 P-U
#6602H Hillside (to 45% slope)	1971-	SP 135 HP dsl	44	18 or 20
#7700	1970-	SP 128/145 HP dsl	55	13 to 24
#7701	1977-	PT, PTO	55	9, 11

*PT = Pull-Type   SP = Self Propelled   GD = Ground Driven   EF = Auxiliary Engine Functioned   PTO = Power Take-Off Driven

This innovation had an additional benefit. Attaching header to separator was greatly simplified and hitch-up time correspondingly reduced. This gave the Detachable Platform system its name—"Quik-Tatch". The system was introduced in 1964 and subsequently licensed to the majority of North American combine manufacturers and to some in other parts of the world.

John Deere Header Combine-ations

Five specialized header types are available for the "Long Green Line" of Deere separators:

- The 200 Series rigid cutterbar platform for grain harvesting, sizes 13 to 24 feet. The rigid cutterbar is adjustable four ways. Auger size is 24-in. diameter.
- The 200 Series flexible cutterbar for soybeans and small grains, sizes 13 to 20 ft.
- The 40 Series low profile corn heads in 2- through 8-row sizes.
- The 50 Series Row-Crop Heads made in 4- to 8-row sizes for maize sorghum, sunflowers and row crops other than corn.
- Pickup platforms made especially for windrow harvesting.

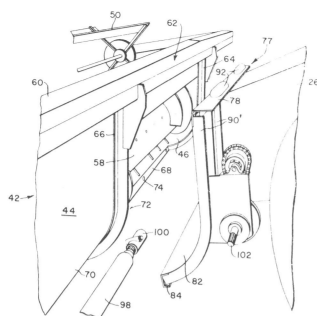

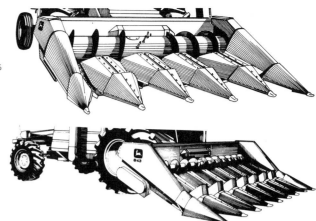

Above: Deere 50 Series Row Crop Head on 6600 combine. Head is available for 4- to 8-row crops other than corn; Below: 7700 shown with 843 8-row corn head and Power Rear Wheel hydrostatic drive option installed – a Deere development for four-wheel traction in rough conditions. Switch on control console permits engagement of rear wheel drive on-the-go.

Above: Patented "Mounting and Dismounting Means for Agricultural Machines" formed the basis for Deere's 1966 patent "Quik-Tatch" header mounting system. The feeder is incorporated with the separator body. A front closure on the feeder house mates with a standard-sized opening on any of the header combinations; Right: Deere's 40 series Corn Head also introduced in 1970 had 25° low profile gatherers to provide better pickup of leaning and tangled stalks and a clearer view for the operator. Variable speed feeder house is standard on the 6600, SideHill 6600 and 7700 corn combines, and an option on the 4400.

Left: Model 7701 pull-type with 55 in. cylinder. This pull-type version of the 7700SP is the largest pull-type combine on the market. Grain capacity is 200 bushels.

"This man's dream has junked every binder, header and threshing machine in use. He has made grain growing a pleasure. He has driven this gypsy band of hobo harvesters from our States and forced them to enter steady pursuits. He has taken the threshing drudgery from the farmer's wives, made farming a joy, and increased the price of a bushel of wheat at least twenty cents."—from a Kansas News Editor's eulogy on Curtis Baldwin, written in 1916.

ALLIS-CHALMERS

17

Rumely, Baldwin and Allis-Chalmers Corporation

Allis-Chalmers traces its history to a Milwaukee business begun in 1847 to make millstones. By 1900, this firm, Edward P. Allis and Co., was a leading manufacturer of steam engines and industrial machinery. A year later, following mergers with other industrial equipment manufacturers, the name became Allis-Chalmers Company. Succeeding years saw the company further expand by acquisitions into the production of hydraulic and electrical equipment, airbrakes, generators, and turbines for heavy industry. (Phillips 1956). In 1914, Allis-Chalmers introduced its first farm tractor, marking the beginning of its development into a full-line farm equipment manufacturer.

The contributions of Allis-Chalmers to the history of the combine originated from two sources: first, the Advance-Rumely Company, acquired by Allis-Chalmers in 1931. It is now the LaPorte, Indiana, division. The second, the Gleaner Harvester Corporation, was acquired in 1955 and is now the Allis-Chalmers division at Independence, Missouri.

The Advance-Rumely Company

German born Meinrad Rumely emigrated to the US in 1848 and settled in LaPorte. Together with his brother John, he set up a machine shop and foundry in 1852 to make products for the thriving railroad construction industry as well as machinery for sugar cane crushing and sorghum-molasses extraction. By 1855, the brothers were building corn

Meinrad Rumely, 1823-1904. His career began in the German army, but ended abruptly when he was unjustifiably hit on the head with a pistol by an officer for a minor breach of discipline. He emigrated to the US in 1848 seeking a fresh start.

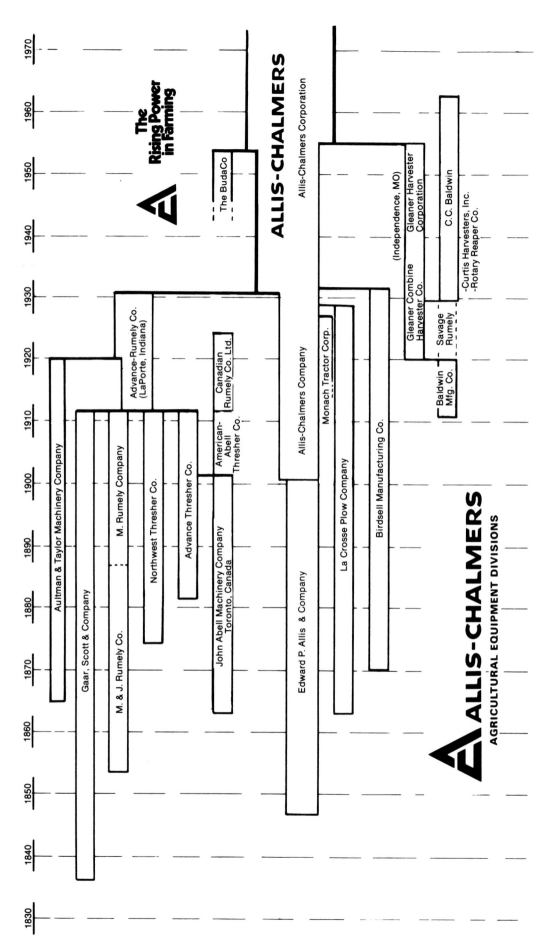

Chart of development of the Allis-Chalmers Corporation. (Harvester, and major farm equipment divisions).

Above, Below and Right: Rumely OilPull tractor, Advance-Rumely combine and Advance-Rumely threshing machine, ca 1927 vintage.

shellers, and in 1857, turned out their first threshing outfit, which two years later took first place at Chicago's Illinois State Fair. In 1861, two medals were awarded the Rumely Company by the US Agricultural Society; one for best horse power and the other for best separator. Before long, Rumely threshers were being shipped to other parts of the US and to foreign countries as well. Rumely later added steam engines and clover hullers, and was one of the earliest US manufacturers to adopt steam power to threshers. In 1909, the company developed an internal combustion traction engine that was claimed to be the first oil-burning tractor equipped with throttle governor and first to use water injection under control of the governor. This was the renowned Rumely OilPull, "Kerosene Annie".

When Allis-Chalmers purchased Advance-Rumely Co., in 1931, they secured a worldwide business, including 24 branch sales offices and about 2500 North American dealers. Advance-Rumely was also selling pull-type combines, but this was not the machine on which Allis-Chalmers established their grain harvester business prior to WW2. That impetus was to come out of the West.

The Fleming-Hall Baby

In 1930, Californians Robert Fleming, a farmer-inventor, and promoter Guy H. Hall came East to sell manufacturers on a combine that was smaller than any ever built in North America. Distinguishing features of this "baby combine" were its small size, lightweight "scoop shovel" front end, and full-width

*Above: Harry C. Merritt in 1937. Allis-Chalmers chief engineer who made the decision to commercially develop the ''baby combine''; Below: The Fleming-Hall combine of 1930; 5-foot cut, 5-foot wide wire brush threshing cylinder (**Farm Implement News**, 1930).*

wire brush cylinder, designed to provide a resilient flailing action on the grain without chewing up the straw.

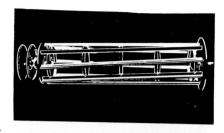

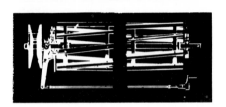

Charles J. Scranton, Jr. in 1952. Scranton was awarded the Cyrus Hall McCormick medal by the American Society of Agricultural Engineers for his work on the ALL-CROP combines; Right: The full width cylinder design of the ALL-CROP Harvester was unique, a foot of width for each foot of cut.

They demonstrated the machine during the summer of 1930, in Kenosha County, Wisconsin. Executives and engineers of Eastern farm machinery companies came and looked, but it was Harry Merritt of Allis-Chalmers who acted first. He paid the intrepid Californians $25,000 for a shop license to produce the small combine. Twenty-five were built in 1931 and test marketed in Illinois and Indiana. The "corn-belt combine" had found its place.

The wire brush cylinder proved a problem from the outset. It was not durable and did not work well in soybeans. The soybean crop had been combine harvested for the first time only seven years earlier and was just coming into prominence.

Harry Merritt assigned the job of redesigning the machine to Walter Dray who headed up engineering at Allis-Chalmers' newly acquired LaPorte plant in 1932. Dray replaced the wire brushes, but retained the principle of resiliency on the threshing cylinder by using rubber-covered angle bars on the 5-ft wide cylinder and on the closed concave. V-belt drives were installed where previously link chain had been the rule. The modified machine was called the ALL-CROP Harvester, an Allis-Chalmers registered trade name. The Model 60 ALL-CROP made its debut in 1934. It weighed 2800 pounds, cut a 5½-ft swath and was mounted on pneumatic tires to reduce draft (Allis-Chalmers had pioneered the use of rubber tires on tractors in 1932). The Model 60 could be pulled and powered with ease from the pto of a two-plow tractor. In 1935, 550 ALL-CROP Harvesters were sold. The one-man harvester, designed for the family farm of under 100 acres, was a commercial success. In 1936, 8000 Model 60's were sold for only $595 each. It could harvest a diversity of seeds—from birdseed to beans. "Harvest the soil builders" said 1936 ads, referring to legume and grass seed production. The ALL-CROP Harvester's versatility gave added scope to the new soil conservation movement at the time.

The merit of the ALL-CROP Harvester was recognized by an award of the Royal Silver Medal at the 1936 British Agricultural Exposition, Bristol, England. The Royal awards committee judged the machine the most notable advancement of the year. This was the first such award to an American manufacturer in 12 years. The credit was Harry Merritt's. He had the vision to promote the machine at a time when many, even among his associates, thought he was playing with a toy.

Charles J. Scranton, an engineer from the financially-troubled Avery Company was hired for the LaPorte team. In 1935, he was assigned to reduce manufacturing costs of the ALL-CROP design in preparation for high volume production. At the same time, Scranton began to work on a smaller version, the Model 40 ALL—CROP, for the under-50 acre crop farm.

In 1937, 11,500 units of the Model 60 ALL-CROP were built. Sales of the "successor to the binder" began to overhaul binder sales. Binder production fell from 66,000 units in 1936 to just 15,000 in 1939. There were improvements to the ALL-CROP Harvester, including the variable-speed V-belt cylinder drive, an idea which came to Scranton during a church service. Scranton reported:

"We had some trouble for a time with this drive and I have often wondered if the Lord was punishing me for my inattention to the sermon."

In 1938, a total of 28,000 ALL-CROP units were produced, or 60 percent of the production of all small harvesters in the United States that year. One hundred 40-in.-cut model 40's were also sold in 1938.

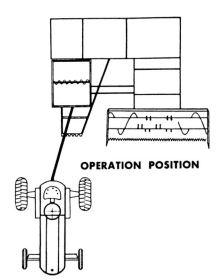

OPERATION POSITION

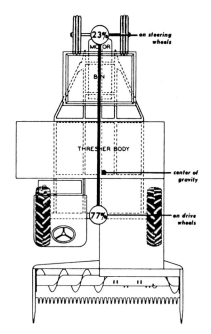

BETTER TRACTION AND FLOTATION
with Balanced Weight Distribution

Above: Model 90 ALL-CROP configuration; Right: Model 100 SP ALL-CROP Harvester configuration. The machine was marketed with 9- or 12-foot cut, 66 in. cylinder.

The Model 40 ALL-CROP Harvester was produced from 1938 to 1942. This 40 in. cut, 36 in. rub-bar cylinder pull-type combine sold for just $345 when it first came on the market. Grain tank capacity was 11 bushels.

Automotive body manufacturing techniques were used on the streamlined Model 40. The sheet metalwork on the body also served as the frame, with the "turret top" formed in a giant press. Sheet metal fabrication techniques were used almost exclusively. The Model 60 assembly line could turn out 150 machines per day. The Model 40 assembly line had a capacity of 250 machines a day. The Model 40 sold for the price of a power binder—$345. This was a price that farmers, still shaky from the Great Depression, could afford.

By 1939, about 80 percent of all combines sold in the US cut swaths under six feet. The little Model 40 alone captured nearly 10 percent of the entire US combine market between 1938 and 1942. (Baker 1960). Allis-Chalmers built 320,000 ALL-CROP units between 1935 and 1958. For their work on the ALL-CROP and other engineering achievements, including the ubiquitous ROTO-BALER hay-baling machine, Merritt and Scranton were awarded Cyrus Hall McCormick gold medals by the American Society of Agricultural Engineers. Merritt's recognition came in 1941, Scranton's in 1952.

During 1941, the LaPorte plant released a low cost, two row tractor-mounted corn-picker-husker, a lightweight and economically priced machine. Production was interrupted by critical material shortages during WW2, but resumed along with the ALL-CROP in 1945.

Larger ALL-CROP Models were produced in the 'fifties', culminating in the design of an SP ALL-CROP Harvester, the Model 100, released in 1951. The company sales department decreed that the SP 100 had to carry the same general design layout as the popular pull-type Model 66. In the next four years, 5000 ALL-CROP 100's were built before the ALL-CROP was gradually phased out of production, following the acquisition of the Gleaner Harvester Corporation in February 1955. Thus passed a colorful era of combine history at LaPorte, Indiana. Today, the LaPorte plant makes corn heads, cotton strippers and implements, but no combines.

The Model 100 SP ALL-CROP combine, 1951.

Gleaner Combines

All Allis-Chalmers combines now carry the GLEANER trademark. The history of Gleaner combines is essentially the story of Curtis C. Baldwin, founder of seven companies, and one of the most audacious farm machinery entrepreneurs of recent times.

The Baldwin brothers (Curt the oldest, George and Ernest) were born and reared in a sod house in Western Kansas. Curt described his family as "stingy Scotch" extraction. The long hot summer days of his Kansas boyhood were spent pitching wheat sheaves into threshing machines. Starting at dawn, Baldwin was so tired by nightfall he recalled collapsing into bed, often the straw pile. He realized that:

> "a machine that could make this drudgery unnecessary would be a blessing and I was fired with ambition to build such a machine". (Baldwin 1931).

In his teen years he gained valuable experience as he graduated to operating his own custom threshing rigs from the Kansas plains northwards to the Canadian prairies. In the beginning, building harvesters was an avocation, then it became a livelihood. His first enterprise, The Baldwin Company, began in 1911 at Nickerson, Kansas.

At the age of 23 he was awarded his first patent and advertised his company's "Standing Grain Thresher". The Standing Grain Thresher was first tested in 1910. It was a pusher, that is, four horses were hitched to the draft pole which extended rearwards, in the manner of the grain "headers" of the day.

Baldwin had accomplished his objective: whereas his stationary thresher rigs had to transport the "field to the thresher", now he took the thresher to the field. As Baldwin described it:

> "With the ordinary (threshing) machine, the grain passes the cylinder at the rate of a mile a minute, with my machine the

The Standing Grain Thresher was built like a vacuum sweeper, shown here at work near Wichita, Kansas in 1911.

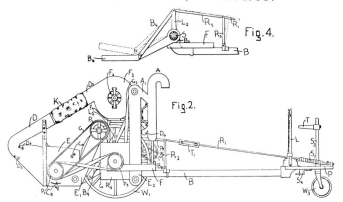

Curtis C. Baldwin, 1888-1960, taken in 1931.

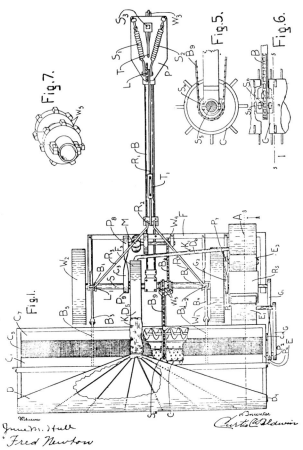

Baldwin's US Patent No. 1,004,134, September 26, 1911 for "Standing Grain Thresher."

the grain passes the cylinder as fast as the horses can walk, or about three and one half miles an hour. You can see that with my machine the cylinder strikes the heads seventeen times where the ordinary machine only strikes them once, I only use the mile a minute cylinder as a rethreshing device . . ."

"The saving to the grain farmer consists not only in dispensing with the many high waged threshers but also in the time formerly lost by cutting the grain, hauling it to the stack, then hauling the threshed straw back to land, where, as a rule, it is ultimately burned up." (Morse 1913).

The Standing Grain Thresher, by using a head-stripping action, dispensed with the need to cut the grain. A small gasoline engine powered the machine. Two men were required, one to drive the draft horses and the other to tie the sacks and look after the engine. Capacity was 25 to 30 acres per 10 hour day.

Baldwin claimed that the machine would reduce the cost of the harvest from 14 cents to 2 cents a bushel in 1913. Although Baldwin persevered with this stripper-type harvester concept for several years, there were difficulties with lodged crops, weeds and gathering losses at the stripper-drum. In 1914, Baldwin replaced the power-hungry air blast over the stripping drum with a slatted mechanical gathering canvas.

Baldwin next tried to overcome the drawbacks of the head-stripping method of harvesting, a method not always suitable to the humid conditions of mid-America. In 1917, he filed for a patent on a unit using the traditional reel, sickle and canvas apron. In this patent is a glimpse of the tractor-mounted combines that eventually followed.

In 1918, the Standing Grain Thresher Company, which had been formed by Baldwin to build these unusual machines, went out of business and a new company was formed.

Clarence A. (Cal) Stevens had become a partner of the Baldwin Brothers in 1915. When Curt Baldwin set off alone to develop the Savage (tractor-mounted) combine in Denver, Colorado, Stevens replaced Baldwin as president.

The Savage Harvester Company's combine of 1921 was side-mounted on a 44 HP Savage tractor. This harvester cut a 16½-ft swath, or two acres for every mile traveled. Thus, at 2½ mph its capacity was 5 acres an hour. A novel feature of the Savage combine was the "Baldwin patented chaff separator". The 1921 sales literature claimed that this device halved the size and weight of the separator. Baldwin retained the overshot threshing cylinder down on the front.

During his sojourn in Colorado, Baldwin enlisted the services of Norman E. (Pop) Bunting to work with him on the Savage as superintendent of engineering. Bunting had considerable experience with combines

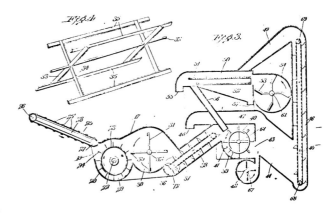

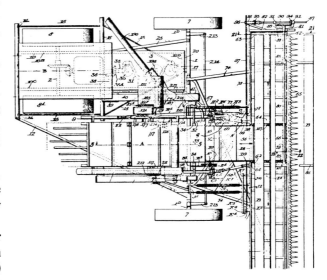

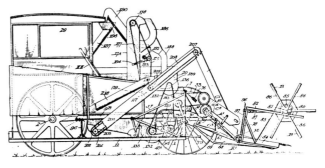

Top: Baldwin's US Patent 1,290,484 filed July 20, 1914 for Standing Grain Harvester; Center and Bottom: Baldwin's US Patent No. 1,598,234 "Automatically Operating Power Actuated Combine Grain Cutting and Threshing Harvester," filed October 18, 1920. It was advertised in the Savage Harvester Company's literature, but issued in 1926 on consignment to Advance-Rumely Co. Note the "T"-shaped platform configuration.

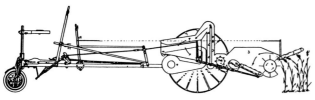

The 1916 Model Standing Grain Harvester Thresher in cross-section.

and had secured patents, while working for International Harvester, on the Deering Reaper-Thresher.

Twenty-five Savage harvesters were built and capacity was said to have been a respectable 7 acres per hour in average wheat. But the tide of fortune ran against Baldwin and the design was acquired in the late 1920's by Advance-Rumely Co., of LaPorte, Indiana.

The Wrap-Around Combine

Meanwhile, Stevens and the younger Baldwins, working independently of Curt, designed and built their own combine around the ubiquitous Fordson tractor. Henry Ford's tractor, manufactured along automobile assembly methods and designed to undersell competition, had been an instant success among smaller farmers. Within one year after production began in 1917, Ford had usurped IH's place as the leader in US farm tractor sales. (Phillips 1956).

The first Fordson-mounted combine was built in a machine shop at Wichita and demonstrated in 1923. After a successful harvest season, the group

US Patent 1,702,323 "Combined Harvester, Thresher and Separator" was filed on December 3, 1924 by C.A. Stevens, George D. Baldwin and John I. Michaels. The Fordson tractor is illustrated. Patent issued February 19, 1929 on assignment to the Gleaner Combine Harvester Corporation of Independence, Missouri.

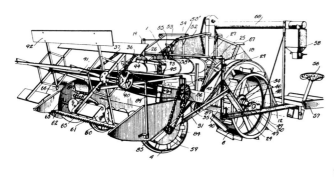

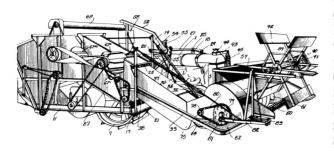

contacted W. J. Herman of Wichita who, in turn, interested Stephen H. Hale of Kansas City in financing the building of five combines for the following season. The Gleaner trademark dates from April 22, 1924, when the Gleaner Manufacturing Company was incorporated as a Delaware corporation.

One of the five original Gleaner combines was used as a demonstrator in Oklahoma, Kansas and Colorado. The others were sold through dealers. The Gleaner combine cut an 8 ft. 3 in. swath—just enough to cover an acre in a mile's travel. The Butler Manufacturing Company of Kansas City built 100 units on contract in 1924. At the Wichita Tractor Show in February, 1925, so many orders were received with cash deposits that Gleaner doubled their Butler order; 200 machines were sold for the 1925 harvest. Encouraged, Gleaner purchased their own plant at Independence, Missouri in August 1925.

Then in 1926, Curt Baldwin decided to rejoin his brothers. His reception is not a matter of public record! Gleaner built 641 units for the 1926 season, but the year ended in a net loss of $33,619, and with total inventory valued at $33,029. Hale, who had advanced $160,000, was not satisfied with the operation and demanded a reorganization. The name of the company was changed to the Gleaner Combine Harvester Corporation which assumed the assets of the Gleaner Manufacturing Company on December 30, 1926, in exchange for 15,000 shares of common stock and 1600 shares of preferred stock issued to Hale to cover his $160,000 loan. Hale also became president.

Meanwhile, Curt Baldwin had designed and built a 10-ft pull-type combine of similar design to the tractor-mounted Gleaner combine, but with a larger separator. He named it "Baldwin Gleaner" and insisted on the Corporation building this machine as well. The directors said no. Baldwin resigned. He then contracted with Standard Steel Company of Kansas City to build his pull-type combine and, in 1927, formed the Baldwin Harvester Company, reusing the old name. Gleaner reached an agreement with the maverick Baldwin and received a royalty on all combines sold by his company.

Through the 1920's IHC had fought the Fordson tractor by responding with newer and lighter tractors, the McCormick-Deering 10-20 and 15-30. IHC also more closely integrated implements with their tractors. This step was to prove decisive as IHC tractor sales climbed steadily at Ford's expense. By 1927, confidence in the Fordson had evaporated and in 1928 Henry Ford discontinued US tractor production. It was obvious that without the Fordson, sales for the pull-type Baldwin combines would exceed those of Gleaner tractor-mounted combines. However, Standard Steel was having difficulty manufacturing the Baldwin. In 1927, when the Fordson seemed doomed, an agreement was reached between Gleaner, Baldwin and Standard Steel allowing Gleaner to take over the manufacture and sales of the Baldwin combine. Five hundred Gleaner combines and 1050 of the Baldwin pull-type units were built. Business expanded rapidly, and the 1927 sales exceeded $1.2 million. Gleaner ceased Fordson-mounted combine production in 1928 and increased the size of the pull-type Model P to 12-ft cut. By improvements to plant and design changes, 3000 combines were built in 1929 for a record gross of $2.8 million, with net earnings after taxes of $532,240. Contracts were let to further expand the factory. Then Curt Baldwin moved again.

During the Fall of 1928, Baldwin, C. B. Ruble and G. T. O'Mally of Universal Equipment Company formed the Baldwin Harvester Corporation and started building a machine they dubbed "The New Baldwin".

Gleaner promptly filed suit against the new firm, which resulted in an injunction in May 1929 that prevented the firm from using the name "Baldwin" or "New Baldwin". Curt Baldwin switched to the "Curtis" name and some 1928 literature was overprinted to reflect the change. The Baldwin Harvester Corporation was shortlived. In 1929, Baldwin Harvester became Curtis Harvesters, Incorporated, of Ottawa, Kansas, and began building Curtis combines in the plant of the Ottawa Manufacturing Company.

These Curtis harvesters were sold directly by factory agents. Baldwin actively discouraged middlemen and parts storage, in the interest of cutting costs for the farmer. This time the rift between Gleaner and Curt Baldwin was permanent.

For the Gleaner Combine Harvester Corporation, 1929 was a record year. Close to 5000 Gleaner Baldwin pull-type combines were sold and after-tax profits exceeded the million dollar mark for the first time. Plans were begun for the sale of 7000 combines and a corn harvesting attachment was developed.

The Gleaner Baldwin Corn Combine became the first combine in North America to offer a corn head.

This was a whole-plant harvester, distinctly different from the ear-snapping corn pickers then available. The first models used a heavy-duty reciprocating sickle, an oversize reel and ear lifters. An ensuing development eliminated the reel, substituted rotary cutters for the sickle, and had rigid snouts to guide the two 40-in. crop rows into the pan.

The Great Depression and Gleaner

The manufacturer of the 7000 Gleaner Baldwin

Above: the 1929 Gleaner-Baldwin 12-foot cut Model R produced by the Gleaner Combine Harvester Corporation; Below: 1927 "Gleaner" Harvester on the Fordson tractor. Note the direct feed from header auger to down-front threshing cylinder, the closed concave, and the use of a slinger-unloader on the grain bin. The operator's instruction manual was quite specific about keeping straw input down.

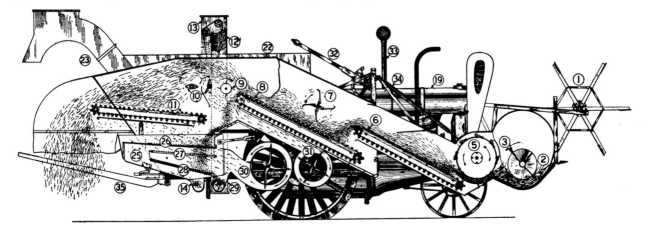

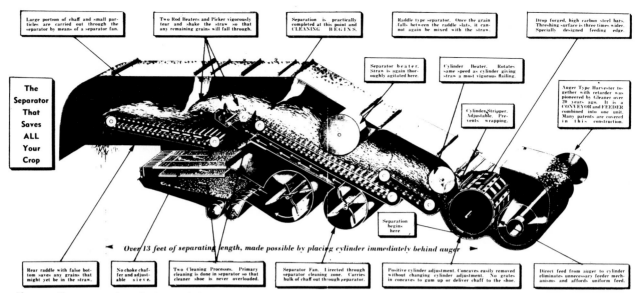

The Separator That Saves ALL Your Crop

Large portion of chaff and small particles are carried out through the separator by means of a separator fan.

Two Rod Beaters and Picker vigorously tear and shake the straw so that any remaining grains will fall through.

Separation is practically completed at this point and CLEANING BEGINS.

Raddle type separator. Once the grain falls between the raddle slats, it cannot again be mixed with the straw.

Drop forged, high carbon steel bars. Threshing surface is three times wider. Specially designed feeding edge.

Separator beater. Straw is again thoroughly agitated here.

Cylinder beater. Rotates same speed as cylinder giving straw a most vigorous flailing.

Cylinder Stripper. Adjustable. Prevents wrapping.

Auger Type Harvester together with retarder was pioneered by Gleaner over 20 years ago. It is a CONVEYOR and FEEDER combined into one unit. Many patents are covered in this construction.

Separation begins here

◄ Over 13 feet of separating length, made possible by placing cylinder immediately behind auger ►

Rear raddle with false bottom saves any grains that might yet be in the straw.

No choke chaffer and adjustable sieve.

Two Cleaning Processes. Primary cleaning is done in separator so that cleaner shoe is never overloaded.

Separator Fan. Directed through separator cleaning zone. Carries bulk of chaff out through separator.

Positive cylinder adjustment. Concaves easily removed without changing cylinder adjustment. No grates in concaves to gum up or deliver chaff to the shoe.

Direct feed from auger to cylinder eliminates unnecessary feeder mechanisms and affords uniform feed.

Above: Cross-sectional view of the working components of the Gleaner Baldwin pull-type combine; Lower Left: the Model 30 Curtis combine sold and serviced directly from Curtis Harvesters, Inc. factory in Ottawa, Kansas, (from 1930 Curtis Harvesters sales bulletin "Inside Facts").

combines was nearly completed in 1930, but sales to dealers amounted to only 5,425 units. In addition, 1000 corn harvesting attachments were built, but only 500 shipped to dealers by the end of the fiscal year. Gleaner dealers could neither sell combines nor corn attachments, nor pay off debts owed Gleaner. With bank credit cut, the company ran short of cash and, in February, 1931 was placed in receivership. In April 1932, the assets of the Gleaner Combine Harvester Corporation were purchased by Gleaner Harvester Corporation, a company formed by Commer-

cial Credit Company, W. J. Brace (Acting Receiver), several Kansas City banks, and the creditors of Gleaner Combine Harvester. The Company was purchased for $270,000 and the assumption of liabilities. In 1932 and 1933 Company efforts were mostly directed toward liquidating machines and parts inventories. But in March 1934, production resumed with the building of 217 combines. In the succeeding years, the company continued to concentrate on the pull-type combines which were sold primarily in the wheat belts of North America and on a limited export basis.

For Curt Baldwin, the depression years were fertile, if lean ones. Curtis Harvesters Inc., went under. The patent record indicates that he had another goal: centrifugal threshing and separation, and the use of all-rotary components. He applied his axial rotary separation principle, developed on grain harvesting equipment, to a gold separator. Next, in 1936, he incorporated a new business, the Rotary Reaper Corporation, with a factory at Ottawa, Kansas. Baldwin developed a "Rotary Reaper"—a tractor-mounted combine with 9-ft swath. It employed the Australian comb/auger/sickle front. He acknowledged that he was acquainted with H. V. McKay, pioneer of the Australian

Thirteen Gleaner Baldwin combines at work on the 6500 acre Carter wheat ranch at Adrian, Texas in 1929. Note that some of the combines are being pulled in tandem, one tractor pulling two auxiliary engine-functioned machines.

Left: The 6-foot cut Model T equipped with swath pickup attachment, 1937; Right: The 1947 Gleaner Baldwin 12-foot cut Model E. This machine was produced at Independence from 1939 to 1951.

stripper-harvester. The machine had no reel and claims for harvesting speeds as high as 12 mph were made in advertising. The only non-rotary processing elements in this machine were the cutterbar and the cleaning shoe.

In 1955, Curt Baldwin, his restless mind still pursuing the dream of a high-speed harvester, demonstrated his 4½ ft cut "Bearcat" combine harvesting at 14 mph in Wichita wheat. In 1956, "The Reflector" of *Farm Implement News*, asked Curt to record his story for posterity. Curt agreed, but he became so absorbed in the development of an alfalfa wafering machine in California that his personal saga remained unwritten upon his death April 19, 1960 at the age of 72.

Gleaner Harvester After the War

The Gleaner Works at Independence had begun to experiment with walkers on their smaller combine in place of raddle separators. The first self-propelled Gleaner Baldwin Combine, the 30 in. cylinder, walker-type Model A, was produced in 1951. The contemporary Model R was the last with raddle-rake separation.

Allis-Chalmers Takes Over at Independence

Allis-Chalmers began development of a two-row corn head (snapping-roll type) at LaPorte in 1955 to fit their Model 60 combine. Offered as an option on the 1957 SP Model 100 ALL-CROP Harvesters following the acquisition of the Gleaner Independence works, the corn head was also adapted to the Gleaner Baldwin Model A of 1957. The LaPorte plant also developed a 3-row, 30 in. corn head in 1963 and, in 1966, 4-, 6-, and 8-row 20 in. corn heads were offered on the Models E, A and C Gleaner combines.

Allis-Chalmers' Hillside Venture

Gleaner Harvester was not in the hillside business at the time of the 1955 merger. When Wes Davis became general manager of the Allis-Chalmers Farm

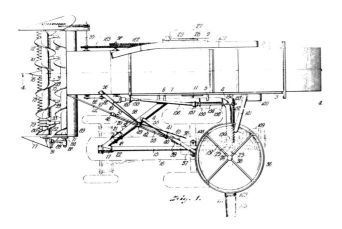

Above: The SP Gleaner Baldwin Model T, 1954 with 23 in. cylinder was designed for the smaller farm. The Model T2 had a 10-foot cut, Model T3 a 7-foot cut; Left: Curt Baldwin's "Rotary Reaper" of 1936, a tractor side mounted combine with 9-foot cut swath. Claims of harvesting speeds of over 12 mph were made in advertising. It used an Australian comb-type front. The only non-rotating processing elements were the cutterbar and the cleaning shoe.

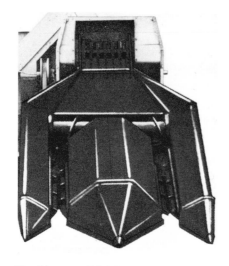

Left: The earliest Allis-Chalmers corn head attachments were developed at LaPorte for the Model 60 ALL-CROP Harvester and released in 1957. The head also matched the Model 90 and Model A Gleaner combine; Above: On display at an Allis-Chalmers dealers' convention (right to left) 1927 tractor-mounted "Gleaner," 1947 Model E pull-type, 1951 Model R SP and 1959 Model C SP.

Equipment Division he brought the experience of trying to sell tractors and equipment in the hillside areas on the Pacific coast without having a hillside combine to 'round out' the line.

At his insistence, Allis-Chalmers investigated the "Gleaners" that farmer-engineers were modifying in Washington and Idaho. The Model AH combine was developed in 1959 for the Pacific hillside market. The Model CH, GH and MH followed, still retaining the down-front cylinder concept, but encircling the thresher infeed with a feeder-head self-levelling ring for hillside operation.

New Concepts Released On The Model L In 1972

The "Top of the line" 48 in. cylinder Model L Gleaner combine was released in 1972 with a number of new design features:

- Electro-hydraulic "rocker-switch" push button controls
- Torque-sensing traction drive
- Power-fold closed bin unloader system
- Open concave with raddle conveyor under the down-front cylinder, and
- Transverse flow dual-outlet fan cleaning system.

Transverse Flow Fan

A recurring problem with combine designs for greater capacity and increased width is that of obtaining uniform air-flow over the cleaning section

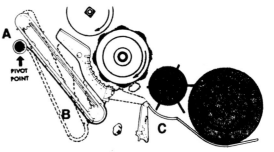

Top: Model GH Hillside, 1969 with 44 in. cylinder, 53 in. separator; Below: Among the features of the Allis-Chalmers Models L2 and M2 down-front cylinders are the automatic rock ejector door and swing-down concave.

Left: Allis-Chalmers has adhered to the idea of providing the operator with means for visually checking tailings. 1977 Model F2 shown in cross-sectional view. Allis-Chalmers has retained the down-front cylinder design consistently for over 50 years.

ALLIS-CHALMERS
"Gleaner" Combines (Independence Works)
and "ALL-CROP" Harvesters (La Porte)
NORTH AMERICAN COMBINE PRODUCTION YEARS

Model Identification	Production years	SP or PT, GD, EF or PTO*	Cylinder or body width (in.)	Header sizes (ft.)
Independence Works				
Gleaner; mounted	1924-1927	SP (tractor mounted)	24 Raddle	8.25 (tractor-mounted wrap around)
Gleaner Baldwin P	1928	PT	Raddle	10, 12
Gleaner Baldwin A	1929-1935	PT	Walker	12
Gleaner Baldwin R	1929-1935	PT	Raddle	12
Gleaner Baldwin NR	1936-1938	PT	Raddle	12
Gleaner Baldwin T	1937-1941	PT	Raddle	6
Gleaner Baldwin E	1939-51	PT, EF	Raddle	12 (Walker in 1950 only)
Gleaner Baldwin F	1939-49	PT	Walker	12
Gleaner Baldwin J	1939 only	PT	Raddle	9
Gleaner Baldwin S	1942, 1945-1953	PT, EF	22 Walker	6
Gleaner Baldwin H	1943, 1948-1950	PT	Walker	9
Gleaner Baldwin A .... (first SP)	1951-1963 1964-1967 (Bagger)	SP	30 Walker	10, 14 First SP largerheads — 1963
Gleaner Baldwin R	1951-1961	SP	30 Raddle	12, 14 (Last raddle type)
(All walker separation from here on)				
Gleaner AB	1953 only	SP (Rice m/c)		14 Rice
Gleaner P 80	1953-1955	PT		80″ cutterbar (2 rows of beans)
Gleaner T2, T3	1954	SP	23	7, 10 resp.
Gleaner B	1955-1963	SP	30	12 (rice special)
(1957: Corn heads introduced for SP's)				
Gleaner AH	1959-1962	SP	30	14 (First AC Hillside)
Gleaner C	1960-1963	SP	40	12, 14, 16, 18 20, (24 in 1963)
Gleaner E	1962-1967	SP (G 226)	27	8, 10, 12, 13
Gleaner A-II	1964-1967	SP	30	10, 12, 13, 14, 16
Gleaner C-II	1964-1967	SP	40	12, 13, 14, 18, 20, 24
Gleaner CH, CKS, CR	1967	SP (G262)	40	16, 18
Gleaner E-III	1968 only	SP (G 226)	27	10, 12, 13
Gleaner G (Variants GKS, G-Hy, GH, etc.)	1968-1972	SP (LP 2800)	44	13, 15, 17, 20 23, (12 and 16 in 1970-71)
La Porte ALL-CROP Harvester Models				
ALL-CROP 60	1935-1951	PT, PTO, EF	60	60″
ALL-CROP 40	1938-1941	PT, PTO	36	40″ sickle
ALL-CROP 66	1952-1958	PT, EF/PTO	60	66″ sickle
ALL-CROP 72	1959-1968	PT, PTO	60	72″
ALL-CROP 90	1956-1960	PT, PTO	42	7.5
ALL-CROP 100	1953-1957	SP (WD45 50HP)	48	9, 12
Production Models — Independence Works				
Gleaner F2 (and variants, FKS, etc.)	1968-	SP	37.5	13, 14, 15, 16, 17, 18, 20
Gleaner K2 (and variants KKS, KR etc.)	1969-	SP	27	10, 12, 13, (14 and 15 from 1975)
Gleaner L2 (and variants LKS, L-Hy, etc.)	1972-	SP	48	13, 15, 16, 18, 20, 22, 24
Gleaner M2 (and variants MH2 MKS, M-Hy, MKS-Hy, MH-Hy)	1974-	SP	40	13, 15, 16, 18, 20, 22

*PT = Pull-Type,   SP = Self-Propelled,   GD = Ground Driven,   EF = Auxiliary Engine Functioned   PTO = Power Take-Off Driven

Above Left: Allis-Chalmers pioneered the use of the electro-hydraulic controls in 1972 with the push button console on the Model L; Rocker switch controls were offered on: electric header and separator clutches, electro-hydraulic header lift and lower, variable speed reel drive, torque sensing ground-speed drive and hydraulic swivel on bin unloader; Above Right: Floor-to-ceiling curved windshield and glass side panels reflect the engineering effort given to cab design. Cab, which is mounted on four rubber vibration isolators, is available with tilt-forward option for transport.

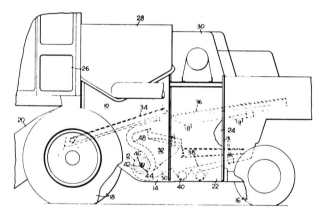

Allis-Chalmers Transverse-Flow cleaning fan with dual outlets. The upper outlet sends a blast of air through the crop stream leaving the separator raddle and grain pan. The lower fan outlet directs a controlled volume of air to the cleaning shoe.

(Quick 1969). Allis-Chalmers resolved the problem on the Model L, released in 1972, and on subsequent models, by using the transverse flow fan principle. With this concept, air flows through the rotor twice. There is no axial component of airflow. A uniform air-flow is achieved with a smaller diameter fan, reducing head room on the machine. Allis-Chalmers engineered their transverse-flow fan with a shaft running through the rotor, at a small sacrifice in airflow, in order to increase the structural rigidity of the 12-bladed fan. They controlled airflow by means of a simple choke vane which could be adjusted from the cab without engaging the separator. Variable speed fan adjustment was dispensed with. Three-stage separation was retained by using dual fan outlets (Shaver and Temple 1972).

> **"Man minus the machine is a slave. Man plus the machine is a free man."—Henry Ford (1929).**

18

The Other Slice

Over 90 percent of the North American combine market of around 35,000 machines a year is sold by the "big four combine manufacturers"— Deere, Massey-Ferguson, International Harvester and Allis-Chalmers. The "other slice" of the market was shared by four other North American manufacturers and the distributors of imported models.

White Motor Corporation, a leading builder of heavy duty trucks and onetime car manufacturer, entered the agricultural machinery market in 1959 with the acquisition of the Oliver Corporation. Three years later, White acquired Cockshutt Plow Company and Minneapolis-Moline Inc. Farm equipment lines manufactured by all three were then marketed under the White Farm Equipment Company logo. Operations were streamlined by the assignment of specific product lines to certain plants. All White "Field Boss" tractors were produced at Charles City, Iowa; tillage implements and planters at South Bend, Indiana, home of the famed Oliver Chilled Plow; and "Harvest Boss" combines at the Brantford, Ontario plant in Canada. Before the merger, each of these three companies made their own combines.

Oliver's entry into the harvesting business began with the 1929 acquisition of the Nichols, Shepard and Company's "Red River Special" threshers and combines, built at Battle Creek, Michigan.

The Cockshutt Plow Company

James G. Cockshutt founded the Brantford Plow Works in 1877 to build tillage implements. The business flourished, product lines were increased and the company renamed the Cockshutt Plow Company. In between, Cockshutt found time to serve as mayor of Brantford.

The 1933 merger with the Frost and Wood Company, founded in 1839 at Smiths Falls, Ontario,

John Nichols

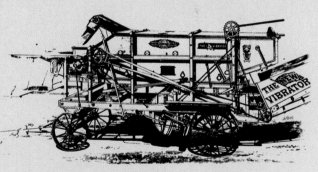

Nichols & Shepard "Vibrator" thresher, one of the early models of the line established by John Nichols at Battle Creek, Michigan in 1858.

brought to Cockshutt a range of harvesting equipment to round out Cockshutt's existing lines.

Cockshutt's 1924 tractors were the first to incorporate live power take-off, providing power for trailed equipment that was not ground speed dependent but could be operated at constant speed. Cockshutt's first combine, the pull-type Model 6, was not introduced until 1941, although Frost and Wood had tested several pre-production combines in the early 1930's.

Combine production continued—war years excepted— at the Brantford plant. During the period follow-

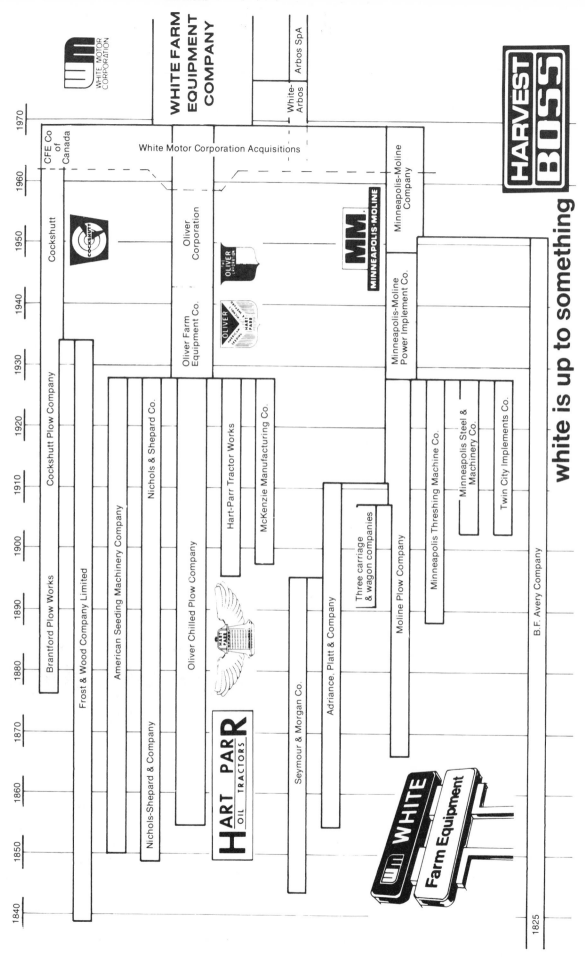

Development of White Farm Equipment Company, Oak Brook, Illinois.

The Vibrator gave way to the Red River Special in 1902.

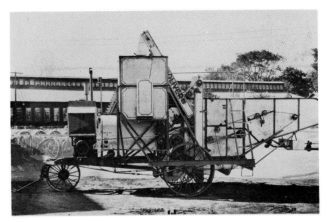

Above Left: Nichols & Shepard's first combine, the Model 22 x 30 Prairie type, appeared in 1925; Above Right: Oliver received a very substantial order for tractors and combines from the Soviet Union in 1929. This pull-type unit is being readied for shipment to the USSR from Battle Creek. Over 2,000 units were believed to have been shipped; Below: Oliver acquired the Red River Specials in 1929 when Nichols & Shepard became one of four firms that merged to form Oliver Farm Equipment Company. The Oliver row crop tractor is shown driving a 28 x 46 thresher in 1937.

Right: Oliver "Grain Master" combine Model 30, 1931. Note "tip-toe" wheels on Oliver 80 tractor. Oliver continued to produce combines in Battle Creek for 30 years after taking over Nichols & Shepard.

Top Left: Unloading the grain bin of the Grain Master Model 30 combine; Left Center: Oliver Model 2 combine in soybeans in 1946 – a 5 foot cut straight through type, first produced in 1941; Bottom Left: Cross-sectional view of Oliver Model 33 SP combine; Above: The Model 33 was Oliver's first SP combine. It was released in 1949.

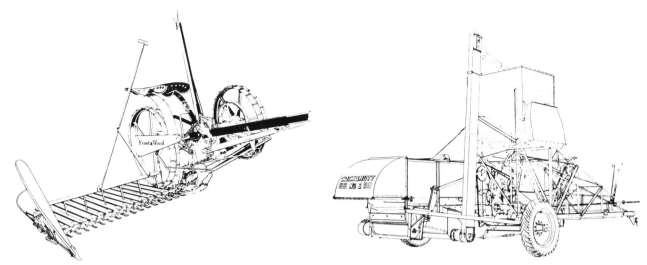

Upper Left: Frost & Wood equipment sold by Cockshutt in the 1930's included the No. 3 hand-rake reaper and the "Simplex" self-raking reaper; Upper Right: Cockshutt's first combine, the pull-type Model No. 6 produced in 1941. It had a 6 in. stroke, right hand drive 3 in. sickle; Lower Left: Cockshutt's Model 525 combine with 27.6 in. cylinder, released in 1964, enjoyed a sound reputation in both Canada and overseas markets; Lower Right: Canada's Cockshutt Model 555SP with 52 in. cylinder was also offered on the US market as the Oliver Model 5555 in 1969.

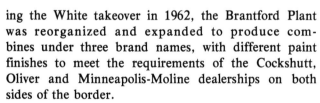

ing the White takeover in 1962, the Brantford Plant was reorganized and expanded to produce combines under three brand names, with different paint finishes to meet the requirements of the Cockshutt, Oliver and Minneapolis-Moline dealerships on both sides of the border.

Minneapolis-Moline

The Minneapolis-Moline Power Implement Company was formed in 1929 by the merger of the Moline Plow Company, Minneapolis Threshing Machine Company, Twin City Implements and the Minneapolis Steel & Machinery Company. The corporate name changed several times before it became a wholly-owned subsidiary of White in January 1963. The most significant contribution of Minneapolis-Moline to harvesting technology was the development in 1950 of the Uni-Farmor, a self-propelled "universal" chassis for powering "every harvest job at less cost than pull-behinds". The one basic propulsion unit

with interchangeable harvesting machines provided self-propelled operation at an obvious cost saving over using individual self-propelling machines for each job. The machines offered were the Uni-Harvester, Uni-Forager, Uni-Huskor, Uni-Picker-Sheller and Uni-Windrower.

White "Harvest Boss" Combines

The consolidation of manufacturing and service facilities within the restructured White Farm Equipment Company began in 1962. One result was the development at Brantford of a unified series of combines, culminating in the release of the 7300, 8600, 8900 SP series and 8650 pull-type "Harvest Boss" combines.

New Idea Farm Equipment, Division of Avco Corporation

Avco Corporation's New Idea Division entered the combine harvester market in 1964 by purchas-

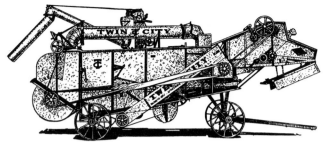

Twin City threshing machine

Minneapolis thresher of 1910

Above: The Minneapolis-Moline W4 Uni-Combine;
Right: Brochure page promoting
Minneapolis-Moline's "Universal" chassis system.

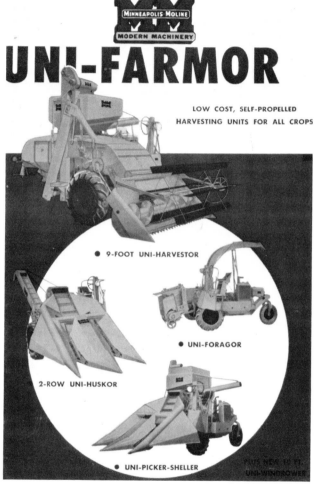

ing the rights to the Uni-Combine concept from Minneapolis-Moline. New Idea had been in the farm equipment business since 1899. Their corn shellers had established a particularly solid reputation. New Idea's corn picker-sheller, the 2-row Model 345, was marketed in 1962. This machine was suitably modified to match the New Idea Uni-System when the Model 722 Uni-sheller was offered in 1965.

The uni-system was designed with the objective of extending the use of a basic tractor unit by mounting different integrated tools on the Uni-Power chassis. New Idea offered three power units, the 706D and 708 with Perkins 6-354 diesel engines, 95 and 125 pto HP ratings respectively; and the 709, with 6-426 diesel engine at 160 pto HP.

Units available to mount on the hydrostatic Uni-Power units included the Uni-Combine with grain and corn heads, Uni-Picker, Uni-Sheller, Uni-Forager, Uni-Tool Carrier and Uni-Rotary Snow Plow. All offered considerable savings compared with the purchase of individual self-propelled machines.

Sperry New Holland,
Division of Sperry Rand Corporation

Before the introduction in 1975 of the revolutionary walkerless TR70 combine, Sperry New Holland had been marketing their 1400 and 1500 series combines, built in their Belgian Clayson Plant. This plant had been purchased from the Belgian combine manufacturer, Leon Claeys N.V. in 1965, and the renamed Clayson machines were introduced to the North American market by New Holland in 1968.

Encouraged by the success of the TR70 when it was first marketed in 1975, the company opened a new factory at Lexington, Nebraska, for serial production of the axial combine. Corn heads and grain headers for all three combine models sold by New Holland in North America were built in the Grand Island, Nebraska plant. More specific details of the TR70 axial combine appear in a later chapter.

Upper Left: The SP 7300, the smallest machine in the White combine line. Introduced in 1972 with a 34 in. cylinder, it was available with either gas or diesel power and a choice of White Kwik-Switch grain, corn and pickup heads; Above: The White 5542 combine with 42 in. cylinder was first released by Cockshutt in 1969 and found a steady market as a lower cost machine, particularly on the Canadian prairies; Below:8600 White Harvest Boss. Standard machine was diesel powered with torque-sensitive drive. Dual-speed gearbox at the cylinder enabled matching of harvesting conditions to crop load.

Above: White 8650 pull-type combine was released in 1976. Requiring the P.T.O. power of a 100-plus HP tractor for the 44.5 in. cylinder, the 8650 featured a 150 bushel bin. A control box installed in the tractor cab was connected to the combine via a wiring harness for tachometers, 9-channel monitoring of key functions and included three switches for feeder and header, unloader clutch and lighting systems.

Western Roto Thresh Limited

Western Roto Thresh Ltd. of Saskatchewan, Canada, traced its origins back to 1963 when Roto Thresh Limited was formed to manufacture a combine using rotary-drum grain/straw separation instead of the traditional straw walkers. The rotary drum concept originated with farmers Frank McBain and Bill Streich of Clandeboye, Manitoba, in 1951. The reorganized company began selling combines in 1974 with the financial backing of the Saskatchewan Government in 49 percent partnership.

Advantages claimed for the Roto Thresh combine, which has a 48.5 in. conventional cylinder,

5.5 foot diameter rotary separator drum and Cat 3208 160HP diesel engine, are:
- Low losses at throughputs up to 500 bushels/hour in wheat;
- Less loss-sensitivity on gentle sideslopes;
- Air aspiration cleaning system provided a clean sample at higher throughputs. (Lipsit et al. 1977).

Ford Motor Company

Ford Motor Company's Ford Tractor Operations have never built combines in North America. Ford's farm equipment line of the 1970's included one model of the West German Claas combines.

Assembled at Troy, Michigan and designated by Ford as the Model 642, it is equipped with a 52 in. cylinder and 127 HP Ford diesel engine.

Upper Left: Avco New Idea Model 329 pull-type picker-sheller used a 15 in. diameter sheller, 2 row head; Upper Right: New Idea Model 717 Uni-System combine with 36 in. wide cylinder, 4 row corn head. New Idea assumed manufacturing rights on the Uni-System from Minneapolis-Moline in 1964; Above: New Idea Model 315 tractor-mounted picker-sheller used 12 in. diameter cage sheller, 2 row head.

The Ford combine name has also been linked over the years with Wood Brothers of Des Moines, Iowa, whose pull-type combine and corn pickers were marketed by Ford for a brief time. Also mar-

keted by Ford was one model of the Oliver combine line and at another time, a Canadian-built Cockshutt combine.

Long Manufacturing Company

The "Other Blue Line"—Long Manufacturing Company of Tarboro, North Carolina, offered the Model 5000 combine, with 49.4 in. cylinder in two versions, mechanical or hydrostatic drive, and a choice of two engine sizes. This combine design was of Hungarian origin. Long also manufactured peanut combines for the Southern States.

J I Case Company

No history of combines could be complete without at least a passing mention of the last of the Case combines, which the Company discontinued in 1972. At that time Case was manufacturing in Bettendorf, Iowa four basic combine sizes: the 660 series with 40 in. cylinder, rack type separation; 960 series, 40 in. cylinder, walker type; 1060 and 1160 series, 42 in. cylinder; and the 1660 series with 52 in. cylinder width. Case entered the combine business in 1923, the same year that the 100,000th Case thresher came off the line (the last Case thresher was produced in 1953).

J I Case offered the first combines with cylinder speed controlled from the cab and a tailings sample check facility right at the operator's elbow.

The Mathews Company

The Mathews Company of Crystal Lake, Illinois, a manufacturer for 35 years of grain driers, entered the combine business when it acquired the Harris hillside combine plant in California. In 1969, a Mathews "Harvest King" combine was demonstrated at the Farm Progress Show. The machine was designed to reduce kernel damage to corn, long a bone of contention in the corn-drying trade.

Left: Sperry New Holland's TR70 Turbo Thresh axial combine released in the Fall of 1975. The twin 17 in. axial rotor design eliminated the need for walkers and provides compact high-capacity threshing and separation. Note that the glass wraparound cab windshield has no corner posts to interfere with visibility. TR70 was equipped with quick-attach grain heads to 22 feet, 13-foot pickup or 4-and 6-row corn heads.

Right: Cross-sectional schematic of the Western Roto Thresh combine shows high cylinder location. The rotary drum separators and four fans removed and expelled the majority of chaff, while the remainder was discharged mechanically over the end of the shoe.

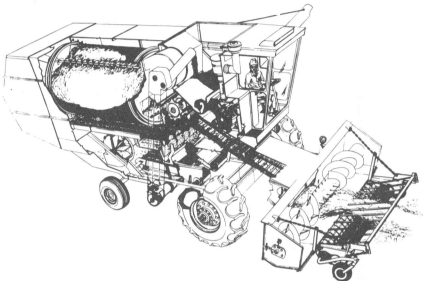

*Left: Roto Thresh combine shown equipped with swath pick up at work on Canadian prairie farm.
This combine had a 48.5 in. cylinder, 5.5-ft. corrugated-perforated rotary drum separator and 150 bushel grain tank; Right: Ford Model 642 combine, 52 in. cylinder, 127 HP Ford diesel engine. The 130 bushel bin unloaded by a top-driven closed auger system which could be positioned hydraulically.*

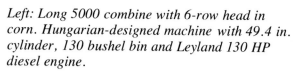

ESTABLISHED 1842

CASE

Above and Right: Case combines enjoyed a sound reputation before they were phased out of production in 1972. The Case 660 combine with 40 in. cylinder had a one-piece straw rack separation system.

THIS CROP IS SAFFLOWER.

Left: The Mathews Harvest King combine demonstrated in 1970. It was among the largest combines of its day.

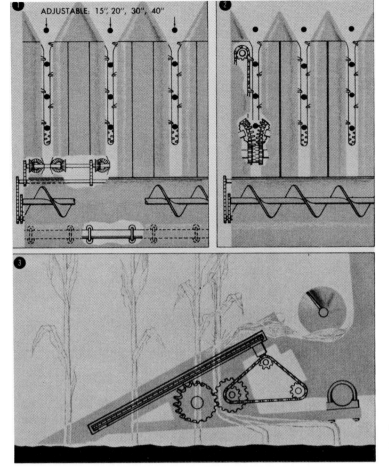

Unique features of the Harvest King combine were the easily removed cylinder and concave design. These illustrations from a company brochure show the gear-type ear-snapping heads and adjustable row spacing. The corn head could harvest up to twelve 15 in. rows.

Among the largest of combines available at the time, the Harvest King had a 54 in. wide cylinder, 14½ feet long walkers, 140 bushel bin and a 135 HP Hercules gasoline engine. Features included a unique cylinder and concave arrangement allowing these components to be serviced or changed in 30 minutes by leaving the cylinder attached to the feeder when the combine was backed away from the header. A 12-row 15 in. spacing corn head was also possible using the Mathews compact geartype snapping wheels and tool-bar style of mounting for the row heads.

Unfortunately, a slump in the combine market, together with the high cost of tooling, forced Mathews out of the market.

Lilliston Corporation's Peanut and Edible Bean Specials

The Lilliston Corporation of Albany, Georgia, began as the National Machine Company, founded in 1911, in Suffolk, Virginia by Thomas M. Lilliston. After moving operations to Albany, Georgia in 1919, the Company adopted the Lilliston name. The First Lilliston combine, based on the Lilliston peanut-picker, was developed by John T. Phillips Jr. in 1949. In 1964, the 400 Series Peanut Combines were replaced by the 1500 Series of machines developed under the engineering supervision of William G. Moore, who was president of the American Society of Agricultural Engineers in 1977. In 1972, Lilliston's

6000 Series HI-CAP Peanut Combines were introduced. An edible bean version, the Model 6200, followed in 1973.

Peanuts and edible beans have a number of similarities, from the crop harvesting standpoint. Both are quite vulnerable to harvesting damage—visible cracks and split seeds are objectionable as well as detrimental to the food value of these crops. Both crops are dug up and windrowed for curing before being threshed. Sticks, stones, trash, etc. may be gathered in the windrow and picked up by the combine. Thus the combine must digest the foreign matter in the windrow, yet gently separate the peanuts and seeds from the vines (Frushour & Johnson 1974). Gentle crop handling: was the primary constraint in the design of these specialized pull-type combines and the spring tooth cylinder and concaves introduced by Lilliston are now practically standard for harvesting these crops.

Grain Swathing

Peanuts and edible beans are only recent beneficiaries of the art of crop harvesting from the swath or windrow. The story really began around 1904. Two brothers, August and Ole Hovland of Ortley, South Dakota, noticed that when a sheaf of wheat was tossed off a binder with a broken band, the loose sheaf lay on top of the stubble and in several days

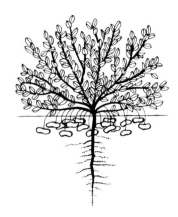

Left: A ''pusher-type'' mounted four-row bean cutter on a Caterpillar Thiry tractor illustrates one type of machine developed for digging and cutting low-growing crops such as peanuts or edible beans. This photo was taken in a Camorillo, California edible bean field in 1929 (Caterpillar Tractor Co.).

Left: The Lilliston HI-CAP 6000 Peanut combine used a series of three 50 in. spring tooth cylinders and concaves to thresh the crop after gathering by the pickup head. Peanuts grow underground and after lifting have to undergo a field curing stage in the windrow before threshing. The Model 6000 was the first peanut combine to use walker-type separation; Right: Lilliston 1580 Peanut combine had 40.5 in. cylinder behind 5.5-foot standard Lilliston windrow pickup and paddle-type separation. Bin on Lilliston combines lifted and tilted to unload.

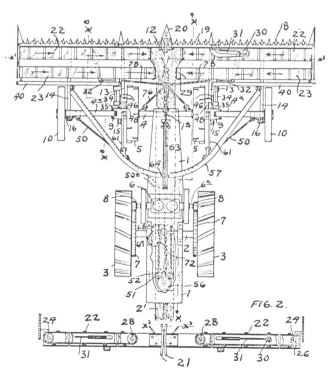

FIG. 2.

Left: Original patent on ''Central Delivery Reaper'' by A. Hoveland of Waubay, South Dakota. The machine was a pusher type, self-propelled with ''an explosive engine'' to provide power to the rear traction wheels. It deposited the cut grain in a windrow with the grain heads at the top of the row ready to cure, thus eliminating the need for binding or stacking; Below: Companion patent on a traveling thresher awarded to Hovland in 1908, reveals the earliest windrow pickup device. A traction engine was provided to propel the pickup and combine or ''traveling thresher.'' The ingenious Hovland apparently planned a self-levelling cleaning shoe, complete with hydraulic damping.

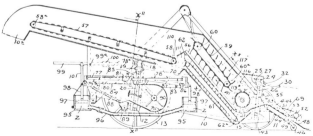

would be dry and ready to thresh. In this simple observation lay the germ of the idea of swathing or windrowing grain for harvesting. Once started on this line of thought the brothers went down a field of sheaves cut by the binder, breaking bands and spreading the loose crop back to ripen on top of the stubble. A few days later the wheat was picked up with forks and threshed by a conventional threshing machine.

For a year or so the brothers pondered the idea of using some kind of machine to accomplish the same results. On February 21, 1907 they applied for separate patents on a central delivery reaper and on a traveling thresher.

Shortly thereafter, they formed a company and by 1910 were testing their first center-delivery grain windrower and traveling thresher. The Hovlands claimed that the trials were successful and that they could handle loose grain and sheaves with equal ease. But these men were ahead of their time. They were unable to attract the interest of the farmers or of the eight manufacturers they contacted. The Hovland Harvesting Company of South Dakota disbanded. The original machines, now restored, can be seen in the Western Development Museum at Saskatoon, Saskatchewan.

The Windrowing Method

It seems fitting that the original Hovland windrower should now be located in Saskatchewan, where the Canadians refer to it as a swather. Today, the majority of the grain crop on the Canadian Prairies is harvested by "swather" and pickup.

Although the Hovland brothers had been discouraged by the indifference shown their windrow method of harvesting, the idea was revived by Helmer Hanson of Saskatchewan. Hanson and a brother built and successfully used similar machines in 1926 and 1927. In 1927, International Harvester Company sent an engineer from Chicago to inspect the Hanson models (Vicas 1970). The outcome was:

"The International Harvester Company built and sold the first (commercial) swathers. They were center delivery and were cut back as suggested. They after went to inside end delivery". (Hanson 1966).

The windrow method gained steadily in Canada, until the Great Depression followed by war, curtailed machine production (Hardy 1928).

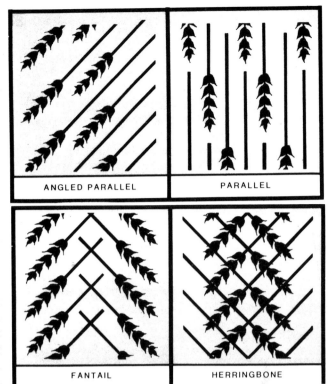

ANGLED PARALLEL   PARALLEL

FANTAIL   HERRINGBONE

Swath Formation. Four possible patterns are illustrated. The ideal is the formation of "Fluffy" swaths with the grain heads above and in line (Deere & Company).

At first glance, windrowing, which involves two entirely different operations, appears at a great disadvantage to direct combining. But there are certain geographical areas and crop conditions where the combine is useless without the windrower, for example:

- Where grain crops have extensive green weed infestations, the crop can be cured in the windrow. The spread of weed seeds is reduced by cutting before weeds can produce mature seed.
- Where the crop ripens unevenly or there is considerable moisture at harvest time, cutting the crop early with the windrower, and allowing it all to ripen and cure before threshing can save up to two weeks. This is especially critical where the season is short, as on the Canadian Prairies (Dodds 1967).

The original Hovland swather has been restored at Saskatoon's Western Development Museum. The traction engine had 24 in. wide, 8-foot high drive wheels. Hovland used the same tractor to power his "Traveling Thresher."

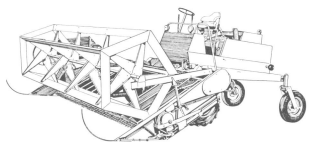

Above: Deere's SP grain swather (Deere & Company); Upper Right: Canadian Co-operative Implements Limited (CCIL) 30 foot swather equipped with foldup ''wings'' for transport between fields. Transport width was 14 feet.

Swathing on the Canadian Prairies. Cutting usually begins when the wheat crop is at the late-dough stage (about 35 percent moisture).

- Where delays may increase risk of crop damage due to hailstorms, wind, diseases or insects, windrowed grain may be better able to stand these depredations.
- Finally, in certain regions, as in Northern US and Southern Canada, the growing season is usually too short for complete ripening. Therefore the crop is usually windrowed to speed drying so that it can be threshed and stored safely (Griffin 1973).

Self-Propelled Windrowers

Windrowing as a harvest procedure had almost ceased in the early 1930's because of the Depression, but then severe infestations of the wheat stem sawfly, which nipped off the ripening heads of grain, posed a threat to straight combining. Investigators at Agriculture Canada's Swift Current Experiment Station confirmed that the windrow could be used to avoid damage caused by sawfly. Wheat cut at a grain moisture content of about 35 percent avoided the crop stage that coincided with the sawfly's most active cycle and yet produced a threshed sample that graded and yielded very well. Sales of swathers increased rapidly after the experimental results were published (Dodds 1967). While binders could still be seen, they were often modified to perform the swathing oper-

ation, particularly when wheat harvesting loss comparisons such as the following were publicised:

Binding and threshing method 3.58%
Straight combining. 1.16%
Windrow and combine method 1.23%

The first self-propelled windrowers were made by Owatonna Company of Minnesota and by Killberry Industries, Winnipeg around 1951 (Vicas 1970).

In 1966, Versatile bought out a 20 foot pull-type swather with movable canvasses. This meant the swather could deliver either to the center or, in the case of a lighter crop, could deliver to either end, therefore making a 40 foot double windrow. Versatile Manufacturing Limited of Winnipeg, Manitoba, had produced their first swather in 1954. One of the largest swathers produced by the several manufacturers in the business was the 47 foot swath version manufactured by Canadian Co-operative Implements Limited (CCIL) of Winnipeg. In 1976, CCIL produced a 30 foot unit that could double swath up to 60 feet. The potential for the self-propelled grain swathers was readily appreciated by hay growers and after some considerable strengthening of the grain models, self-propelled machines became widely used by alfalfa growers in the 1960's. In the US hay regions, the process is more generally called "windrowing".

"As the steel plow made possible the tilling of the vast Central States prairie lands in the middle 1800's so the development of hillside combines made possible the conversion of the Western and Northwestern States into productive grain lands."—Witzel and Vogelaar 1955.

19

Hillsides and the SideHill

The major hillside combine market regions.

A vast area of the US Pacific Northwest has fertile soil and an equable climate, yet was considered unsuitable for grain growing because the rolling hills were too steep to harvest. Specialized hillside equipment was needed. The Holt Brothers of Stockton were one of the first with an answer. They built three hillside combines in 1891 that were widely acclaimed. Twenty more were made the following year, principally for the Oakdale, California, area. Best Manufacturing Co., began producing hillside combines soon after. The huge steam traction engines that had been developed to pull the early California level-land leviathons were unsafe on hillsides. They were top heavy and there was the ever present risk of an explosion if the boiler crownsheet was exposed on side-slopes. The first commercial hillside combines were wood-framed machines hauled by 24- to 44-horse or mule teams. They were ground wheel driven and required a crew of five or more.

The 1892 Holt had a 30 in. cylinder, 44 in. separator and cut 24 ft of crop. On machines built after 1892, the main or ground wheel drove the separator; while the right hand or "grain" wheel powered the header.

The largest hillside market developed in Washington, areas of Oregon and Idaho, and a fair percentage of California's wheat and barley acreage were also harvested with hillside combines. Grades on the hillside regions vary from level to as steep as 65 percent. The field outlines are usually irregular, often following the contour, and provide spectacular harvest scenes in season.

Five or six men were required on the early Holt California hillside combines. A driver or "mule skinner" handled the jerklines. Other crew members were the "leveler," the "header puncher," the "separator tender," or "combine man," the "sack jigger" and the sack sewer (F. Hal Higgins Library).

These five Holt hillsides were on George Drumheller's wheatfield near Walla Walla, Washington in 1910. There were 165 head of horses and mules on these outfits. Drumheller was reported to have paid as much as $600 each for some of his lead mules. He was also reported to have received $25,000 for that year's wheat crop. Machines like these made possible the sevenfold boost in wheat production in Washington between 1893 and 1920 (Caterpillar Tractor Co.).

The first hillside combines in Washington—10 Holts—were shipped in 1893. In the next 30 years the State's wheat production increased sevenfold.

The Hillside Combine Body-Leveler

Without a body-leveler, a combine on a sideslope does a very inefficient job of threshing and separating. The crop works its way to the downhill side of the machine, where it piles up to overload the thresher or "walks out" with the straw or chaff. Lateral leveling systems corrected this deficiency by enabling the combine body to remain level irrespective of header position. The body-leveler also provided an adjustable center of gravity to reduce the risk of toppling.

Lower Left: This "notched post" or rack and pinion leveler mechanism was part of a Holt power lift system that enmeshed the appropriate pinion to shift the ground wheel relative to the separator body by means of the rack; Below: J. Jonas and A. Phariss of Sonoma County, California where issued this patent in 1880 for developing a combine brake. The wheel brake became standard on horse-drawn hillsides combines "to discourage run-a-way teams when bumble bees, rattlesnakes or other 'acts of God' scared the horses or mules" (F. Hal Higgins Library).

Bulking wheat. No sacks were needed since the grain was handled in bulk in this 1927 scene. The hillside combine was engine-powered.

Early levelers were manually operated and included a carpenter's spirit level taped to a transverse handrail to indicate when the machine was really level. "A person's sense of what is level becomes greatly distorted on those large steep hills"—observed Jerry C. Boone, Allis-Chalmers combine engineer.

To the casual observer it appeared that the driver had a most glamorous job. There he sat—perched high on the end of a ladder slanting out over the draft animals. Kirkby Brumfield wrote:

"It was like riding in the prow of a ship. The hills were the waves. There were times just before reaching the peak of a hill that the driver was so far up and at a backward slant that he couldn't see the horses in front of him. They were already over the hill and out of sight. Even the most experienced drivers had a feeling of relief as the combine pulled over the hill and the horses came back into view. The comparison was the same at the bottom of a hill. It was the trough of a wave. With the horses already starting up the next rise and the combine coming down the last, the driver was suddenly thrust right down in and among the draft animals just a few feet above the ground." . . .

Right: Cross-sectional illustration of the Harris Side Hill Combined Harvester. Note the use of three separating fans. The basic design remained unchanged from 1930 to 1954, although the wooden framed machine gave way to all steel construction in 1937; Below: Harris Gas Side Hill No. 6. The Harris Manufacturing Company was located in Stockton, California.

" . . . For the driver, . . . a runaway accident was scarier than any roller coaster ride and considerably more dangerous. As the combine began careening down the hill the teamster was no doubt wondering why he hadn't chosen a more genteel occupation." (Brumfield 1968).

Considerable skill was needed not only to drive large teams but to set them in motion as well. Drivers used various methods. One method involved carrying a tubful of green apples or clods beside the seat so the driver could give the lead team a signal—provided he could throw straight. Where there were neither apples nor clods, the driver might fire a pistol—using blanks for an old lead team, but aiming a light charge of buckshot at new leaders the first few times.

Gasoline Engine Powered Hillside Combines

In 1904, the first successful gasoline engine-powered combine harvester—a hillside—was sold by Holt Mfg.

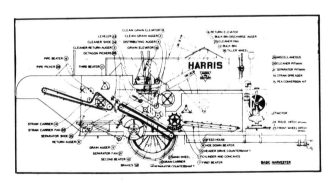

A 1928 Case Model W Hillside combine at work near Thornton, Washington. JI Case built their 100,000th thresher and their first combine – a prairie type – in 1923. They entered the hillside business shortly thereafter.

Co., to Pat Tute of Oceanside, California. That same year a 40 in. cylinder width, 60 in. separator level-land harvester was tested with 38 ft cut and rasp-type cylinder and concave bars. The users were not satisfied with the "English-style" thrasher and the machine was refitted with a peg-tooth cylinder and concave.

It was still deemed economical to use draft animals for pulling hillside combines as late as 1929. In 1926, at a Wheat Growers Economic Conference at Moro, Oregon, it was reported that on farms of less than a thousand acres, wheat could be produced at lower cost using horses rather than tractors and horses. In 1929, "The Combine Yearbook" reported that many horsedrawn harvesters were at work in the wheat fields of Palouse Country in the Northwest. But the days of the "four footed hayburners" were numbered.

Self-Propelled Hillside Machines

The first self-propelled hillside combine was developed in 1912, shortly after Holt produced the first gasoline-powered SP level-land machine. Steel-framed machines were tested by Holt in 1913 and went into production in 1921.

The first self-propelled hillside combines were slow to gain popularity because the crawler tractor was very effective and economical for pulling the cheaper pull-type hillside combines during the period from 1910 to 1930.

Harris Manufacturing Co.

George F. Harris had been manager of Matteson & Williamson Mfg. Co., where he gained experience on Shippee, Houser, Harvest Queen and Holt combines before forming his own company in Stockton in 1902. By 1904, he had become a leader in building pull-type combines with auxiliary gasoline engine power. Business was booming in 1912 as sales of the Harris Combined Gas Harvester soared.

Harris prices in 1912 were:

42.5 in. separator, 10 ft head, level land
pull-type with 4-cylinder 45 hp engine $3,975

54.5 in. separator, 24 ft head, level land
pull-type with 4-cylinder 55 hp engine $4,900

and for the "Harris Gas Side Hill Combined Harvester" 26 in. separator, 14 ft head, hillside pull-type (18 ft model also available) $3,725

Although Harris built his first self-propelled hillside combine at about the same time as Holt, around 1912,

Above: A Caterpillar hillside combine crabbing through a Washington grainfield in the 1930's (Deere and Company). On the steepest of slopes, the combine might crab at such an angle that a 20 foot combine might be cutting only a four foot swath!;
Below: JI Case of the Model K pull-type hillside about 1936. Hilly fields were usually opened so that most cutting would follow the contour, starting from the bottom of the hill and working up. That way, if the combine started sliding, it would slide away from the uncut grain (USDA).

Right: Advance-Rumely hillside combine at work with Best 60 Crawler on a 60 percent grade near Cheney, Washington in 1921. The rig harvested 1,600 acres that season for a reported $1.25 an acre.

The Harris SPH 908 combine with 30 in. cylinder, 1965. The machine was manufactured by Harris Harvester Co., Fresno, California.

he decided that the greatest market potential lay in his pull-type hillside developments. He pioneered the power-assisted leveling separator body. The Harris grain separation system was designed to steadily increase the crop flow by maintaining each processing element at a slightly higher speed than the previous one. This basic principle remained unchanged in the Harris Hillsides for 25 years. Harris was also adamant that spiketooth threshing elements were superior. His spiketooth cylinder was approximately three-fourths the width of the separator. By the Great Depression most major combine manufacturers were competing for pull-type hillside business.

Post-War Developments

In the late 1930's and early 1940's, a few prairie type (level-land) self-propelled combines from the East began to appear on the hillside farms of the Northwest. Their use was limited to opening up fields for pull-type hillside machines or cutting out level patches, but farmers took notice of their effectiveness, economy and one-man operation.

Production of pull-type harvesters for the Northwest came to a halt during the war. The Case V2 was the last hillside pull-type combine built. In 1946, Harris and others began developing one-man SP hillside machines. Harris placed 10 pre-production machines in the Pacific

region in 1950 and produced 150 more units the following year. Production on the SPH 88 increased to 225 in 1953, but stopped abruptly with the untimely demise of the original Harris Company in 1953.

The "Easterners" Join the Fray

The pioneering developmental work on the Harris SPH 88 up to 1953, hydraulic control systems, rubber tires, uniform feeding and general SP versatility, rapidly displaced the older pull-type hillside machines for hillside work. The design met a demand. Soon several major Eastern companies were "cooperating" with "Inland Empire" machine shops and farmer-innovators in projects aimed at converting their level-land SP combines into hillside machines.

As for Harris, its assets passed into the hands of Harvesters-Implements Co., a division of Wilkerson & Nutwell Inc., Fresno, California. The SPH 88 reappeared in 1956. In 1959, J. I. Case Co., marketed the Harris line as the Case-Harris, with the models 88 (34 in separator) and 98 (38 in separator). Case did not renew the Harvesters-Implements agreement in 1964, and in 1967, the Harvesters Implement Division was sold to the Mathews Company of Crystal Lake, Illinois. Shortly after Harris hillside combine production was discontinued.

hillside regions faithfully since 1904, and the acceptance of self-propelled combines practically marked the end of California combine manufacturing. Since 1954, the hillside market has become the province of Easterners Deere, International Harvester, Allis-Chalmers and briefly, Massey-Harris, Case, Oliver, Mathews, and Minneapolis-Moline.

Automatic Levelers

On a hillside machine the cutterbar must follow the ground slope while the separator stays level. Lateral leveling was not a problem on the pull-types, since the outer end of the header platform followed the ground contour through a gage wheel. On SP machines however, there is no gage wheel on the cutterbar since the gathering head is supported entirely at the feeder. The problem of providing for relative motion between header and feeder was generally overcome by a pivot circle or "fifth wheel", large enough to encircle the feeder housing and usually located between the end of the feeder and the header platform. Two methods have been employed on SP's to keep the header parallel to the ground slope: (1) mechanical connection (linkages or cables) between the header platform and the fifth wheel mechanism; or (2) by the displacement of oil between two hydraulic cylinders.

The separator body is kept level by means of pivoting drive axles which adjust to changing slope. The rear axle is mounted on a center pivot and adjusts freely to the ground contour. In both instances, however, all wheels remain vertical to better hug the hillslope. When the axle swings, the point of contact between the uphill wheel and the ground moves closer to the combine body so that a greater proportion of the weight is carried on the upper wheels. The lower the axle pivot point, the greater the effect. The earliest successful automatic hillside leveler was developed by a Palouse, Washington farmer, Raymond A. Hanson, in the early 1950's. Many of his electro-hydraulic sensors were installed on pull-type hillside combines (Boone 1977). Another Washingtonian, David Neal, developed a solid business in Garfield, Washington, converting International Harvester hillside pull-type machines into self-propelled units—some equipped with four-wheel drive as well. He then began making conversion kits for Deere level-land SP combines. Before long, a flourishing but modest business developed for a number of small machine shops in the Northwest. The revamped machines worked so well that the full-line companies purchased manufacturing rights and began to build automatic hillside levelers in their Eastern plants.

In 1955, International Harvester offered the world's first fully self-leveling hillside machine, the 141H, with fore/aft as well as lateral leveling. In later years, the successor, IH's Model 453, was the only machine of this type on the US market, since the main demand was for lateral leveling only. The demand for automatic lateral leveling made this feature standard on the self-propelled hillside combines.

There were many problems connected with the design of automatic levelers. A combine is subject to shock and inertia forces as it rolls over rough terrain, making it necessary to reduce the effect of these inertia forces on the slope-sensing device. Closely related to this system-damping problem is the need to provide a time delay in the sensor so that the combine will not compensate at every minor pothole. The "time delay", built into the automatic leveler may be a few seconds slower than manual control. To overcome the gap, the controls may be designed with an integral fast initial response and slower-finish electro-hydraulic system. The system may be manually over-ridden at any time. Skilled operators have taken advantage of this for servicing and to maneuver hillside combines under or between obstacles that would have been impossible otherwise.

Side views of International Harvester's Model 141H, the world's first fully self-leveling hillside combine with longitudinal as well as lateral leveling (International Harvester).

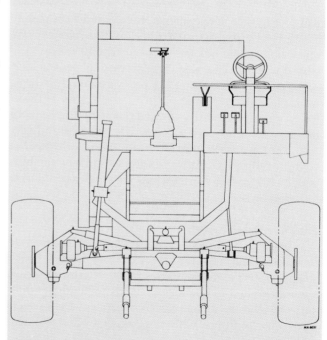

The IH automatic leveling system, operating on the right-hand and rear axles, maintains the separator level on sideslopes up to a maximum of 32 percent; uphill to a maximum of 31 percent; and downhill to a maximum of 11 percent (International Harvester).

202

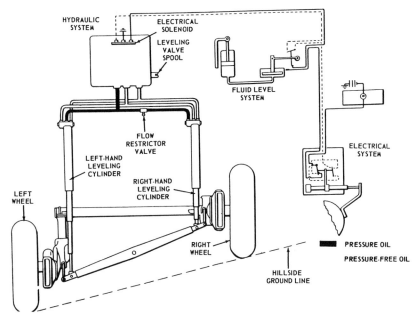

HYDRAULIC SYSTEM · ELECTRICAL SOLENOID · LEVELING VALVE SPOOL · FLUID LEVEL SYSTEM · ELECTRICAL SYSTEM · FLOW RESTRICTOR VALVE · LEFT-HAND LEVELING CYLINDER · RIGHT-HAND LEVELING CYLINDER · LEFT WHEEL · RIGHT WHEEL · HILLSIDE GROUND LINE · PRESSURE OIL · PRESSURE-FREE OIL

Left: Schematic diagram of hydraulic leveler mechanism on the Deere hillside combine. The master cylinder operates by relative motion between the separator body and wheel mechanism to automatically keep the separator level (Deere and Company); Below: Hillside combines have pivoting axles front and rear. The linkage geometry ensures that wheels remain vertical for better grip on hillsides and ensures a better weight distribution for greater safety (Deere and Company).

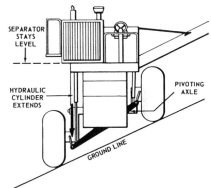

SEPARATOR STAYS LEVEL · HYDRAULIC CYLINDER EXTENDS · PIVOTING AXLE · GROUND LINE

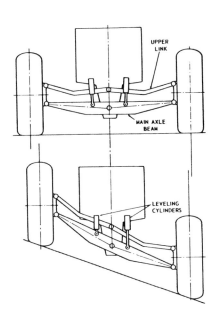

UPPER LINK · MAIN AXLE BEAM · LEVELING CYLINDERS

Left: Schematic showing uphill weight shift with swinging axles. The center pin on each axle beam of the Allis-Chalmers MH2 Hillside is located at a lower point than the end pin when the combine is on the level. As the axle tilts, the uphill wheels swing closer to the combine body, shifting a greater proportion of the combine weight onto the uphill wheels.

Below Left: The IH 453 fore/aft leveler cylinders automatically actuate the rear axle support and keep the separator level up and down hillslopes.

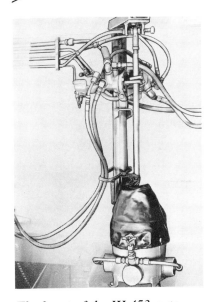

The heart of the IH 453 automatic leveling system is the damped-pendulum slope sensor. The electrohydraulic control system has feedback and two-speed capability, but can be manually over-ridden by the operator at will. Speed of actuation is adjustable. The pendulum is located between the dampers.

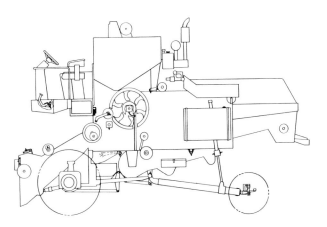

The John Deere SideHill Combine

Forty-five percent of US crop land is classed as sloping 6 to 18 percent. This includes some prime corn and soybean growing regions. In these intensively-managed regions, harvesting losses and productivity are closely monitored. The inherent sensitivity of harvesters to slopes has long been recognized:

"part of the ritual of positioning the old stationary threshing machine was the process of digging it in to ensure a level separator . . . today's modern turbo powered combines have little in common with the old stationary threshing machine. In terms of function, however, the same basic acceleration, aerodynamic and gravitational forces are used to thresh, separate and clean in today's combines as in their stationary ancestors. In both cases *maximum capacity is obtained when the machines are level.*" (Bichel and Cornish 1974).

Other companies have recognized this and Allis-Chalmers were possibly the first to make a tentative probe in 1964 into the market potential for hillside combines in the cornbelt region. Two hillside combines were sent to England in 1963.

In 1974, Deere & Company announced a new combine designed specifically for gently sloping cropland, the SideHill 6600. This self-leveling machine was designed specifically for small grains or row-crops grown on slopes of 20 percent or less.

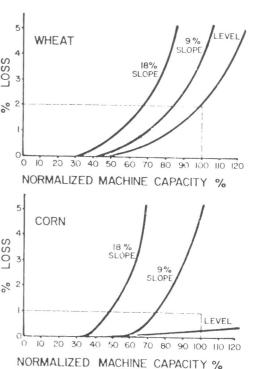

Above: Effect of sideslopes on machine capacity. A level machine can harvest faster at the same loss level. Separator losses in corn harvesting become more severe as sideslope increases (Bichel & Cornish 1974).

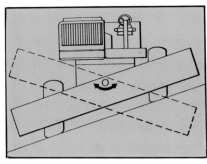

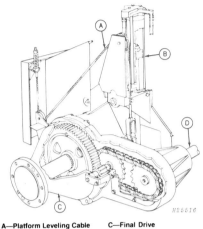

A—Platform Leveling Cable   C—Final Drive
B—Left-Hand Leveling Cylinder D—Drive Shaft

Upper Left: The Deere SideHill 6600 automatic side-leveling system for sideslopes up to 18 percent. A leveling control mounted on the front axle contains a small pendulum which senses deviations from level; Lower Left: The final drive leveling arm assemblies are actuated by hydraulic cylinders which support and propel the 6600 combine. The leveling arms pivot around the center line of the transmission output shaft. The cylinders positioning the two leveling arms are slaved together. The pivot-mounting feeder house allows the gathering head to follow ground contours while the rest of the combine remains level; Right: Released in 1974, the 6600 was designed for row crop conditions and the cornbelt market (photos Deere and company).

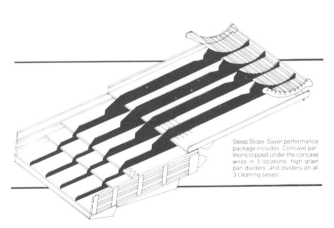

Steep Slope Saver performance package includes: Concave partitions clipped under the concave wires in 3 locations; high grain pan dividers; and dividers on all 3 cleaning sieves

Above: Vertical dividers extending from the concave to the grain pan and shoe are recommended options from Massey-Ferguson. This ''slope-saver'' kit is claimed to cut grain losses by up to 50 percent on a 10 percent sideslope; Upper and Lower Right: A pendulum controlled device developed by Iowa farmer Vernon Sietman provides lateral air blasts across the front of the cleaning shoe to reduce grain loss on gentle sideslopes. An Allis-Chalmers option.

Auxiliary Devices for Sloping Ground—Apart from the Self-Leveling Combines

Over the years designers and ingenious farmers have developed many ways to reduce grain loss on side hills. Grain flow dividers on reciprocating surfaces, mechanical deflectors, and pendulum-controlled air blasts, are examples of devices that have been added to combines with varying degrees of success. Such devices can never be completely successful because they only deal with one or two machine components. Every process of the machine, starting with the gathering head, is affected by sideslopes. High grain losses occur at relatively low feed rates. In corn for example, Deere engineers found that the capacity of an unmodified level-land combine was reduced 55 percent for the same loss level at 18 percent sideslope. Hence the effort and expense devoted to self-leveling combine improvements.

"The staff of life in the West can be described with laudable brevity: corn . . ."—Otto Bettmann, in "The Good Old Days—They Were Terrible"! Random House 1974.

20

Gwaiakowe: King Corn

Corn is the most valuable crop grown in North America. In 1976, 6 billion bushels of corn were produced on 70 million acres. By comparison, there were 54 million acres producing 1.4 billion bushels of soybeans, the nearest competitor, and 40 million acres of wheat produced 900 million bushels.

Corn, native to Latin America, has been under cultivation in North America for centuries. Hardy, disease resistant and nourishing, it was the principle crop of the Indians long before the Europeans crossed the Atlantic. (Mangelsdorf 1974).

The European colonists first planted, cultivated, fertilized and harvested wild maize using methods taught by the Indians. It is doubtful whether the Jamestown or Plymouth settlements could have survived if the colonists had not learned about "Indian corn", which could be planted even before the stumps were cleared from the land.

If corn husbandry had failed to progress beyond the skill of the Indians, it would never have become America's foremost grain crop. Corn had the inherent capacity to outyield smaller grains—there seemed no limit to the response to fertilizer and water. But the characteristics that made it the preferred crop in Mid-America were unique. It yielded well in rows and could be cultivated after planting, making it possible to fight weeds while the crop was growing. It also has a curious flowering habit. The male part of the blossom is in the tassel, at the very top of the plant, while the female part is well down on the stalk, in the silk of the ear. Here was a unique opportunity for large-scale crossing and hybridization (Johnson 1976). Hybrid corn became a reality by 1930 and seed production of the king of grains soon turned into a vast industry. Hybrid corns frequently produce 200 bushel/acre on well managed farms. The national average approached the 100 bushel mark in the early 1970's.

The frontier family's steady diet of corn and hog meat – even if there were over 30 ways to prepare it – would cause riots in modern penitentiaries. Corn was grown by Indians long before the white man arrived in North America. Corn could be boiled, roasted, parched, mashed or popped, and served as bread, hominy cake or pudding (From "The Good Old Days – They Were Terrible!" by Otto L. Bettmann. Copyright© 1974 by Otto L. Bettmann. Reprinted by permission of Random House, Inc.)

Pioneer Corn Harvesting

The pioneers soon found that corn made a nourishing livestock feed. They departed from the Indian method of stripping the ripened ears from the standing stalk, preferring to cut the stalk close to the ground to save the fodder for winter feed. Their first tools were sharpened hoes and long corn knives or machetes. Some even converted old scythe blades into short handled corn knives. The next improvement, the corn "hook", allowed a man to cut the stalk close to the ground without stooping. (Jackson 1950).

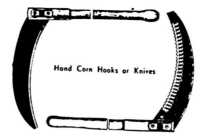

Upper Left and Center: Corn knives or corn hooks; Upper Right: Leg knife for cutting corn stalks (Wooley and Miller 1955); Lower Left: Wooden horse to support the corn shock (Zintheo 1907).

Another form of corn cutter developed by the pioneers was the leg knife. A sharp blade was strapped to the boot and braced to the knee, in the same way a telephone lineman's climbing irons are fastened. The laborer kicked vigorously at each stalk to sever it. (Horine 1924).

"Shocking" Corn

Corn stalks cut early retain their saccharine juices with minimal setback to the ripening ears. They can be cured in the shock and gathered after fall plowing. Getting the corn into the shock, which might contain from three to twelve dozen stalks, was hard work.

The ear-laden stalks were chopped down one at a time and piled around a wooden horse, or the tops of four "hills" were tied together to support the shock. The hills of corn nearby were cut and carefully stacked around the support, then tied with twine near the top so they would resist wind and weather. A field of shocks was a magnificent sight, "like hundreds of tepees ranged in straight rows". (Johnson 1976).

Corn Shucking

The process of removing the ears from the stalks was known as "picking", "shucking" or "husking". When corn was shucked in the field the total workload was reduced, but the intensity of effort was increased as the farmer, his hired men and boys worked from dawn to dusk every decent day to get the crop in before the snow began to fly.

Corn in the shock was either shucked in the field and the stalks reshocked, or the cured plants were loaded onto wagons or sleighs, to be hauled to the farmstead. Either way the ears ended up in the corncrib and the stalks were kept for winter feed.

"The American cornbelt was practically built on the husking pin, a hook or pin strapped to the hand to permit quick opening of the sheath that holds the precious ear of corn. There were dozens of models and adaptations. Plain farmers as well as corn husking champions had their favorites and swore by them. Even as late as 1915 most corn husking—more often called corn picking—was done by hand." (Johnson 1976).

A man using a corn husking hook could husk about an acre of corn in the shock or 1½ acres of standing ears a day. It was a good day's work for three men to cut and shock ¾ of an acre of corn by hand. A bushel of corn could be harvested from 150 to 175 corn stalks. By comparison, it took 40,000 to 50,000 stalks of wheat or oats to reap a bushel of grain.

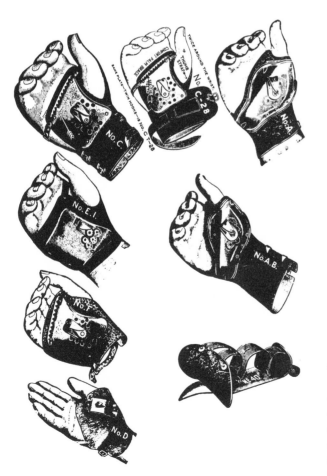

Hand corn "huskers." The husking palm hook reduced wear on thumb and fingers and increased the rate of hand harvesting ear corn. Still available in 1977 and used in corn picking competitions and on experiment station plant breeding plots (Johnson 1976).

Stalk bundles being moved to storages for winter feed in Goderich, Ontario, 1908 (Ontario Ministry of Agriculture and Food).

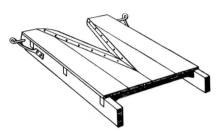

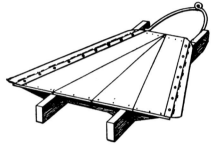

Far Left: This one-row sled harvester was pulled by two horses. A patent was taken out on the Vee notch design by J.C. Peterson of Ohio in 1886; Left: A two-row harvester pulled by one horse. A nervous horse made these sharp knives a real danger (Zintheo 1907).

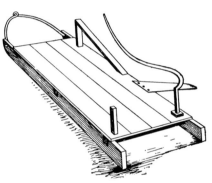

Left: Improved one-row corn sled harvester. The knife had to be kept razor sharp or it would ride down the stalk instead of cutting. The guide arm collected the stalks on the platform; Right: Platform corn harvester with knives on hinges so that either side could be lifted to pass obstructions. The wheels reduced the draft of the sled of this two-man machine (Zintheo 1907).

Right: Another two-man corn sled. Each man perched on a stool, guided the stalks against the knife with one hand and collected them against his leg to be dumped when there was enough for a shock (Johnson 1976); Far Right: Two-row corn bundler had stalk dividers for lifting lodged plants. The cut stalks were deposited on the platform. The width of the singletree governed row spacing (Zintheo 1907).

Mechanizing the Corn Harvest

Corn harvesting machines appeared later than machinery developed for cereal crops, for good reason. Corn was a hardy plant that could be left on the stalk for months after ripening with less deterioration than cereals. The snapped ear could be carried directly to storage where it was picked, husked and shelled at a convenient time to better use manpower throughout the year. Much of the corn crop was (and still is) used on the farm. As early as 1820, attempts were made to construct a mechanical corn harvester. Those early inventors who tried to follow the style of the mower or reaper were deterred at first by the tough, thick stalks and turned instead to horizontal knives fixed to horse-drawn sleds.

Sled Harvesters

On most early sled-harvesters the driver stood on the

*Above: Another type of early two-row corn harvester, ca 1888. The machine was stopped when there were enough stalks for a bundle. The corn was then tied and set up in the shock as shown (Zintheo 1907); Right: Mason's corn shock loader tackled another problem, probably with limited success. Some of the hand built shocks practically needed a field crane to handle them (**The Prairie Farmer** 1865).*

Left: Corn shocker based on the 1888 patent of A.N. Hadley. A five foot diameter rotary table assembled about 100 plants around the central shock form. The shock was hand tied, then lifted by rope and tackle high enough to clear the retainers and deposited on the ground by the revolving crane. The whole operation took about five minutes (Zintheo 1907); Right: Elevating corn stalk harvester, ca 1888. The stalks were elevated into the wagon pulled alongside. This Osborne machine had circular cutters and was one of the first commercial machines with power-driven stalk lifters using "traveling chains," as they were called (Zintheo 1907).

platform, gathering the stalks in his arms before they were cut, to prevent them falling in a random fashion. The work was exhausting. Lodged or tangled corn was the greatest problem encountered by the platform-type harvester. At best, the horse had to walk at a fast pace for clean cutting, thus the work was slowed down considerably and was more difficult if the crop was down. Soon, ingenious stalk lifters and gatherers appeared. A two-man sled harvester could cut 4 to 7 acres in a day, or about 300 shocks.

Corn Binders

The vertical corn binder patented by A. S. Peck of Geneva, Illinois in 1892 was a singularly important development. It contained most of the features incorporated into later commercial corn binders. Peck demonstrated the machine in his own fields and then built others for his neighbors. Success attracted manufacturers to Peck's patent, some were licensed, while others evaded the patent with slight variations in design. By the turn of the century, vertical, horizontal and

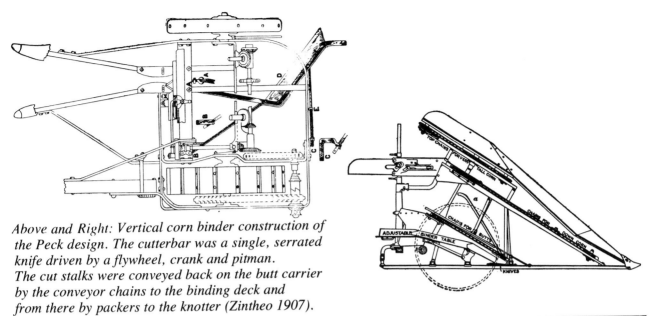

Above and Right: Vertical corn binder construction of the Peck design. The cutterbar was a single, serrated knife driven by a flywheel, crank and pitman. The cut stalks were conveyed back on the butt carrier by the conveyor chains to the binding deck and from there by packers to the knotter (Zintheo 1907).

McCormick vertical corn binder advertisement, 1903 (Johnson 1976).

*''Gwaiakowe''-in the language of the Indian – Corn is king. That was part of the copy advertising the McCormick vertical corn binder in another 1903 ad (**The Threshermen's Review**, June 1903).*

inclined binders (referring to the bundle position during collection and tying) had been developed.

Inventors claimed the vertical binder broke off fewer ears because fewer stalks were knocked aside and it worked better in tall corn. A stubble cutter was often attached to the binder to cut the remaining stubble close to ground level. This left shorter stalks on the ground to work under while preparing the field for the next crop. The average corn binder weighed up to 1,800 lbs, had a draft with stubble cutter of over 420 lbs and required three horses. A corn shocking machine cost about the same as a corn binder, but required only one man, where a binder required a driver and two or three men following to shock the corn. In 1907, the shockers had a price advantage: $1.06 per acre compared with $1.18 per acre for the sled and $1.50 per

211

Upper Left: John Deere vertical
corn binder (Deere & Company);
Above: Plano Company corn
binder advertisement.

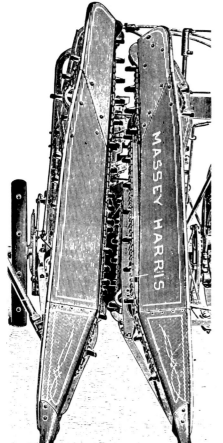

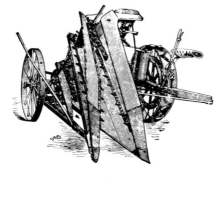

Far Left: Massey-Harris vertical
corn binder, 1911; Center: An
inclined corn binder showing
guide rod and tilting lever for the
shock (Zintheo 1907);
Below: Frost & Wood's vertical
corn binder.

*Two early machines to speed husking and save hands from chapping; Far Left: Crude husking bench built in Connecticut was advertised in **Country Gentleman**, 1858; Left: The Rosenthal horse powered husker invented in 1882 by August Rosenthal of Reedsburg, Wisconsin (Johnson 1976).*

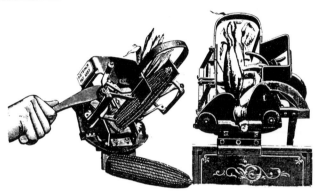

*The Champion corn husker, designed by A. Steiner of Illinois, 1888. The unit was bolted to the side of the wagon along with a step for the operator and a collecting box under the machine. One person operating the machine could keep up with two persons "snapping" corn (**Farm Implement News**, 1888).*

acre for the binder and even more by hand. (Zintheo 1907).

The bundles made by the binder were easier to handle, especially if the corn was to be cut and stored as silage. The corn shocker left a shock that practically needed a field crane to lift it.

Corn Huskers

Ears develop inside a tightly-wrapped leaf envelope, or husk, which loosens at maturity, but still impedes shelling. The first patent on a device for removing the husk from snapped ears mechanically was issued in 1837. It consisted of a pair of counter-rotating roughened rollers which tore the husks off the "snapped" ears. Various hand-powered huskers were also developed. One person with a hand-powered husking machine mounted on the side of a wagon could usually keep up with two people "snapping" corn in the field. The first successful field huskers were developed around 1880— they incorporated snapping and husking in the one machine.

Mechanical Corn Pickers

When stalk fodder was not needed, ears were snapped from the standing corn by men walking down the rows. The ears were thrown into piles then the piles were put into baskets to be dumped into a wagon.

When a wagon with a "bangboard" and a trained team of horses was available, it moved along the rows of corn while two men on foot tore the ears from the stalk and tossed them into the box. The wagon needed no driver, a trained team of horses automatically moved ahead as the husker approached the front of the wagon. An experienced picker could pick 100 bushels of ear corn from dawn to dusk. This work was generally one of the last tasks of Autumn; afterwards cattle and hogs were turned out to browse on the trampled fodder.

Edmund W. Quincy of Peoria, Illinois, obtained the first patent on a corn picking machine in October, 1850. His first machine picked the ears by means of wooden rollers studded with iron pegs. As the stalks passed between the rollers, the pegs tore off the ears and dropped them on a canvas belt, which conveyed them up to a trough where they slid into a wagon driven alongside. Traveling from farm to farm living in abject poverty, often without food or shelter for days at a time, he constructed his crude machines with the money given by sympathetic but skeptical farmers of the area, where he became known as "Old Father Quincy". (Ardrey 1894).

A score of inventors followed Quincy, devising

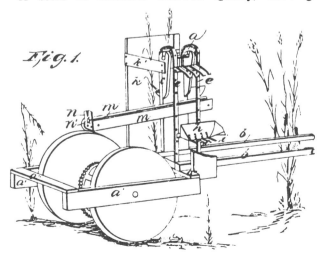

One of "Old Father Quincy's" corn picker patents which emerged from a lifetime spent trying to perfect a corn picking machine.

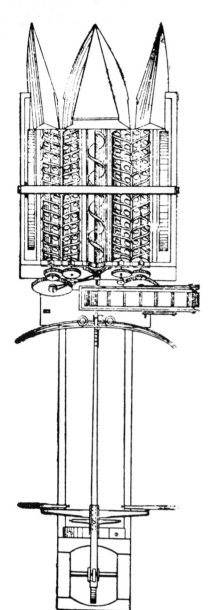

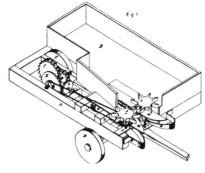

Corn picker patented by Richardson, White and Weed of Illinois, June, 1859 was the first with spiral ("terete") snapping rolls. These inventors anticipated the cantilevered ear snapping rolls by a century.

Left: Practically all the earlier commercial corn pickers consisted of a sloping pair of rollers. The mechanism was inclined to allow the front end of the rollers to pass below the lowermost ears and rake the stem from bottom to top (Zintheo 1897); Above: A belt-driven husker and shredder was placed on the market in 1896 by the Rosenthal brothers who established a thriving business in Milwaukee for making corn handling equipment (Johnson 1976).

machines for husking ears, for snapping ears from the stalk, and for shredding the stalk. The shredding machines of the 1880's were stationary units so that farmers did not have to pick ears in the field. They could cut the stalks with hooks and knives in the same manner their forefathers had. Or they used sleds to carry the stalks to the shredder.

Among the many devices tried on field pickers were gathering prongs, cutters, roller and breaker devices, parallel vibrating bars, etc., but it was the inclined, roughened rollers with snapping plates which prevailed and captured the interest of manufacturers after 1874. Then pusher-type machines were produced, with dividers to guide the corn into the snapping rolls, which at the front were low enough to pass under the ears. Since the ears were larger than the stalks, they were pinched off as the stalks were pulled down between the rollers of the advancing machine. There was a conflict between function and strength. Large diameter rollers butt-shelled the ears, resulting in high losses. Small

rollers did the snapping without shelling, but were not rugged enough. Corn pickers steadily improved in design, but until the turn of the century, were eclipsed by the popularity of corn binders and shockers.

There was also a difference of opinion among farmers on whether or not to husk ears clean. In the North, it was common to husk the ears clean before they were cribbed. In the South, the common practice was to leave the husks on the ears. It was claimed this prevented insect damage. The objections to using corn snappers which left the husk on the ear was that more crib space was required; the husks would serve to attract and harbor rodents; the ears would not dry, but would become moldy; the husks interfered with shelling; and, that while the husks were good cattle and hog feed, horses would toss the ears trying to remove the husks, thus losing ear and all.

Corn for shelling needed to be husked clean to command the best market price. Farmers also took pride in adjusting a picker that delivered fully-husked corn to the wagon. Later, cornpicker contest rules would dock

WESTERN
CORN SHELLERS

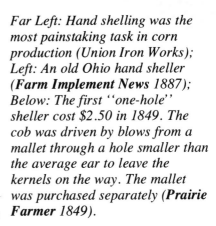

*Far Left: Hand shelling was the most painstaking task in corn production (Union Iron Works); Left: An old Ohio hand sheller (***Farm Implement News** *1887); Below: The first ''one-hole'' sheller cost $2.50 in 1849. The cob was driven by blows from a mallet through a hole smaller than the average ear to leave the kernels on the way. The mallet was purchased separately (***Prairie Farmer** *1849).*

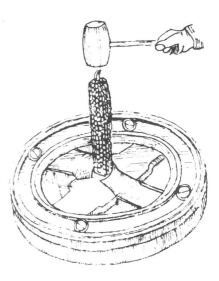

Left and Above: Two early English picker-wheel maize shellers.

Early American hand sheller of the picker-wheel type popular in the closing decades of the 19th century (Johnson 1976).

Burrall's 1882 patented one-hole sheller (Johnson 1976).

Patch's Patent Corn Shellers.

These Shellers are Strong, Simple and Durable.

PRICE $3.00.

Will Shell Easily and Rapidly.

They also separate the cobs from shelled corn. Twelve

packed in a barrel for shipment   Dealers write for Circulars and Prices. ☞ Agents wanted in all States and Territories.

A. H. PATCH. · **Clarksville, Tenn.**
Dealers in New England and Canada please address   DAN'L A. PATCH, Boston, Mass.

*Patch's ''Black Hawk'' corn sheller sold for $3 in 1888 (***Farm Implement News** *1888).*

The Iron Star Corn Sheller

Left: Burrall's 1845 US patent 4300 was one of the earliest picker-wheel shellers designed to specifically expel the cobs separately from the kernels; Above: F.N. Smith's "cannon" sheller, 1843. This was the earliest power-driven commercial sheller. The cobs were claimed to pass unbroken from the end of the machine and separated from the shelled corn, which fell through the middle. The rotor was 14 in. in diameter, 6 foot long and could be driven by water, steam or two-horse power. Cost was $45 (Collections of Greenfield Village and Henry Ford Museum, Dearborn, Michigan).

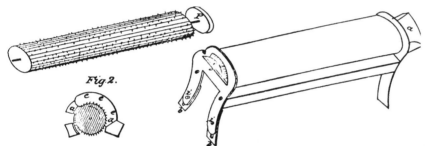

US patent 3114, June 1, 1843 to F.N. Smith of Kinderhook, New York. First cylinder-type corn sheller. The ears were confined within the concavity (designated eee) over the rotor.

hand-pickers as well as mechanical pickers for any husks left on the ears.

Corn Shellers

Whenever there was a demand for shelled corn in colonial days, shelling was done by hand. Flails were used, or horse treading, or the ear might be rubbed along a sharp projecting edge—a bayonet was said to be ideal. Corn for domestic use was hand-shelled into a tub after baking or drying. Shelling was the most difficult and time-consuming task in corn production. The earliest shelling machines, of the picker-wheel type, originated in England during the 1820's. Someone over there must have wasted little time in developing a sheller, since Indian corn had only been brought to England a few years earlier.

Power and Cylinder-Type Shellers

The first power-driven corn sheller, patented by F. N. Smith of Kinderhook, New York in 1843, was also the forerunner of the cylinder-sheller. Smith's appropriately named "Cannon" was a cast iron assembly 6 ft long, had 14 in. diameter rotor and could shell 200 bushels of ears an hour with two horse-power. The de-

velopment of the corn sheller occurred about the same time as innovators were at work on other corn implements. The first effort was directed toward shelling only, the next to separating or removing the cobs, then the chaff and the litter, and lastly to increasing capacity and perfecting the operation. Burrall's 1845 patent was one of the earliest picker-wheel shellers to simultaneously dispose of the cobs. Cob carriers and fans were sold as attachments to power shellers before the Civil War. The next step was the addition of an elevator to carry the shelled corn from under the machine to bags. The capacity of power shellers was limited by the dexterity of the attendants and their ability to place the ears endwise into the sheller. The first self-feeding power sheller was credited to Augustus Adams in 1859. His four-hole machine was capable of putting through about 800 bushels a day when operated by two experienced men and driven by a two-horse-power. (Marsh 1887).

"Shuck" Shellers

A "shuck" sheller was one used for shelling and separating "snapped" corn, i.e. ears with the husk on. The process of removing the kernels from the cob was similar to separating them from the envelope of

The Rustler Corn Sheller

Left: The Rustler corn sheller was sold in 1903 by the Buckeye Syndicate for $73; Above: At the same time, Buckeye offered a one-hole hand sheller for $10.

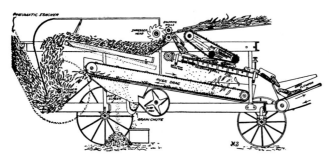

Above: One of the earliest combined husker-shredders was patented by J.F. Hurd of Minnesota in 1890. A more sophisticated version was produced in 1903 by Corn King Husker Co., Chicago; Right: The Western Corn Sheller, 1897 made by the Union Iron Works.

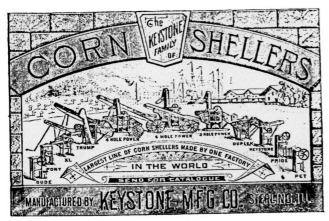

*Keystone advertising in the November, 1885 issue of the **Farm Implement News**.*

husks, cob and chaff. The need for shellers originated in places "where farmers have the slovenly habit of cribbing unhusked corn" or where Southern corn was cribbed with the husks on for insect protection (Marsh 1887).

The First Mobile Field Sheller Originated in Australia

Field shelling of corn on the farm was not feasible in the US without adequate drying facilities and the associated equipment for handling and storing shelled grain (Lounsberry 1957). These facilities did not become widely available until the 1950's.

In Australia, on Queensland's Darling Downs, harvest mechanization and a drier climate provided just the environment for a farmer-inventor to come up with a practical corn combine in the 1920's.

George Iland of Toowoomba, Queensland suffered the fate of many an inventor who was ahead of his time. Iland had been raised on the farm and as a boy of 11 he could repair any piece of farm machinery, including binder knotters. He recalled that many knotting troubles were caused by a lack of uniformity of the binder-twine then in use.

When corn production started in Queensland, young Iland had to pitch in to help with the harvest. He hated the job of "pulling corn" by hand, and resolved to build a machine to do the job mechanically. In time, he developed a set of plans and in 1921 convinced the Eclipse Foundry of Toowoomba to build a prototype which was assembled from old grain harvester parts and fabricated components. It was field tested the same year.

The "Eclipse Maize Reaper Thresher", as it was called, used three pairs of gathering chains on the single row gathering head to lift loose ears and crowd the standing plants past a series of rotating cutters which chopped the plants into 18 in. lengths. The chopped material and ears were then conveyed into the threshing drum. The heavy peg-tooth iron drum, locally cast at the foundry, shelled the ears and discharged the crop onto a four-walker separator. The grain was cleaned in a conventional winnower and chaffer on the machine. The clean kernels were then elevated to a five bushel grain tank for bagging. That first year, over 100 acres were harvested with the ground-driven prototype which was pulled by three horses. (Quodling 1924). Seven more modified machines were sold, including an engine-powered unit. A brand new car was purchased just for that engine. The rest of the car was pushed into a junkyard. Before the end of 1924, Iland had gathered 42 additional orders. Then the blow fell. The foundry directors failed to see a future for the machine and decided to drop the project. Iland was outraged and didn't hesitate to let certain directors know what he thought of them—but to no avail. He had burned his bridges too.

An offer from the US had been rejected earlier. He believed his machine belonged in Australia. Iland's spirit was broken for some time, but he eventually found other interests and the first corn combine was abandoned. About six years later, the Gleaner Combine Harvester Corporation of Independance, Missouri, started independently down the same road.

Above: The world's first corn combine, designed by Australian George Iland of Toowoomba, Queensland. The Eclipse Maize Reaper Thresher, a one row pull type whole plant harvester was first tested in 1921. The stalks and ears were crowded past a series of rotary cutters in the head. Chopped material was then threshed, separated and cleaned. The shelled kernels were conveyed into a 5 bushel bin. A ground wheel drove the first machine, but this 1924 model was engine driven; Below: Gleaner corn combine first sold in January, 1930. The header had a heavy-duty reciprocating sickle, oversize reel and five vertical pickup chains within the row dividers. The unit was designed to fit the 1929 Model R Gleaner combine.

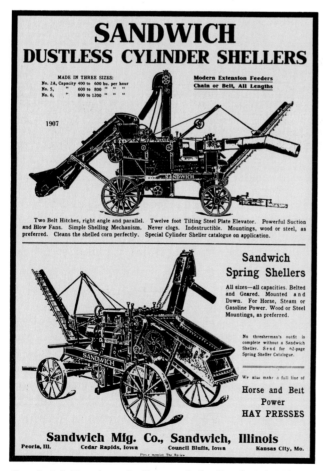
*Sandwich Dustless sheller made by Sandwich Mfg. Co., Sandwich, Illinois (**Threshermen's Review**, 1907).*

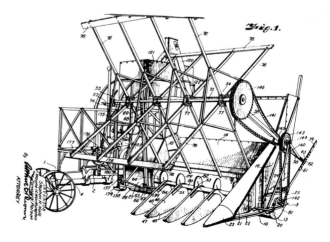

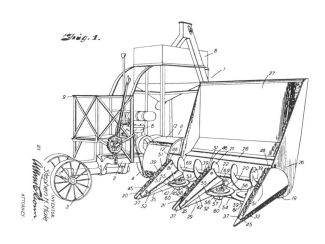

Right: Gleaner Combine Harvester Company was granted patents for corn heads: US patent 1,882,823 "Apparatus for harvesting corn," issued to S.H. Hale, et al, October 18, 1932 and 1,901,099 "Cutter mechanism for corn harvester" issued to Hale on March 14, 1933; Below: 1930 sales literature on the Gleaner Baldwin shows the second developmental stage of their whole plant corn head attachment with rotary cutters for two 40-in. rows. The project was dropped and not resumed after the Company plunged into financial difficulties in 1931

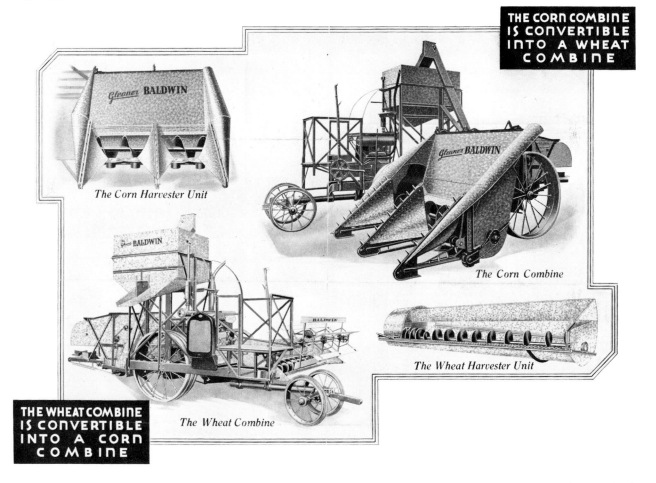

THE CORN COMBINE IS CONVERTIBLE INTO A WHEAT COMBINE

The Corn Harvester Unit

The Corn Combine

The Wheat Harvester Unit

The Wheat Combine

THE WHEAT COMBINE IS CONVERTIBLE INTO A CORN COMBINE

The Gleaner Combine Corn Attachment

The Gleaner Baldwin Corn Combine was the first commercially-available combine with a corn head in North America. The Gleaner corn attachments were not pickers, but whole-plant harvesters. The first development, released in 1930 on the pull-type Model R combine at the Wichita show, used a heavy duty reciprocating cutterbar, extra large corn reel, and lifters. Five lifter chains operated inside the four row-dividers to lift the ears of down and tangled corn. The next model had no reel although it used similar row dividers, with a series of rotary cutters for the two 40 in. rows. The corn head would fit on the regular Gleaner left-hand cut combine models. The ears were shelled by the standard rasp-bar cylinder and separator. A thousand corn attachments were built at Independence by August 1930, but the project was abruptly and permanently shelved during the Great Depression. It took several years for the holding company that purchased Gleaner to liquidate the corn attachments in stock.

Allis-Chalmers' LaPorte Corn Heads

In the fall of 1936, Allis-Chalmers began developing a corn head attachment for their ALL-CROP Model 60 Harvester that saw the whole corn stalk cut by the regular 5 ft sickle and conveyed to the thresher. Dissatisfaction with the excessive power required for processing the entire plant led Allis-Chalmers engineers to begin development of a two-row snapping and shelling attachment. They were still experimenting when the war restrictions of 1941 caused the project to be shelved. (Lounsbury 1957). The project was revived in 1957 and the LaPorte corn heads finally marketed the same year on the Model 60 ALL-CROP Harvester. In addition to the corn heads for the Gleaner at Independence and Allis-Chalmers developments at LaPorte, there were other developments during the 1930's. Collins and Shedd at Iowa State University worked on a corn head for a Caterpillar combine, and J I Case worked on a pull-type field sheller which eventually appeared in 1950.

The Farmall No. 10 power-driven corn picker had a 25 bushel bin. It could harvest 12 acres a day at 3 mph.

John Deere pto-driven picker with wagon hitched to tractor was "a strickly one-man outfit" according to 1925 advertising (Johnson 1976).

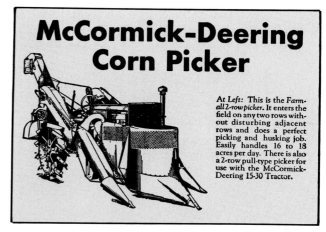

The McCormick-Deering two-row picker-husker of 1939 could harvest 16-18 acres a day. Over 113,000 units of the IH 2-row tractor mounted picker were produced between 1939 and 1952 (Johnson 1976).

Corn Picker Sales Overhaul Corn Binder Sales

These sporadic developments in field shellers took place during a period of steady sales growth for corn pickers. Before WW1, one-row corn pickers, both horse and tractor drawn, were sold by several companies. USDA agricultural statistics for 1925 showed corn binder sales peaking and field picker sales too inconsequential and hardly worth reporting. The development of the tractor power take-off in 1924 provided the catalyst that sent corn picker sales soaring (Davidson 1931). By the 1930's, the tables were turned. Corn picker sales proceeded to climb to nearly 89,000 units by 1950, as corn binders disappeared from the market (Johnson & Lamp 1966). The earliest two-row picker-husker for power take-off operation, a McCormick-Deering, was sold in 1928. One-row tractor-mounted pickers came out the same year. In 1929, two-row mounted pickers became available.

Hand methods still persisted, in spite of mechanization. USDA surveys showed that in 1939, 42 percent of the US corn acreage was still harvested by hand (Brodell 1940).

The first self-propelled corn picker was introduced by Massey-Harris in 1946. In 1949, portable batch driers

The Farmall No. 10 made full use of the versatility of the Farmall row crop tractor. It could pick and husk about 8 acres of corn in a 10 hour day at 2 mph.

were marketed with capacity to match that of harvesting techniques by picker-husker and stationary sheller. Sheller attachments for the field corn picker were first marketed in 1958. Sales of corn picker units reached their all time high in 1959, when there were an estimated 792,000 pickers on US farms.

Corn Pickers Injured Many Farmers

As corn picker sales rose to their peak by 1959, so did picker-related injuries. The cornfields became a veritable battleground during harvest. Farmers lost fingers, hands, arms and feet in the snapping rolls and husking bed. A gathering of farmers at county extension meetings in the mid-1950's looked like a group of war veterans—there were men with steel hooks for hands and some with artificial legs.

The aggressiveness of the snapping and husking rolls had to be limited to prevent excessive shelling loss, a loss which in earlier days reached as much as 10 percent. On the other hand, when the stalks were damp and

tough, the snapping rolls slipped and clogged. Farmers risked injury when they tried to clear the stoppage with the machine running. Vigorous educational campaigns were mounted by safety specialists throughout the corn-belt States in the 1950's. The solution to the safety problem came with safety education and the redesign of

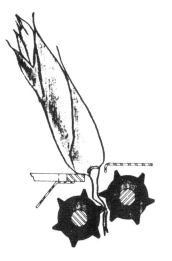

The ear snapping action of modern corn heads.

the snapping mechanism. An aggressive set of stalk rolls was mounted below a set of snapping bars spaced ¾ to 1 in. apart. The new snapping mechanism did not clog as often and it was virtually impossible to get a limb into the stalk rolls. The combine corn head marketed in 1954 eliminated the husking bed problem by doing away with the need for a husking bed. Corn picker sales fell as the combine took command of the corn fields.

Combine Corn Heads Introduced in 1954

In 1954, the corn head was successfully adapted to match the combine (Hurlbut 1955). The release of John

The JI Case two-row pull-type corn picker, 1932 (USDA).

221

Deere's Model 45 and No. 10 corn head in 1954 served notice on the corn picker. Soon the corn head was offered on all North American combines, but with it came the need to redesign the combine itself. Corn harvesting places far higher loads on machine components than normal small grain harvesting. The corn head, for example, may require from three to five times the average power of the grain head for the same cylinder size, and instantaneous loads can be an even larger multiple. Field conditions may be even worse—the corn crop is harvested later in the year than cereals. Bin loadings of wet corn cause heavier strains on chassis and transmissions. Combine designers found it prudent to drive corn heads from both sides of the feeder and head.

The French Corn-Only Combines

At least three European companies, all French, had a

corner on the market for a combine built exclusively for corn. The ABM Rivierre-Casalis, Burgoine and Braud combines fed the crop from the corn head into twin overshot cage shellers. The compact size of the shellers and absence of walkers allowed these machines to have large bin sizes, and they were compact in relation to their capacity.

Field Shelling Takes the Upper Hand

Corn directly harvested by combines exceeded that harvested by pickers in 1965 and the sales curves of corn heads versus cornpickers have widened ever since.

Combine corn heads were introduced in 1954 and reached a sales plateau of 26,000 units a year in 1966. Lately, corn head sales have followed combine sales, with a sales volume equalling about two-thirds of the number of combines sold.

The acreage of corn harvested for grain has declined gradually since 1930 as technological advances have steadily increased productivity. For instance the acreage

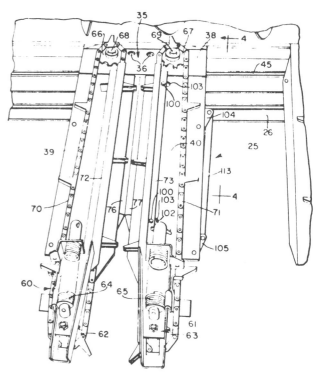

Above Left: Both Deere's No. 45 and No. 55 combines were equipped with the No. 10 corn head, introduced in 1945; Lower Left: In 1957, the first post-war Allis-Chalmers corn heads were developed at LaPorte, Indiana as an attachment for the Model 60 ALL-CROP harvester and the Model A Gleaner combine; Right: Deere's patented corn head design with through-shaft hexagonal shaft drive gearboxes and adjustable row-spacing proved a basic design in the industry. Modern corn heads have dual drives with power taken from jackshafts on both sides of the feeder to meet the heavy power demands of ear snapping, especially under weedy and tough conditions.

222

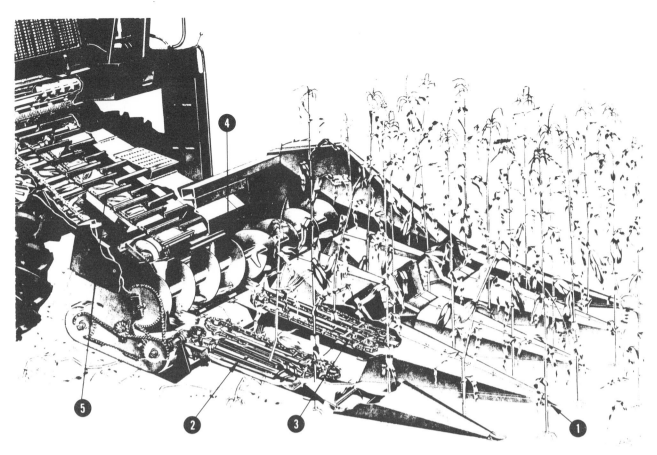

Cross-sectional view of Deere's 40 series corn head. The 40 series, introduced in 1970 with the New Generation combines, featured cantilevered stalk rolls, full length trash knives and 25° low profile gathering snouts. A variable speed feeder house with dual drives was standard on the Deere corn combines (Deere and Company).

The IH 874 eight row cornhead on a 915 combine. Hinged gathering shields could be swung up out of the way for easy access to gathering chains and for servicing.

Above: Avco New Idea's Model 729 Uni-Sheller is an interchangeable unit for the Uni-Tractor chassis; Below: the 15 in. cage sheller is the heart of the Uni-Sheller.

Left: New Idea's Superpickers were available as pull-type (Model 314) or tractor-mounted (315) models; Lower Left: snapping rolls and husking bed components of the Superpicker series.

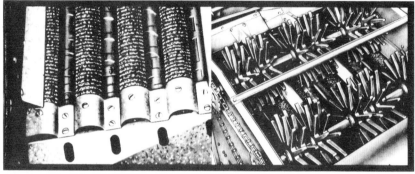

of corn harvested for grain was slightly less in 1960 than in 1950, but 40 percent more corn was produced. The 1940 corn farmer, using hybrid corn, could expect to harvest about 45 bushels an acre using 6 gallons of fuel. The 1955 farmer, using abundant fertilizer (equivalent to 30 gallons of fuel) could expect to harvest 90 bushels an acre. By 1960, using pesticides equivalent to 4 gallons of fuel, an average yield increase of 10 bushels an acre could be expected. The 120 bushel barrier was breached with longer season hybrids, combine harvesting at high moisture content and with the use of grain driers. And if energy was not a cost constraint, yields would continue to escalate.

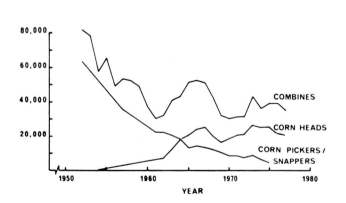

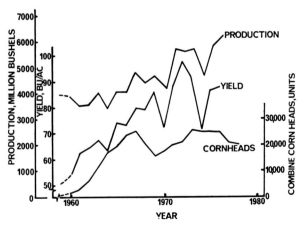

US corn production and cornhead sales in relation to combine sales.

> "The harvest truly is great, but the labourers are few" . . .—Luke 10:2 (King James Version).

21
Cinderella Soybean

During recent decades the rising need for protein has made the soybean a staple in many parts of the world. Yet, at the turn of this century, the plant was an oddity in the US. Today the crop is the top US dollar earner abroad.

Soybeans are by no means new to agriculture. The first written record of soybean cultivation was penned by Chinese Emperor Shen-nung in 2838 BC, and the crop was certainly cultivated for a long time before that. The plant was introduced into Europe by way of Japan in 1712. The ancient oriental "meat of the field" of antiquity was grown in the Royal Botanical Gardens in Kew, England as a botanical curiosity in 1790.

In 1804, the captain of a Yankee clipper ship in a Chinese port stocked a few bags of soybeans as a reserve food supply. Little did he realize that they were the beginnings of an industry which a century and a half later would produce more soybeans in the US, than two-thirds of the entire world's supply. Yet as late as 1920, the primary use of soybeans was a legume-hay for stock feed and for soil enrichment.

Mechanized Soybean Harvesting

The mower, binder and thresher were the first machines used to harvest soybeans for seed. Seed losses were high as the repeated handling shattered open the brittle pods.

Soybeans were first harvested by combine in Illinois in 1924, and were the crop that brought the combine east of the Mississippi. Elmer J. Baker Jr., "The Reflector", of *Farm Implement News*, was instrumental in procuring a Massey-Harris No. 5 combine for one of his Illinois subscribers. The subscriber just happened to be an International Har-

The soybean, Glycine max. (L) Merrill, one of the five sacred grains of the ancient orient. Soybean seed is among the richest protein sources and soybean oil finds a multitude of uses (Quick, 1972).

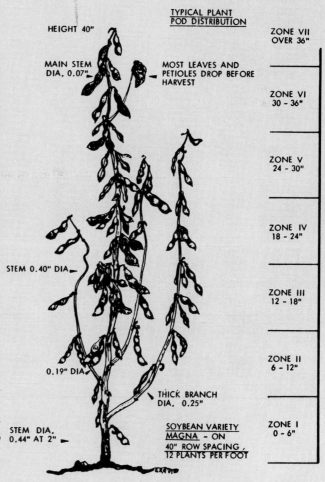

TYPICAL PLANT
POD DISTRIBUTION

HEIGHT 40"

MAIN STEM DIA. 0.07"

MOST LEAVES AND PETIOLES DROP BEFORE HARVEST

STEM 0.40" DIA.

0.19" DIA.

THICK BRANCH DIA. 0.25"

STEM DIA. 0.44" AT 2"

SOYBEAN VARIETY MAGNA - ON 40" ROW SPACING, 12 PLANTS PER FOOT

ZONE VII OVER 36"

ZONE VI 30 - 36"

ZONE V 24 - 30"

ZONE IV 18 - 24"

ZONE III 12 - 18"

ZONE II 6 - 12"

ZONE I 0 - 6"

vester dealer who was disgruntled because the company would not ship him an IH combine that he could sell to the Garwood Brothers—customers who had decided to go into soybean seed production in

Above: Soybeans – the crop that in 1924 brought the combine East of the Mississippi (Baker 1961); Below: R.D. Browning of Orange, Virginia built this 5 foot harvester especially for broadcast-sown soybeans (USDA).

9 percent of total yield and changed very little until the 1970's. (Quick 1974). About 84 percent of the loss was found to be at the header—primarily due to the action of the cutterbar on the low-podding crop. The first real breakthrough in significantly reducing harvesting loss was the floating cutterbar.

Pea Soup and the Floating Cutterbar

Restauranteur Joe Zeb, fed up with Chicago, moved to Spokane, Washington where he started a new restaurant. His house specialty remained a delicious

pea soup. Zeb was an entrepreneur who not only managed the business, but as a ventriloquist and clown provided the entertainment as well. As business florished, Zeb found he needed more and more dried peas.

He contracted with farmers to grow peas on a commercial scale in the "Inland Empire" around Spokane in the early 1920's. But there were problems with the pea harvesting—big problems. Losses as high as 90 percent of the pea seed could occur and a 50 percent harvest loss was not uncommon. By harvest time, the vines dry out completely and collapse onto the ground. The harvest method at first involved mowing early, windrowing, then handloading onto wagons for transport to a thresher. Each handling operation multiplied losses. Where the pull-type combine was used, people were employed to walk ahead of the combine and pitchfork the vines into

a big way on their farms at Stonington, Illinois. The Company's reluctance can be understood since there was no evidence of any combine having been tested on a soybean crop in Illinois before 1924. The Reflector referred his reader to Massey-Harris in Toronto, with full knowledge that they had no combine dealers in the US. The ultimate success of the Garwood operation prompted The Reflector to write:
to write:

"The adaptation of the combined harvester to soybeans may open up a market of profitable proportions . . . Heretofore there has been no machinery that harvested soybeans for seed to the satisfaction of the growers . . . With the harvester-thresher it has been shown possible to cut and thresh the beans in one operation with minimum shattering and at low cost. The price received for soybean seed is sufficient to justify the large grower to purchase a machine as expensive even as a combine."—Farm Implement News, Nov. 20, 1924.

The success of the combine in Illinois soybeans was followed by intensive breeding trials on the Garwood farms. The increase in demand for the crop led to a preference for the "scoop type" combine for soybeans, although several other harvesters were developed exclusively for soybeans, but none approached the efficiency of the combine header. (Heitshu 1928). Soybeans have always been comparatively easy to thresh, separate and clean. The problem was excessive gathering loss. (Lamp et al 1961).

Agricultural engineers conducted field assessments of combine performance during the Garwood trials in Illinois. A review of published soybean loss data in the US showed that in 1925, harvesting loss averaged

Gathering losses have been a problem from the very beginning. The four classes of gathering losses are illustrated. 1) Shatter loss – pods and beans shaken free and dropped; 2) Stalk loss – pods cut and dropped or otherwise not collected; 3) Lodged loss – beans in pods on stalks that slipped under the header; 4) Stubble loss – beans in pods attached to the free standing stubble left by the machine. Four beans to the square foot represent a bushel to the acre (Quick 1973).

the header as it moved through the field. (Neal 1977). Conventional rigid cutterbars could not cut low enough to retrieve all the incumbent vines, nor was the operator able to follow ground contours closely enough. Nevertheless, growing peas became a lucrative business. In fact, by 1931, high prices enabled one farmer to actually buy and pay for a farm in one season.

In 1930, Horace D. Hume and J. Edward Love formed a partnership to develop and promote a new concept—a floating cutterbar to adapt the combine for harvesting ripe peas. Hume was a farmer-stockman who in 1920 began a business in Garfield, Washington, selling Pontiacs, Chevrolets and Oliver farm equipment. In 1927, he added the Rumely hillside combine line.

Love was a farmer and commercial pea grower in Garfield who purchased one of Hume's Rumely combines. Thus began a personal association that was to revolutionize the pea harvest and ultimately the harvesting of soybeans. Love's fully floating cutterbar prototype, built in 1929, was successful in harvesting millions of peas that would have been left for the birds. Hume had the vision and wanted to patent the design, but Love felt that the market was too small to justify the patent costs. Nevertheless, Hume assumed all financial responsibility for the patent application in return for a half-interest in the business.

Initially, the partners manufactured the "floaters" on the Love farm. Soon they purchased a shed in Garfield, a shed so small that they had to cut a hole in the wall to poke the bar-backs through as they were drilled. Thus began the Hume-Love Company.

Hume recalled the early battle for acceptance:

"We hauled a sample around on a four-wheel trailer for four years, demonstrating to farmers. People called us crazy. Intelligent men educated in mechanical contrivances were telling everyone our invention couldn't possibly work. We sold it through the use of it. Farmers bought them because they worked." (Hume 1976).

The floating cutterbar worked alright, but lacked a reel suitable for carrying the cut crop back onto

Horace D. Hume of Mendota, Illinois.

Above: Minneapolis-Moline "Harvester" pulled by a Twin City tractor harvesting Ohio soybeans in 1936; Below: The Hume-Love Company's 1931 floating cutterbar attachment for combines.

227

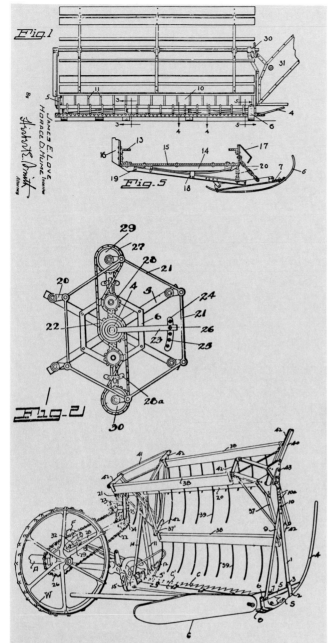

The 1934 Hume-Love Company pickup reel design. This reel design has found universal application, particularly for lodged and tangled crop conditions and for legume harvesting.

Top: "Pea Harvester," US Patent No. 1,881,411 issued to J.E. Love and H.D. Hume, October 4, 1932; the foundation patent on the floating cutterbar; Center: The Hume-Love Company feathering reel, 1932 (US Patent No. 1,926,538, September 12, 1933); Bottom: Polygonal eccentric cam-action pickup reel, one of the original patents (US Patent No. 2,102,711, December 31, 1937). The reel tines had a feathering action and adjustable tine pitch.

the canvas apron. A reel was needed that would lift the crop high enough for the bar to slide underneath and move the vines onto the apron. The first step was simple; belting was tacked onto the wooden bats of the existing reel. This device, said to resemble a Dutch windmill, did the job of sweeping the cutterbar clean, but it was too aggressive. Its slapping action threshed a lot of peas out of the pods before they

could be conveyed to the thresher.

Horace Hume had another good idea. Some years earlier he had seen a pickup reel with "feathering" arrangement to keep the reel tines in motion in a vertical plane. The company adopted this reel, and mounted the tines by wrapping steel wire in a coil around a steel pipe, fastening them in place on the wooden reel bats. Their first reel drive feathering mechanism was a length of link chain running over planetary sprockets. It was clever and effective, but undependable. Next a simpler polygonal linkage and cam drive was designed that was immediately successful. Almost universal now, this type of "pickup" reel has since been produced for over 200 different makes and models of combines.

In 1933, Hume successfully demonstrated the floating cutterbar and reel in soybeans at Champaign, Illinois, but acceptance in the soybean region was slow to come. So slow, that of the 12 flexible floating cutterbars and 12 pickup reels delivered to farms in Illinois in 1934, the 12 cutterbars were returned as having no apparent value. Back in Washington, however, orders were coming in from other areas of the US as farmers realized the possibilities of the reel design, particularly in down and tangled crops.

Some growers attempted to build their own units and, in one two-month period in 1935, 26 "cease and desist" orders were issued by the company's lawyers. The floater and reel had also caught the interest of a group of Western vegetable canners, who suggested that the fledgling Hume-Love Company adapt the design to green-crop harvesting. After several years of experimentation, the "Tractor-Rower"— a windrower—was introduced in 1938 specifically for the canning trade. This innovation was followed a year later by a heavy duty green crop loader. Thus began the "two-stage" method of green crop harvesting.

Love remained an inventor, not a promoter. It was Hume who foresaw the promotional possibilities for Hume-Love products. The partnership had worked well for a decade, but Love was reluctant to expand. He wanted to maintain close control through local operations. In 1941, the two men agreed to an amicable split.

In 1929, when they formed their partnership, 25,000 acres of peas were grown in and around Spokane and Western Idaho. By 1945, 800,000 acres of dried peas were being harvested annually in the Pacific Northwest. This valuable source of protein became a significant contribution to the war effort.

Love continued to produce equipment at Garfield, Washington, but the younger Hume was willing to take calculated risks. He wrote:

"I took a compass and a map of the United States and drew a circle of the area I felt would use my business the most. Mendota (Illinois) was right in the center, and it had three railroads."

Hume established the H. D. Hume Company in Mendota, Illinois, in 1941. The war disrupted manufacture of Hume products but business expanded rapidly after the conflict ended and by the mid 1950's, several branch plants were in operation. The Hume name appeared on a wide range of harvesting equipment as well as on floating cutterbars and pickup reels. Horace Hume retired in 1965, about the time that the floating cutterbar was gaining some acceptance in soybeans. The Hume Company's plants were sold to Hart-Carter Company, the largest manufacturer of combine sieves and chaffers, which had been developed from the original Closz "No-Choke" design.

The Hart-Carter Flexible Floating cutterbar. Unit production at one time reached 27,000 annually. Hart-Carter, Agri-Division, Mendota, Illinois.

Hart-Carter had acquired the Charles Closz Company of Webster City, Iowa, in 1928.

Love's work continued at the J. E. Love Company, Hart-Carter's chief competitor for the floating cutterbar and pickup reel market. The Love Company has remained a family concern in Garfield, with John E. Love, son of the founder, as president. The company has plants in Garfield and in Mississippi.

In the mid-1970's, practically all the pickup reels and floating cutterbars were built by either Hart-Carter or Love. Recently, there has been surge of demand for these products, caused by the rise in

soybean prices, which at one point in 1973, reached a record $12 a bushel. This demand was heavy enough to have lured several newcomers. Some full-line manu-

Top: Deere 50 Series Row-Crop Head. Unit must be operated on planted rows and guided accurately on rows for optimum performance. Available in 4- to 8-row sizes, spacing down to 27 in.; Center and Bottom: Each row unit of the Deere Series 50 had its own stationary knife and six bladed rotary cutter. A pair of chain-mounted gathering belts convey stalks past a rotary cutter to the platform.

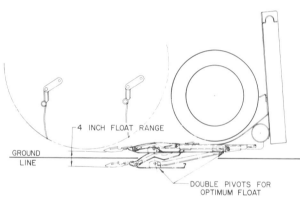

Above: The 200 Series sickle, belt-driven at 565 cycles per minute by an enclosed wobble box drive system. The gearbox floats with the cutterbar; Below: White's "Kwik-Cut" header, available in 13-, 15- and 18-foot widths with integral automatic height control. The 15 foot Kwik-Cut shown here on the 8600 Harvest Boss provided 228,000 cuts per minute.

Top: A parallel-link support system maintained the postion of the 20° "low profile" snout points with respect to the ground, regardless of position of row unit over the 6 in. float range of the Deere Row Crop Head; Center: 20 foot unit Model 220 head on Deere's Turbo 7700 combine with four wheel drive option. The 200 Series heads were released in 1975 and were available in 13 to 20 foot sizes with 24 in. cross-auger. Extra long floating dividers improve crop dividing between adjacent rows; Bottom: 200 Series flexible cutterbar was suspended on a parallel linkage system for a 4 in. floatation range. The cantiliver spring counterbalance member consisted of two layers of 0.030 spring steel sheet running the full width of the head. The spring steel sheeting also closed the gap between the cutterbar and platform (Bichel et. al. 1977).

facturers began making their own floating cutterbars, although others still purchased their pickup reels from the shortliners. And it all started with Joe Zeb's pea soup!

Deere and Company Soybean Heads

Continuing increases in the threshing, separating and cleaning capacity of combines have resulted in a need for improved front-end equipment. Approximately 90 percent of US soybeans are grown as a row crop. In 1974, Deere and Company engineers developed a rowcrop type of head and in 1975, an "open front" head with integral flexible cutterbar. The row-crop head was also suited to other row crops, such as sorghum and sunflowers. Cutting was achieved by a rotary knife on each row; the rotary knives enabling higher forward speeds. There was

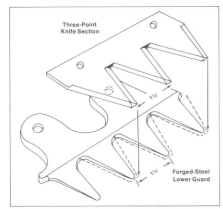

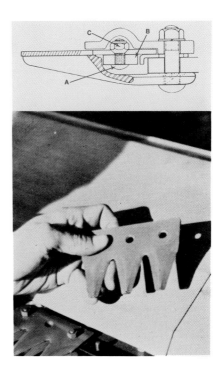

Left: Kwik-Cut ''live knife'' sections protrude marginally ahead of the four-prong forged guards. This design was said to clear trash and mud more readily than conventional guards. The 1½ in. pitch cutterbar had a 3 in. stroke with plate-type mid drive; Right: Triple-point Kwik-Cut knife sections, bolted on instead of the traditional rivetting, could be rapidly changed in the field with a wrench. Knife bolts were held captive in the barback.

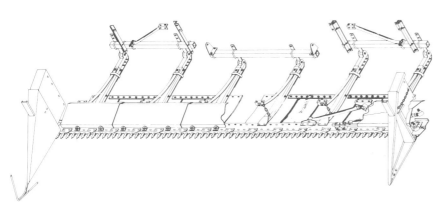

Left: The Hiniker floating cutterbar attachment utilized the conventional knife and guards from the header comb on which it was mounted. Hiniker Mfg. Co., Mankato, Minnesota; Lower Left: Baker ''Bean Saver'' pans. This attachment, produced by Full Vision, Inc., in Newton, Kansas in 1976, was fitted onto the regular fixed platform head to save beans that would otherwise be dropped between the rows.

no reel, giving the operator a better view of the row. Severed plants were gently conveyed to the cross-auger by intermeshing, corrugated, gathering belts. Productivity of the combine with the row-crop head was said to be higher than a regular header since the row-crop head harvested only the material in the row and it was claimed that the gathering belts fed the crop more evenly to the cylinder. (Bichel et al 1976). Typical loss levels for the 50 series Row-Crop head were 4 percent of crop yield, a performance superior to a flexible floating cutterbar under comparable 12 percent moisture soybean conditions.

White's Kwik-Cut Head for Soybeans

The major limitation on header capacity and the primary cause of losses in soybeans is the cutterbar. (Quick 1972). This conclusion was reached after extensive laboratory testing in Iowa and Illinois, and was confirmed in field trials. (Nave & Yoerger 1975). As a result of the research, White released, in 1977, a flexible soybean cutterbar with a narrow-pitch

231

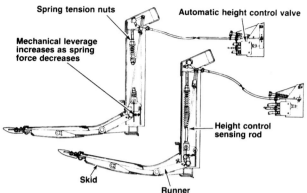

Spring tension nuts

Automatic height control valve

Mechanical leverage
increases as spring
force decreases

Height control
sensing rod

Skid

Runner

Above Left: Lynch verticle drum reel attachment for crops such as soybeans, sorghum, sunflowers and in the illustration seed alfalfa. Extended reel dividers lifted lodged plants into the reel drums. Lynch Mfg. Co., High Plains, Texas; Above Right: The International Harvester 820 Series grain header released in 1977 featured a 6 in. flex range floating cutterbar with regular cutterbar, but driven at 600 cycles per minute, full width skid-shoes, automatic height control and lockup for small grain harvesting (Kerber & Johnson 1977).

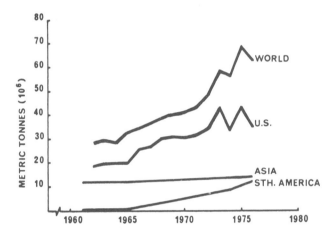

World soybean production. Approximately two-thirds of the world's production in 1977 was grown in the US and one-third of the US soybean crop was exported. USDA projections for 1985 place soybeans at 62 million acres and ahead of corn as the major cultivated crop.

that had been prototyped at Iowa State University. (Quick 1974). The 1½ in. pitch cutterbar was mounted on a flexible barback and had a flat plate knife drive. Sickle sections and guards were manufactured in multiples, with sections bolted on (instead of rivetted) for easy field servicing. Claimed performance was a

50 percent higher forward speed lower losses and a fairly flat speed/loss characteristic, superior to the conventional flexible floating cutterbar. Losses as low as one percent were reported, with narrow row or broadcast beans being more efficiently harvested than those in wide rows. (Quick & Mills 1977).

Soybeans—Biggest Acreage by 1985?

Not until the combine was brought to the Midwest from Western wheat areas in 1924, did the soybean become an important US crop. No other major crop has presented such a challenge. Until the 1970's harvest losses of almost one-tenth of the crop were commonplace. At 1977 prices, each percentage point of the US soybean crop wasted at harvest-time represented a loss to growers of $100 million.

The soybean plant's growth habit, accentuated by uneven ground conditions and the important effect of rapid pod drying on seed shatter, have prevented more efficient harvesting. The challenge of increasing crop yields has brought onto the marketplace header designs that permit higher speed harvesting with losses as low as one to four percent. Automatic controls and further machine refinements promise even better returns as plant scientists and engineers work towards the goal of meeting the increased demand for soybeans—the Cinderella crop that may be planted on more acres than any other in North America by 1985.

"It is a common error to suppose that the British farmer is inherently unprogressive and unenlightened. The real fact of the matter is that, until he had the heart knocked out of him by the years that began in 1879, his skill and leadership were recognized throughout the world. From that time on, and for thirty years, discouragement and poverty were his lot. He watched his fellow farmers in the other old countries, in France and Germany especially, enjoying economic protection and State aid in various forms—treated as men who filled a place of some importance in the scheme of things; but no one seemed much to care what happened to him or his land or his men. During the few brief years that began in 1914 he found himself suddenly again a person of importance but the episode was soon forgotten . . . But these are trifling questions. The important thing is that no people has ever done very much good in the world that did not make a job of farming its land."—Watson and Hobbs 1937.

22

European and Russian Harvest

In the century preceding the Great Exhibition of 1851, British inventors greatly advanced the state of the art of mechanized harvesting. Yet there was a painful lapse before the first combine harvester—of American origin—was tested on an English estate in 1928.

The first combine in Britain was a McCormick-Deering No. 8 pull-type, expressly imported for trials on the Flamsteadbury Estates Limited, Hertfordshire. At almost the same time the Institute of Agricultural Engineering, University of Oxford (forerunner of today's National Institute of Agricultural Engineering at Silsoe) imported a Massey-Harris pull-type combine from Canada. The Flamsteadbury machine cut 53 acres of oats and 137 acres of wheat "in a very satisfactory manner", while the Institute's harvested 89 acres in Wiltshire.

Until that time, the opinion was held in "the best instructed circles" that the combine was "unfitted by weather and other conditions for duty in this country".

The conclusion from the 1928 trials, which was said to represent "true farm opinion", was that:

"costs when compared with the customary methods were reduced by at least half, . . . We must admit our first season's work with a 'combine' in the main has been a complete success; and we feel it is only a question of time when this valuable machine will become an essential part of English farm equipment". (**Implement and Machinery Review** 1929).

The trials and reports continued. In 1929, a Holt windrower and combine imported from California were used to successfully demonstrate the windrow and pickup method of wheat harvesting in England.

In 1929, the first British-built combine—Clayton & Shuttleworth's "Combined Harvester-Thresher"—was produced in Lincoln. Tested by the Institute in 1929, the Clayton showed great promise, prompting a test in Scotland in 1932. (Fenton 1966). The Clayton was equipped with a 12-ft cutterbar and a novel pre-thresher cylinder and concave of the spike-tooth type,

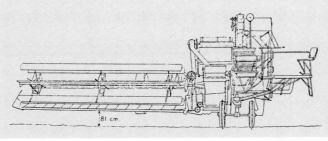

This is how a European draftsman saw the McCormick-Deering No. 8, the first commercial combine tested there in 1928.

located ahead of the main rasp-type threshing cylinder.

In 1932, pull-type combines harvested some 4,400 acres in Great Britain. They were used in many types of crop conditions and on different farm sizes, providing "their practicality in (the British) climate was no longer a matter for discussion". There was, however, still considerable sales resistance and inertia to overcome. Eventually the rising cost of labor made combine harvesting inevitable.

Continental Europe

Someone has said that "an invention, whatever it may be, is never better received in France, than when it is presented under the auspices of a foreigner". Cynicism notwithstanding, it was the French firm of Albert Douilhet at Canderan—Bordeaux which, late in 1928, attempted to market a harvester. But it was not quite a combine. Their object was to start harvesting before the crop was mature, cut low, thresh in a single pass, and bind the threshed straw into sheaves. Douilhet's engine-powered pull-type did not clean the grain. The grain and material-other-than-grain (m.o.g.) was all bagged for cleaning in a subsequent operation. No provision was made for cleaning and it was a right-hand cut machine, as on contemporary European binders.

Right-Hand vs Left-Hand Cut

Douilhet's machine copied European binders which were always right-hand cut. North American self-binders, on the other hand, were left-hand cut. This difference created an additional expense for companies making both types. Elliott Adams has traced the reason for the European right-hand vs left-hand cut binders back to the earliest reaping machines:

"The McCormick Reaper was left-hand cut because a man walked at the right hand end of the platform to hand-rake the grain onto the stubble. Since most men are right handed it was normal to make the Reaper left hand cut. There were reapers of both right hand and left hand cut depending on the design, also push and pull type. The later development of the self-rake reapers were all right hand cut, pull type machines. Many of those machines were exported to Europe. They were so successful in Europe that there was never a European market for hand tie binders.

The first American binders were hand-tie and they were again left hand cut because that design was suitable for the men doing the tying. The front man on the tying platform rode backwards. It was more convenient for him to put the wire or twine around a sheaf with his right hand than his left hand. So the left hand binder design continued when the self tying attachment was developed.

When the self tie binder was considered ready for export, the European market demanded that it be a right hand cut as they were still using right hand cut reapers.

This was a tough situation about 1929 when neither I.H.C. nor Massey-Harris had a right hand cut p.t.o. binder for Europe. Lanz and Fahr came with good right hand cut p.t.o. binders and pretty much ruled the market." (Adams 1976).

Germany's First Combine

In 1929, six American combines were purchased for evaluation in Germany. One of the three imported by the Federal Agricultural Technology establishment (RKTL) received a favorable report, and by 1930, 19 imported machines were in use. One company, Deutsche Industrie Werke, proceeded to develop a machine along similar lines to the French Douilhet, but did not market the unit. Despite the optimism and efforts of Professors Vormfelde and Brenner of the Farm Machinery Institute at Braunschweig—Volkenrode, there was still considerable resistance to use of the new-fangled American "Wundermaschinen". The usual reasons were given—long straw, weeds, wet grain, and so on.

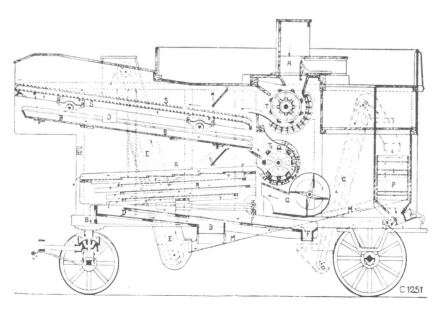

The Lanz stationary thresher was in widespread use during the period combines were being evaluated in Europe. Note that two separate closed concave threshing drums were used on this German machine in 1928 (Firma Heinrich Lanz, Mannheim).

Above Left: Earliest Claas experimental combine, 1931. Machine was mounted around a Lanz tractor (Claas); Above Right: The first locally-produced combine sold in Germany was the Claas M.D.B. (Mäh-Dresch-Binder). No. 1 was sold in 1936. Note the lateral separator body configuration (Querflussmähdrescher) (Firma Gebruder Claas).

Left: For sheaf-threshing, the pto-driven machine was moved to the shock and the crop loaded onto the sheaf table attachment. Production of the MDB was resumed briefly after the war.

Right: The 1947 version of the MDB illustrated towing a chaff-collecting wagon.

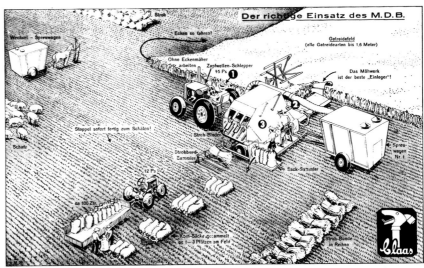

In response to a severe economic recession, the German Government curbed the import of combines. Fears of yet higher unemployment that might be caused by mechanization froze progress. Brenner, as consultant, and the Claas brothers, manufacturers of binders and implements at Harsewinkel, went ahead quietly, and in 1931 tested a tractor-mounted wraparound combine, similar to the 1923 Baldwin Gleaner.

In 1933, Vormfelde & Brenner visited the US, where they were impressed by Allis-Chalmers' ALL-CROP combine developments. Activity at Harsewinkel took a new direction as the Claas brothers and Brenner began to work on the pull-type M.D.B. combine. (Mah-Dresch-Binder). The first combine manufactured in Germany, M.D.B. No. 1, was sold to a farmer in Halle/Saale, in 1936. By 1941, when the war ended production, 1400 Claas combines had been sold and a new design, The Claas "Super", was ready for manufacture. (Isselstein & Schwartz 1965).

Combine harvesting caught on in Europe in the 1950's. A dozen firms produced combines in Germany alone (Brenner 1969). In 1954, there were about 90,000 combines in Western Europe. By 1964 this number had jumped to 420,000, and 80 percent of Western Europe's grain was harvested by combine. (Scott & Smyth 1970).

Claas was the world's largest combine manufacturer for many years, with annual production peaking at 22,500 units in 1965. As the combine market became saturated, sales fell, and Claas placed greater emphasis on other product lines. In the 1970's the majority of the combines produced at Harsewinkel were exported. The Claas Dominator combines were the first to be equipped with grain loss monitors as standard equipment.

Combine usage and Tractor Density relative to cereal production, 1964.								
North America	5.8	Combines	5.5	Combines	9.7	Tractors	90	Tractors
West Europe	4.4	per	4.5	per	17.5	(1966) per	17	per 100
Aust. & NZ	3.2	1,000	6.0	1000 tons	4.2	1000 acres	66	farm
USSR	1.6	acres	2.0	of grain	3.3	arable Land	7	workers

Left: Claas MDB combine hauling a bulk grain tank and chaff-saver wagon. Production of the MDB was resumed by Claas during the Allied occupation in 1945 and the Claas "Super" combine followed in 1946. This machine had an axial threshing drum feeding longitudinal walkers (Quer-Langfluss-maschine). The first self-propelled Claas combine, the Herkules, appeared in 1953.

The Claas Super combine was produced in 1946. Shown here with chaff tank in tow, this machine was ready for manufacture when the war curtailed production.

Claas Combine Models (1977)

Model		Cylinder Width mm in		Typical Header size m ft.		Grain bin capacity l bu.		Engine HP (DIN)	Weight kg lbs	
Dominator	105	1580	62	5.1	17	5500	151	170	8810	19423
Dominator	85	1320	52	4.5	15	4000	93	150	7490	16512
Senator	85	1250	49	3.6	12	3000	83	105	6270	13823
Senator	70	1250	49	3.0	10	2500	69	95	5670	12500
Senator	60	1250	49	2.8	8.5	2500	69	83	5300	11684
Consul		1060	42	2.8	8.5	2000	55	68	4510	9943
Compact	30	960	39	2.4	8	1100	52	50	3240	7143
Compact	25	580	23	2.1	7	1100	30	45	2650	5842
Pull-type Super-Automat S		1250	49	2.4	8	1700	47	pto	2400	5291

Upper Left: The Claas Herkules was the first self-propelled combine produced in Europe, 1953. The Model SF (Shown) was made in 1955. It was equipped with a bagging platform and straw-trussing press. European farmers found plenty of uses for their crop straw for animal bedding, stockfeed, mulches, etc.; Upper Right: The Claas Dominator 105 was the largest in the 1977 range of combines from Harsewinkel. The 105 had a 62 in. wide by 18 in. diameter cylinder, 13 to 25 foot grain heads, 151 bushel bin, 170 HP (DIN) Mercedes diesel engine.

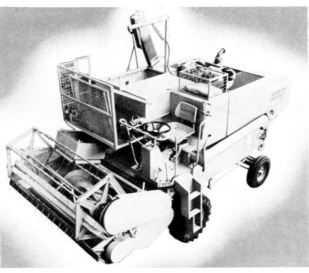

Upper Left: Walter & Wintersteiger's Seedmaster 150 Universal plot (1.5m) combine. Jets of air directed towards the elevator clear the area behind the cutterbar of any crop; Right: Hege plot combine, a 5-foot cut completely self-cleaning combine. (Hege Saatzuchtmaschinen); Lower Left: Claas Compact 25 research version modified by Manns of Saxham, England.

Specialized Plot Research Combines

Plant breeders and grain research scientists need compact harvesting equipment that can be readily cleaned out or self-cleaning, to eliminate contamination between varieties. Small machines are needed because breeders deal with limited batches of valuable seed. Also, the size of research plots is kept as small as possible to reduce yield variations because of changes. in land productivity. During the 1970's, several specialized machines were developed to meet this need. The K.E.M. Company of Haven, Kansas, supplied about 80 percent of the 1977 market for specialized plot combines in North America. West Germany's Hege, and Walter & Wintersteiger in Austria were larger manufacturers of this specialized equipment, with production runs of several hundred machines annually. The Claas Compact 25 combine was modified for research work in England by Manns of Saxham and distributed worldwide.

The American Multi-Nationals in Europe

Deere and Company has manufactured combines in Zweibrucken, West Germany, since 1956. Deere purchased the Lanz enterprise together with its product lines, including combines. Until 1962, Deere continued to produce the German models, and several from the US, selling them under the John Deere LANZ name. In 1959, a development program began on a new line of combines at Zweibrucken, resulting in the introduction of the 300 series combines in 1964. The four self-propelled models were joined by a pull-type model in 1965 and the range marketed worldwide with North America excluded. In 1975, the 300 series was succeeded by the 900 series from Zweibrucken; all with the Deere cross-shaker system for improved walker performance under long-strawed crop conditions.

Massey-Ferguson's first European-built combine, the SP Model 726, was built in 1946 at their Kilmarnock, Scotland, plant. Massey's long-established Marquette Division in France began combine production with the Model 890 in 1953, and in 1954, the Eschwege works in West Germany released the SP Model 630.

By 1976, a total of 37 different combine models, SP and PT, had been produced by Massey-Ferguson's European plants. The average design life between model changes was 5.5 years. In 1969, the Eschwege works produced a novel telescoping grain head on Models 187 and 487 combines, in 10 and 12 ft size heads; these heads retracted under hydraulic power to a road transport width of 8.5 ft. The telescoping design was eventually eclipsed by the Quick-Attach grain heads.

Massey-Ferguson's 1978 combine production in Europe included six models: the MF 206 and 207 from Marquette, and the MF 520/525 and 620/625 from Kilmarnock. A feature of the MF 525/625 Models was Multi-Flow separation—a design which separated a portion of the grain carried out in the straw effluent from the walkers. Massey-Ferguson also sold the Canadian-built 700 Series combines in many European markets. The Power-Flow draper platform was made available for all Massey-Ferguson European models in 1977.

International Harvester Company began building European combines at their Lille, France, works in 1964. First off the line were the Models 8-31 and 8-41 with 32 in. cylinder widths. In 1970, four new models were introduced, with cylinder size increased to 41 and 51.5 in. In 1972, these four versions were updated as the Models 221, 321, 431 and 531. The Model 953 followed in 1977.

Sperry New Holland purchased the Belgian combine company, Leon Claeys N.V. in 1965. Manufacture of the five Claeys combines were continued under the Clayson logo. In 1972, the line was updated with the 1500 series of machines with the same cylinder sizes. In 1977, a new design, the Model 8080, was released. This 61.5 in. cylinder-width machine incorporated a second cylinder, or rotary separator, behind the beater. The company claimed the full-width rotor, with its own adjustable grate-type concave, significantly boosted separation capacity.

Russian Harvesters

In 1975, the Soviet Union was the world's largest producer of wheat, barley, oats, rye, buckwheat, sugar-beet and sunflowers, and it occupied fourth place in soybean, corn, rice and millet production. Recently, Soviet agriculture has placed a heavy emphasis on mechanization. The first combines were imported from the US in 1925. By 1975, around 109,000 combines were being produced by Soviet factories each year. (Tractoroexport 1975). According to information received, there are two main combine plants in the USSR—the Taganrog works, where the "Kolos" models are manufactured, and the Rostov-on-Don works where "Niva" models are produced at the rate of about 100 combines per day. It is not known how many of these machines might be "re-manufactured"; i.e., re-cycled for major overhaul within the same precincts, similar to a proportion of collective farm tractors (Millar 1974).

Above: Deere 955R from Zweibrucken shown with bagger attachment; Below: Deere 965 developed in 1975 was equipped with the Deere cross-shaker over the walkers. The development enhanced separation, particularly in European long-strawed crop conditions.

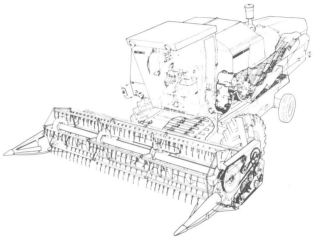

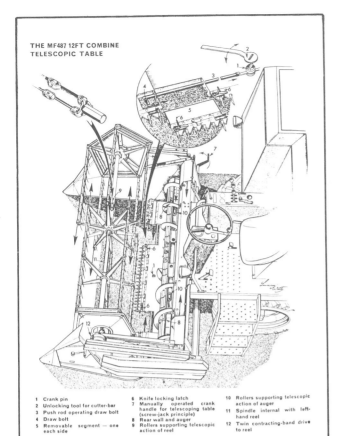

THE MF487 12FT COMBINE
TELESCOPIC TABLE

1 Crank pin
2 Unlocking tool for cutter-bar
3 Push rod operating draw bolt
4 Draw bolt
5 Removable segment — one each side
6 Knife locking latch
7 Manually operated crank handle for telescoping table (screw-jack principle)
8 Rear wall and auger
9 Rollers supporting telescopic action of reel
10 Rollers supporting telescopic action of auger
11 Spindle internal with left-hand reel
12 Twin contracting-band drive to reel

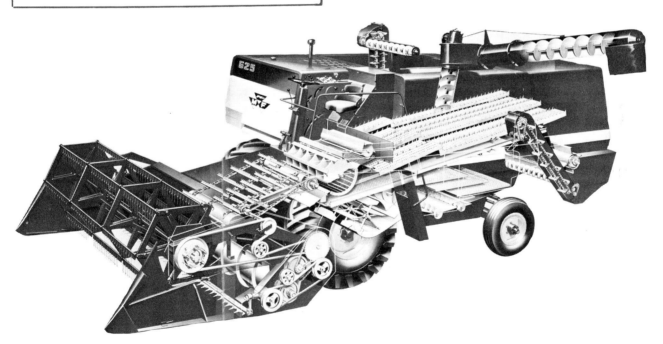

Clockwise from Upper Left: Massey-Ferguson's former combine plant at Eschwege, Germany produced a telescoping grain head in 1969 in 10- and 12-foot sizes for the Models 187 and 487. The operator would remove several latches and filler assemblies before the header was retracted hydraulically to 8.5 foot width; Sperry New Holland Clayson 8080 combine with 61.5 in. cylinder, grain heads to 22 feet, 180 bushel bin, 170 HP Mercedes diesel engine; Semi-sectional view of Clayson 8080. Note the "Total Vision" cab with full glass wrap-around ahead of the operator; Massey-Ferguson Model 625SP combine originally released in 1971 and produced at the Kilmarnock, Scotland, plant. Cylinder width is 54 in. with headers from 10 to 16 feet.

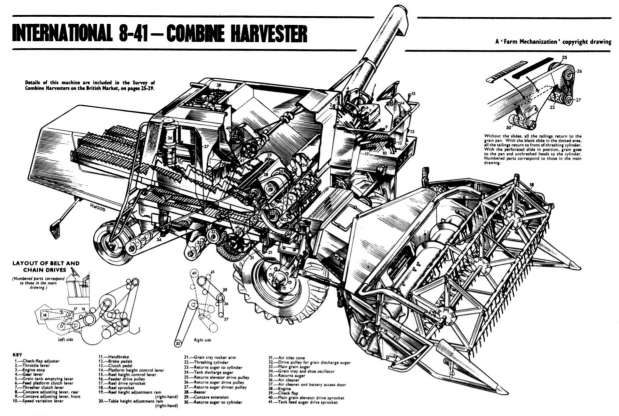

Details of this machine are included in the Survey of Combine Harvesters on the British Market, on pages 25-29.

Without the slides, all the tailings return to the grain pan. With the blank slide in the dotted area, all the tailings return to front of threshing cylinder. With the perforated slide in position, grain goes to the pan and unthreshed heads to the cylinder. Numbered parts correspond to those in the main drawing.

LAYOUT OF BELT AND CHAIN DRIVES
(Numbered parts correspond to those in the main drawing.)

Left side

Right side

KEY

1.—Check-flap adjuster
2.—Throttle lever
3.—Engine stop
4.—Gear lever
5.—Grain tank emptying lever
6.—Feed platform clutch lever
7.—Thresher clutch lever
8.—Concave adjusting lever, rear
9.—Concave adjusting lever, front
10.—Speed variation lever
11.—Handbrake
12.—Brake pedals
13.—Clutch pedal
14.—Platform height control lever
15.—Reel height control lever
16.—Feeder drive pulley
17.—Reel drive sprocket
18.—Reel sprocket
19.—Reel height adjustment ram (right-hand)
20.—Table height adjustment ram (right-hand)
21.—Grain tray rocker arm
22.—Threshing cylinder
23.—Returns auger to cylinder
24.—Tank discharge auger
25.—Returns elevator drive pulley
26.—Returns auger drive pulley
27.—Returns auger driven pulley
28.—Beater
29.—Concave extension
30.—Returns auger to cylinder
31.—Air inlet cone
32.—Drive pulley for grain discharge auger
33.—Main grain auger
34.—Grain tray and shoe oscillator
35.—Returns auger
36.—Air cleaner
37.—Air cleaner and battery access door
38.—Engine
39.—Check flap
40.—Main grain elevator drive sprocket
41.—Tank feed auger drive sprocket

*IH Model 8-41 SP combine with 32 in. cylinder was produced in England from 1964 to 1971 (**British Farm Mechanization**, March 1965).*

A Sampling Of Combines From Continental Europe

Fahr M1600 Hydromat combine with 60 in. cylinder, largest of the Fahr line. This machine was equipped with Linde hydrostatic transmission, 19 foot header, 240 HP, 8 cylinder Deutz air-cooled diesel engine. An unusual feature was the use of two axial flow fans for the cleaning shoe. (Maschinenfabrik Fahr AG).

Above and Left: The Dania D-900 combine of Danish manufacture was introduced in 1958, had a de-awner and secondary cleaning system for graded seed production. The machine came with 8 foot cut, 44 in. cylinder, 51 bushel bin and 62 HP diesel.

Above: The largest grain combine in the Braud line from France, the 801 with 51.2 in. cylinder and 16 foot header; Below: The Arbos MT 125 Giaguaro was the largest model of this level-land and hillside combine line from Piacenza; with 44 in. cylinder, 14 foot head, 102 HP engine. Unit options included rice tracks and dust eliminator fan for operator.

Sequence illustrating folding action of the Fisher-Humphries Lely Victory folding head. Transport width of 9.75 feet. In field operation, the two halves of the grain head were separately driven.

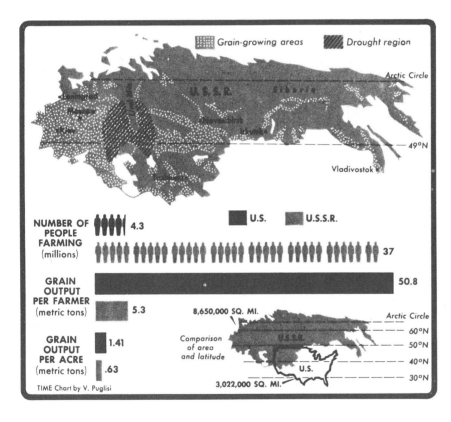

Left: The Laverda Company was founded in 1873 to produce hand-powered farm machinery. The line leader was the Model 152 combine with 14 foot grain head, 42 in. cylinder, 97 bushel bin, 130 HP engine and a secondary cleaning system as standard equipment. Laverda produced two hillside models, the M100AL and the M112AL. Laverda combines could be equipped with capacity enhancing and synchronized "Tridimensional straw shakers" which agitated the straw mat sideways to improve separation on the walkers (Ditta Pietro Laverda S.p.a.); Right: Two versions of the Lova rice combine made in Italy were produced in 1976. The larger Model 2001 (shown) was equipped wih a 72 in. cylinder and separation section. Rice tracks were standard equipment along with 180 HP (DIN) diesel engine, 13.7 foot header (Lova S.p.a.).

Russian-American grain production compared (reprinted by permission from Time, The weekly news magazine; © TIME INC., 1975).

82-Foot Cut Combine

The Soviet's attempts to learn US mechanization techniques may actually have started over a century ago at their Fort Ross settlement north of San Francisco. They became more deeply involved when the 1917 Revolution took them out of WW1 and the decision was made to revolutionize agriculture—from the top down. Oliver, Caterpillar and International Harvester all received substantial orders for tractors and combines. An unconfirmed estimate placed Oliver with an order for 3000 tractors and 2800 combines by 1929. Each of these companies sent over top management people to help train Russian operators in the use of the new machines. A substantial order for Massey-Harris harvesters was to prove most timely for the Canadian company at the depths of the Depression. Curtis Baldwin was also invited and was said to have designed an 82-ft cut combine for the Russians. The machine was rumored to have been built, but did not perform and was never heard of again.

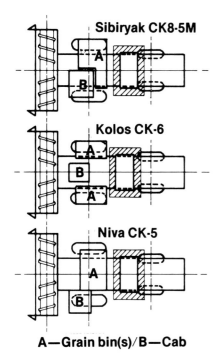

Sibiryak CK8-5M

Kolos CK-6

Niva CK-5

A—Grain bin(s)/B—Cab

Cross-sectional views of Niva, Kolos and Sibiryak combines. Total Soviet production was claimed in 1975 to exceed 109,000 combines annually.

The Soviet Kolos combine Model SK-6 towing a 23 foot grain head on a trailer. Kolos has a 59 in. cylinder, 150 HP V6 turbocharged diesel engine and 132 bushel totally enclosed "saddle" tanks for the grain. Augers were used for tailings and grain conveying. The louvered cab was designed to keep down solar radiant heat and glare in the brief Northern summers where the sun is only 25° above the horizon.

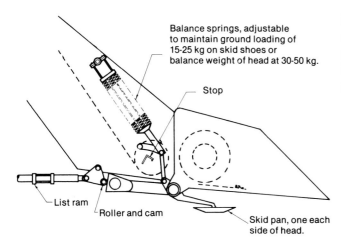

Balance springs, adjustable to maintain ground loading of 15-25 kg on skid shoes or balance weight of head at 30-50 kg.

Stop

List ram

Roller and cam

Skid pan, one each side of head.

Left: Niva and Kolos combines were fitted with a mechanical skid flotation system on the header that automatically retained a predetermined cutting height (50, 100, 130 or 180 mm), independant of ground contour.

East Germany's "Fortschritt" Combines

In 1977, the East Germans produced the Model E516 combine that had the world's biggest threshing cylinder. The largest of their several combine production models, the Model E516, was equipped with a 31.5 in. diameter, 64 in. wide cylinder and walkers and was powered by a 210 HP diesel engine. A feature of the Fortschritt machines was the optional automatic steering control system which, sensing the crop wall, automatically steered the combine, keeping the head cutting a full swath of grain.

Other Eastern-bloc Combines

Top to Bottom: Hungarian combine needs were met by production from the Budapest Mezogazdasagi Gepgyar (BMG) Agricultural Machinery Factory at Budapest. Corn heads built under license to Braud of France were produced by BMG and adapted to fit certain of the Soviet-bloc combines as well as the Hungarian machines; Yugoslavia's Zmaj factory produced four self-propelled models, including the Universal shown; In Rumania, the Gloria C-12 and C-14 self-propelled combines feature a lateral tilting header of Russian origin and a secondary cleaning system located across the grain bin.

From Top to Bottom: The Model E 516 Fortschritt combine, released in 1976, had the biggest cylinder in the world with a width of 64 in. and a diameter of 31.5 in. Machine also has 25 foot header, 210 HP diesel engine (VEB Kombinat Fortschritt Landmaschinen); the Model E 512 Fortschritt front view; Crop wall sensor on the Fortschritt E 516 combine for automatic steering control.

"In my opinion, the changes in methods of farming in the future will be brought about by a wide knowledge and application of scientific principles. I do not think it probable that farm implements will be improved very much, although doubtless on the larger farms means will be devised to perform certain operations by electricity or steam. Nor do I lay any stress upon the possible revolution in methods of farming anticipated by those who think that the rainfall may be controlled at will by explosives, a theory which will, long before the time of which I write, have been itself thoroughly exploded and given a place among the curiosities of so-called scientific investigation, in company WITH ITS TWIN ABSURDITY, THE FLYING MACHINE." —Jeremiah R. Rusk (1830-1893). "American Farming a Hundred Years Hence". Written while US Secretary of Agriculture and published in North American Review 1893.

23

Rotary Separators and Axial Combines

In June 1975, Sperry New Holland announced their "breakthrough combine design"—the TR70 TWIN ROTOR combine. This axial design marked the company's entry into the business of manufacturing combines in North America (Skromme 1977).

The target of Sperry New Holland's marketing thrust was the US cornbelt. The greatest concentration of combines was to be found there and the axial design has been found to perform best in corn.

Comparative tests have shown axial designs to have higher capacity and substantially reduced kernel damage in corn and soybeans than the conventional cylinder (Moss 1977; Saijpaul et al., 1977).

By September 1977, International Harvester had joined the fray. When their AXIAL-FLOW design was unveiled, the North American public was introduced to no less than three new models at once (DePauw, Francis and Snyder 1977).

Not to be overlooked was the Canadian Western Roto Thresh walkerless combine. A limited number have been sold since first introduced in Saskatchewan in 1974.

Solution to an Engineering Problem

This "new generation" of axial combines offered a solution to a combine engineering problem—how to get more capacity out of a given machine size or envelope. One way to increase capacity is to operate at a faster speed, covering more ground in a given time. But sooner or later a machine component becomes overloaded. Usually the straw walkers overload first—provided the crop can be digested by the gathering front. Grain losses climb to unacceptable levels or some part of the machine becomes plugged when the combine is pushed too hard, or crop conditions are unfavorable.

Another way to increase capacity, so that the farmer can get the crop in at the peak of the harvest season, is to increase the body width. With the time-honored tangential threshing cylinder and straw walker design, manufacturers simply produced newer models of wider and still wider cylinders.

As cylinder width is increased, combine wheel tread also increases. Penalties occur once overall machine width exceeds a certain point. The shipping and transportation of oversize machinery requires special permits or even modifications to the machine itself for legal or safe road and rail shipment. Apart from that, there are problems in manufacturing extra-

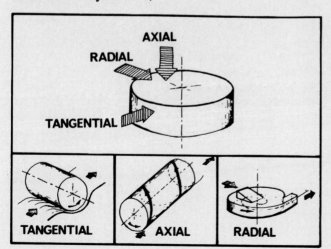

Three basic ways that crop can be fed into a rotating crop processing component.

New Holland's TR70 combine. The two rotor axial design is claimed to have the performance of a 50 in. plus conventional design, yet each TR70 rotor is only 17 in. in diameter.

wide components such as walker cranks. On big machines, grain bin size assumes major importance, as it governs the frequency of unloading, the number of grain trucks needed and their location, etc. The straw walker section takes up space that might otherwise be used to carry more grain. Finally, there are some crops and conditions where the straw walkers need frequent maintenance just to keep them clean and functioning.

What are the alternatives? There are basically three ways that a crop can be presented to a rotary threshing cylinder or crop processing component: tangentially, axially, or radially.

With the axial configuration, the cylinder is turned at right angles, major axis lengthwise along the combine. Much greater cylinder lengths are possible. The walkers are eliminated, and high centrifugal forces can be employed using rotary separation components.

Rotary Separation not so New

Rotary threshing and separation concepts are not new. Once inventors got away from the idea of trying to simulate the motion of the human arm and flail, and began to consider pure rotary motion, a veritable deluge of designs appeared. The now-universal tangential-feed threshing cylinder was not even the earliest thresher design. In 1772, Sven Ljundqvist described a Swedish grain separator which used centrifugal force to separate whole sound grain from cracked seed, weeds and trash. In England, William Winlaw developed a water-powered thresher in 1785. Both these were axial designs, with vertical axes, and helped free agriculture from dependence upon the age-old winnowing fork and jointed flail.

Once Andrew Meikels' 1788 tangential threshing cylinder was proven, rotary separation soon followed. Meikle himself used tangential-fed rotating drums to separate grain from the straw. By the 1800's there were a number of his threshing mills in operation.

Meikle and his contemporaries successfully had combined threshing, separation and winnowing into a single operation.

Multiple tangential cylinder arrangements have been used elsewhere since early Scottish designs. The Ransomes stationary thresher sold in England in the 1860's separated grain from straw with a long series of rotary straw beaters. Several European and Russian combine designs have used a similar principle. The Clayson Model 8080 combine, introduced in 1976, used a rotary drum and walker in a three-stage separation process. Multiple cylinder designs were claimed to increase capacity in crop conditions where grain-straw separation was the main limit on performance. Absolute combine capacity was nevertheless still governed by the width and characteristics of the first cylinder.

Scottish threshing mill, 1805. Andrew Meikle invented the tangential threshing cylinder and sought a patent in 1788. About 1790, he put together the thresher, rotary separators and the winnower into one system for "finishing" grain (Slight & Scott-Burn 1858).

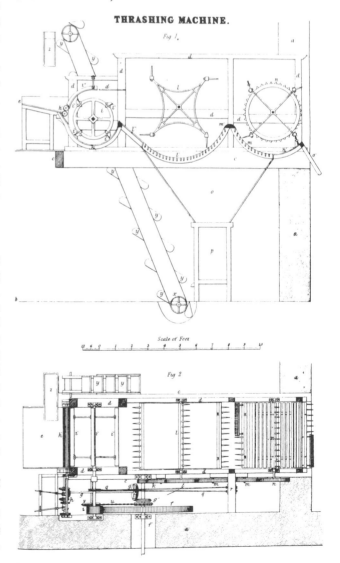

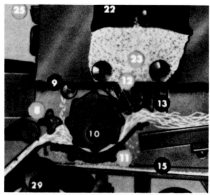

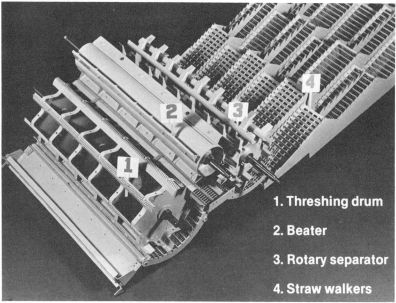

1. Threshing drum

2. Beater

3. Rotary separator

4. Straw walkers

Above: Ransomes Cavalier combine of 1965 had a smaller threshing drum ahead of the main cylinder. It was claimed to have higher capacity and lower grain damage than the conventional design.

Upper Right: New Holland's SP 8080 combine with rotary rake separation. This concept was first introduced by New Holland on their 1550 model in 1973; Below: Several Soviet combines were produced using the multiple drum approach. The Kolos rice machine with twin drums was still being mass produced in 1977.

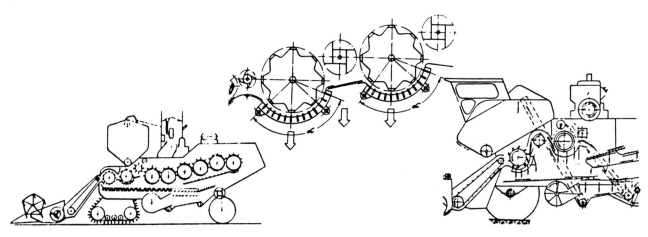

Radial Systems

Some writers claim the 1797 mill of William Spencer Dix in England as a threshing device. If this was the case, then this stone-mill type of machine would be the earliest radial-type thresher. Many radial threshers have been patented, but the reason for their notable lack of commercial success was probably their awkward size, limited intake capability and a tendency to damage grain because of the relatively high rotor tip speeds. A notable exception was the horizontal table thresher developed in the Philippines for hand-fed rice bundle threshing by Amir Khan and associates at the International Rice Research Institute (IRRI) (Khan 1973).

The "Wild" harvesting machine was a radial design that was produced briefly in England around 1950, and was restricted to standing crops that were readily threshed in one pass. It used a series of radial discs mounted on a single axle in the header of the pull-type machine.

The Harvestaire Combine

Harvestaire Incorporated, of Sacramento, California, began work on a pneumatic-centrifugal combine design in 1950 under the direction of H. D. Young. The objective was simplicity. The developers envisaged a compact combine having fewer parts, using all-rotary motion and relatively insensitive to sideslopes. By 1955, the company was testing three units in selected locations in the US, while attempting to interest other manufacturers in producing the combine under license. The combine was massive in appearance, but the bulk was mostly air chambers. With few mechanized parts, weight was low. As *Farm Implement News* editor Baker described it, the machine "fanned its way through the wheat fields with high hopes" (Baker 1955), but failed to impress the manufacturers' representatives. The Harvestaire was "complicated and a great power consumer" according to the Oliver Company's observers.

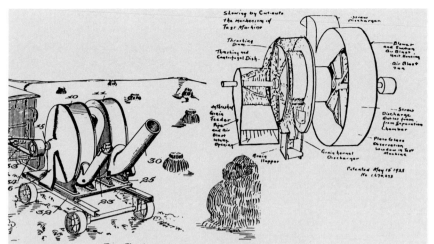

G.F. Nye set up a short lived company in Kansas to produce his centrifugal thresher just before the Great Depression. Typical of a flurry of activity in the US and on the Continent, radial thresher inventors each claimed machine simplicity as the greatest advantage.

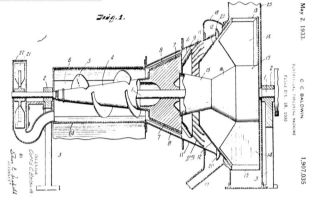

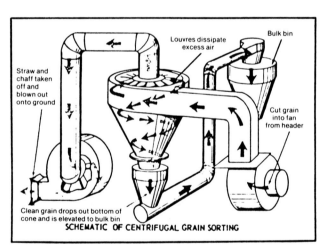

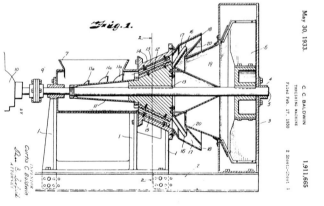

Above: Californian Harvestaire combine, 1955. The crop was processed through centrifugal fans and separating chambers. The massive-looking machine enclosed mostly airspace. Weight was said to be one-half that of conventional combines; Below: Arrangement of the centrifugal grain separation and cleaning systems in the Harvestaire combine. Early accounts of high separating efficiency were negated when it was found by independent consultants Zink and McCuen that the discharge fan tended to grind up any lost grain into powder and discharge it with straw.

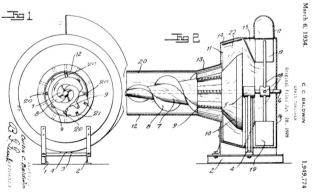

A series of patents were awarded to Curt Baldwin on rotary thresher/separators in the 1930's which displayed the main theme of a through-shaft design with conical surfaces and fans to thresh, separate, clean and convey the crop materials.

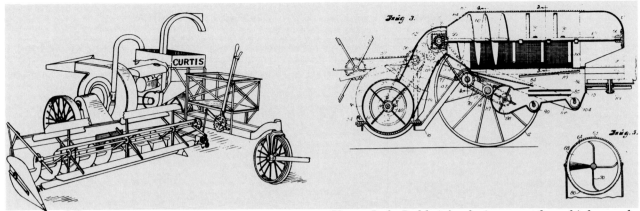

Conical Threshers

The threshing cone principle, known at least since 1860, has tantalized numerous inventors because of the potential use of a progressively-increasing aggressive action as the crop flows over the rotor and the inherent simplicity of clearance adjustment. The Union Iron Works' "Western Corn Sheller"—a conical sheller with an 80 year history—ranks as one of the oldest threshing designs in continuous production. Clearance was adjusted by simply moving the threshing cone axially into or away from the conical outer cage.

One of the earliest patents on the cone principle for cereal grain threshing was the Telschow design, patented in Germany in 1878. A succession of designers have pursued the many combinations of axial, tangential and helical crop paths and rotor orientations that are possible with the cone. Notable designs, some of which were produced for a limited time, included those of Baldwin, Bunting, Wessel, Wise, Buchele and Lalor, Stout and Strohman, Lamp, Mark, and others working for Massey-Ferguson (Caspers 1969).

The attractiveness of the rotating cone principle, which provides an ever-increasing separating force field as the crop passes over the rotor, was probably offset by the complexity and cost of accurately manufacturing conical surfaces.

Upper Left: Baldwin's elusive quest for a high speed harvester led him to start several companies. The Curtis Harvester illustrated here utilized the cone-axial principle in 1931; Left: Bunting's "Aero" cone-axial combine, 1934. For a short time Bunting was an employee of Baldwin, but went into business to make his own combine after they both left Gleaner Combine Harvester works. About 25 Bunting machines were built (USDA); Above: Bunting's combine had a cone-axial thresher mounted co-axially with the header auger, but running at higher speed.

Massey-Ferguson's Conical Thresher Developments

In 1957, the Massey-Harris-Ferguson management set up an advanced engineering group to work on harvester development. Initially they wanted to learn more about the crop processes in the conventional machine. This study later produced work on resilient threshing devices, including inflatable cylinders; transverse flow fans; extensions of the header; and several types of rotary separating and cleaning apparatus.

One of these, a conical rotary combine design, reached the field prototype stage and was evaluated in North American and European crop conditions. There were certain engineering problems and changes

Massey-Ferguson walkerless combine with conical thresher concept was based upon patent issued to Massey-Ferguson in 1969. The combine was built in Europe (Herbsthofer 1974).

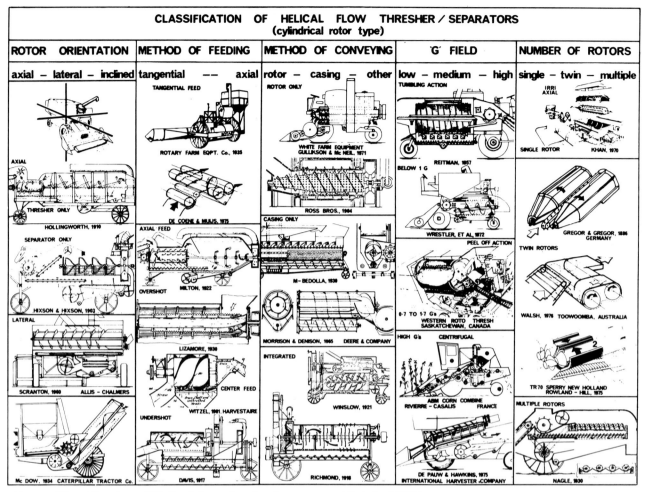

CLASSIFICATION OF HELICAL FLOW THRESHER / SEPARATORS (cylindrical rotor type)				
ROTOR ORIENTATION	METHOD OF FEEDING	METHOD OF CONVEYING	'G' FIELD	NUMBER OF ROTORS
axial – lateral – inclined	tangential –– axial	rotor – casing – other	low – medium – high	single – twin – multiple

Above: Classification of Helical flow thresher/separators (Quick, 1977).

in objectives that prevented this combine from being mass-produced at the time. (Herbsthofer 1974).

Helical Flow Combines

A sampling of some of the axial devices patented, and those machines that have been produced, illustrates that the field of axial thresher/separators has been well explored during the past century. The chief advantages the inventors and manufacturers have sought and claimed for their designs have been: the elimination of reciprocating walkers, and higher capacity within a more compact design. Perhaps the earliest cylindrical axial thresher/separator patented

Twin rotor axial thresher patented by Gregor & Gregor in Germany, 1886.

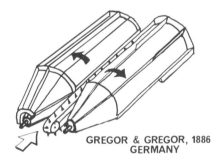

GREGOR & GREGOR, 1886
GERMANY

Letterhead of the National Farm Machinery Mfg. Co., Wichita, Kansas, a company organized in the mid-1920's to market a centrifugal stationary thresher of the axial type.

was a hand-cranked twin rotor design, the Gregor patent, filed in Germany in 1885. Of the many ideas that followed, none were known to have reached the marketplace until the 1920's, when several rotary machines were offered to farmers in the US grain belt. The Great Depression halted the initiative of manufacturers and momentum was not regained until well after the war.

So diverse are the various cylindrical rotary designs that a table has been drawn up to summarize the designs. Those represented are classified according to:

- Rotor orientation
- Method of feeding (tangential, axial, overshot or undershot).
- Method of conveying the crop (by rotor only, by casing only, rotating casing, pneumatic, etc).
- Number of rotors and strength of separating force field.

The classification was restricted to just those machines in which the crop travels in a helical path over a cylindrical rotor.

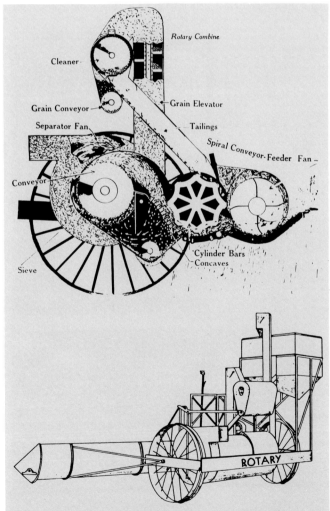

The Rotary Farm Equipment Company, Wichita, Kansas used a conventional "down-front" cylinder but had an auger-type axial separator in a very compact configuration, ca 1929-1935.

Felix Schlayer's Heliaks

Felix Schlayer was born in Schwabia, Germany, like engineer Robert Bosch, his lifelong friend. He adopted a Spanish lifestyle and chose as his vocation the farm machinery business. Married in Madrid, he was the German Consul before WW1. Schlayer began building axial threshers there after hostilities ceased. His first design was of the through-shaft helical-flow type. The crop was flailed as it progressed through the stepped perforate drum. German authorities at the RKTL were favorably impressed by tests of the "Heliaks" in 1928 and 1931 (Schlayer 1928, 1931), although power consumption was high: 40 to 50 HP for a capacity of only 5 TPH. Schlayer accompanied one of his units to the US in the 1930's in an attempt to interest Massey-Harris, J I Case and later Allis-Chalmers. It was during the Allis-Chalmers tests that "The Reflector" of *Farm Implement News* saw it in operation and reported:

> "When ye Ed. observed the machine threshing barley it was doing a clean job. However, a new A-C 24 x 40 threshing the same barley in a comparative test did an equally good job while consuming considerably less power from the Model "U" tractor engine. The Schlayer pneumatic-centrifugal thresher worked, but not as efficiently, power considered, as the standard US designed machine."
>
> "So that was that. A-C did not buy shop rights, and Mr. Schlayer departed for Spain, leaving his experimental machine behind." (Baker 1955).

Schlayer was obliged to flee Madrid at the outbreak of the Spanish Civil War, settling in Switzerland, where he passed the rest of his life. During a wartime scrapdrive, the Allis-Chalmers Heliaks became part of the war effort. The influence of the design may have remained however, as an axial separator project

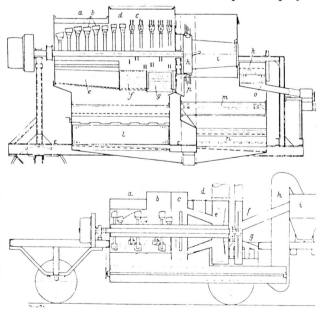

Felix Schlayer's "Heliaks" threshing machine design evolution. Top: First patent in Germany, 1922; Bottom: Versions tested by RKTL in Germany in 1931.

One of the original Spanish Schlayer Heliaks units is held by the Claas works in Germany (Courtesy L. Caspers, Gezb. Claas).

was revived following the war. There was perhaps a hint of the Heliaks in a 1960 Allis-Chalmers rotary separator patent.

An interesting insight into these pre-war developments is contained in the following excerpt from the anecdotes of Elliot Adams, former engineer with Massey-Harris:

"There were two so-called "secret weapons" in the deal when Massey-Harris bought the J. I. Case Plow Works in 1928 after having jobbed the Wallis tractor for two years;
    1. Ed Everett's Centrifugal Combine.
    2. John Rogers' 4-Wheel Drive Tractor.
When at Racine early in 1929 I was shown the Centrifugal Combine ready for test work in the 1929 Harvest.

I was transferred to the European Organization April 1, 1929. I attended the Paris Salon (one of the important Farm Machinery Shows in Europe) in 1929, 1930, and 1931. Mr. Duncan asked me to be sure to have a good talk each year with Mr. Felix Schlayer of Madrid, Spain, who had a very adequate display of his Schlayer-Heliaks threshing machine at the salon all three years. Mr. Schlayer, who as a German by birth, spoke very fluent English. He had been working on his machine for years and had many patents and patent applications. He had only built the machine as a stationary thresher, which he had on display, but he had the layout drawings for a pull type combine. The machine was in the same general category as the Ed Everett machine. I never saw either machine in operation. There were plans for me to see the Schlayer machine doing test work at the University of Leipzig in 1931 but due to problems which arose there in long rye straw plus some rainy weather, I never made the trip. I had put in a negative report to Mr. Duncan each year after talking to Mr. Schlayer as I didn't think that he had a commercial machine. It seemed that he was always running into what he considered minor corrections but they were quite involved in my opinion. I didn't know why Mr. Duncan was questioning me so closely about this development.

I later learned that the company had purchased the Curtis Harvester Company in order to obtain the patent applications which Curt Baldwin had on his Centrifugal machine as it was felt that some of those claims might read into the Ed Everett machine. After Curt was no longer associated with Gleaner Baldwin, he had organized the Curtis

Harvester Company and built quite a number of pull-type machines which were almost identical with the Gleaner Baldwin. However, Curt himself spent most of his time on the Curt Baldwin Centrifugal Combine. In 1931 the Massey Engineering Department in Toronto built what was supposed to be the final edition of Curt's machine. The Ed Everett machine and the Curt Baldwin machine plus a conventional Massey Harris machine were run at Calgary in 1931. The result was that the two Centrifugal machines were shelved by Massey-Harris.

Along about 1935 the grapevine information was that Allis-Chalmers had made some sort of deal with Felix Schlayer and had taken his machine to LaPorte for test work. I was pretty much on a hot seat until it was learned that A.C. had put that machine on the shelf." (Adams 1976).

Asian Axial Threshers

The International Rice Research Institute's Agricultural Engineering section had been involved in the use of power equipment for rice production in Southeast Asia. For rice harvesting, centrifugal threshing and separation was selected for its inherently more compact and potentially simpler design capabilities. One result was the development of axial-flow threshers that have since been produced by a number of companies in Asia. (Khan et al, 1975).

Rotary thresher combine developed by Massey-Harris in the early 1930's but never marketed. Referred to as the Everett machine by E.A. Adams in his "Anecdotes from 43 Years in the Farm Equipment Business," 1976.

Above: Multi-crop pto axial thresher. Spike tooth cylinder with adjustable deflecting baffles to control the degree of threshing. Developed by IRRI agricultural engineers, the machine is one of two axial threshers being manufactured locally by a number of Southeast Asian companies (Khan 1970).

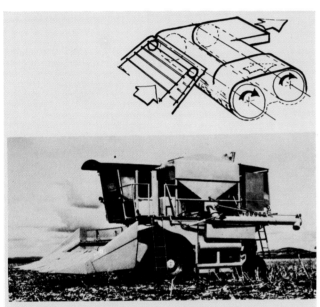

Australian farmer Bill Walsh of Toowoomba, Queensland, developed a custom-built two rotor combine with rotary separation for corn harvesting. The schematic shows the crop flow through the rotors.

European Axial Corn Combines

At least three French combine manufacturers have produced axial flow machines of the sheller-type, specifically for corn harvesting. One of these, the ABM Rivierre-Casalis, has been in production since 1965.

Australian Axial Corn Combine

An Australian contribution to the art of centrifugal threshing, the Walsh Maize Header, has been produced by a Queensland firm for over 20 years.

Both pull-type and self-propelled versions of the specialized corn combine with twin lateral rotors have been produced.

Canada's Western Roto Thresh Combine

Western Roto Thresh Limited of Saskatoon, Saskatchewan sold their first ROTO THRESH combine in 1974. The design used a conventional thresher of 48½ in. width, feeding into a 5½ ft diameter rotating separator drum. The perforated corrugated drum revolved at 30 rpm to create a centrifugal force field slightly greater than gravitational. A stripping auger forced the straw path away from the drum during the crop's motion inside the separator. The straw continued to move towards the end of the drum because of the helical corrugations inside the drum, while the grain penetrated through the drum for delivery to the four-stage cleaner. Although designed originally for harvesting cereals on the Canadian prairies, the ROTO THRESH successfully harvested corn and can be equipped with a six-row corn head option.

Above: ABM Rivierre-Casalis self-propelled corn sheller with twin axial rotors was first marketed in 1965 with a two row corn head. Later models offered with 3- or 4-row heads. Rivierre-Casalis, Orleans France; Below: The Burgoine corn combine with 4-row corn head. This machine was also equipped with a twin rotor axial sheller design. Completely self-propelled and tractor-mounted models were produced. Burgoine Cie., Chantonney, France.

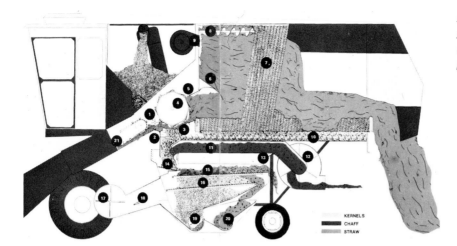

Left: Schematic view of the 1977 ROTO THRESH self-propelled combine, manufactured in Saskatchewan.

Right: ROTO THRESH combine separator principle.

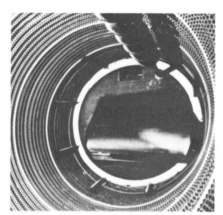

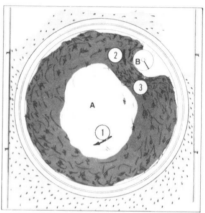

KERNELS
CHAFF
STRAW

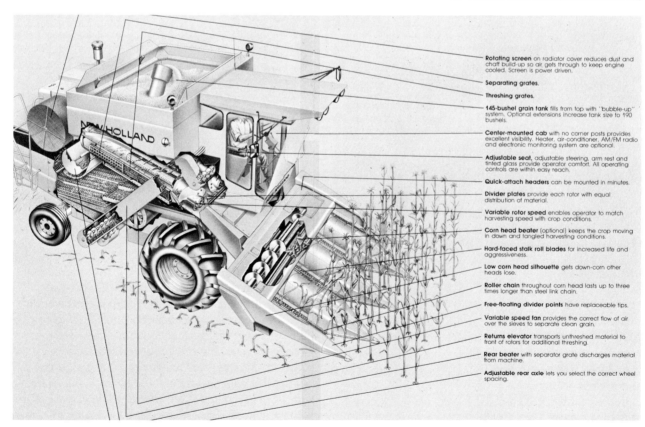

Rotating screen on radiator cover reduces dust and chaff build-up so air gets through to keep engine cooled. Screen is power driven.

Separating grates.

Threshing grates.

145-bushel grain tank fills from top with "bubble-up" system. Optional extensions increase tank size to 190 bushels.

Center-mounted cab with no corner posts provides excellent visibility. Heater, air-conditioner, AM/FM radio and electronic monitoring system are optional.

Adjustable seat, adjustable steering, arm rest and tinted glass provide operator comfort. All operating controls are within easy reach.

Quick-attach headers can be mounted in minutes.

Divider plates provide each rotor with equal distribution of material.

Variable rotor speed enables operator to match harvesting speed with crop conditions.

Corn head beater (optional) keeps the crop moving in down and tangled harvesting conditions.

Hard-faced stalk roll blades for increased life and aggressiveness.

Low corn head silhouette gets down-corn other heads lose.

Roller chain throughout corn head lasts up to three times longer than steel link chain.

Free-floating divider points have replaceable tips.

Variable speed fan provides the correct flow of air over the sieves to separate clean grain.

Returns elevator transports unthreshed material to front of rotors for additional threshing.

Rear beater with separator grate discharges material from machine.

Adjustable rear axle lets you select the correct wheel spacing.

Two rotors are at the heart of the Sperry New Holland TR70 combine. Each rotor was 17 in. diameter, 88 in. long. Rotors revolved in opposite directions to split the incoming crop into two streams.

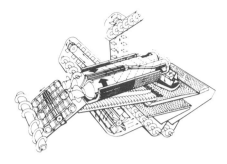

Upper Left: The TR70 feeder that delivered material to the underside of the contra-rotating axial rotors had a stone trap across the elevator floor. The elevating feeder was 40 in. wide; Left: The feed auger flights on the front of the rotors divided and spread the inflowing crop material ready to be threshed by two sets of rasp bars located 180° apart. Crop passed over the adjustable concaves at least three times; Above: Crop leaving the threshing zone was given a final separation through the separator grates in a centrifugal force field as high as several hundred "g's".

Sperry New Holland's TR70

Sperry New Holland's TR70 TWIN ROTOR axial combine released in June 1975, culminated a 10 year effort to develop their own combine in North America. By turning to centrifugal separation, the designers sought higher throughput in a smaller machine envelope. At the same time, lower grain damage and improved separation performance compared with the conventional threshing cylinder was achieved (SaijPaul et al., 1977). The US corn belt market was considered the prime target for machine sales and the machine's performance was accordingly highest in corn. (Rowland-Hill 1975). The export of TR70's began in 1978.

The main features of the TR70, for which Sperry New Holland built a new factory at Lexington, Nebraska, were:

- Two counter-rotating axial rotors for threshing and separation.
- Walkerless operation with low separating losses from the two rotors and discharge beater combination.
- Compact rotors (17 in. diameter, side by side) so that body width was narrow (40 in. feeder width), but with the capacity of a 50 in.-plus conventional cylinder/walkers design.
- Larger thresher clearances than conventional cylinder, since the action depended more on centrifugal force than on rubbing.
- Lower grain damage and lower seed losses.
- Crop spiralled over rotors to pass over concaves three times instead of just once as on a conventional cylinder-concave. The crop may pass over the rotor as many as nine times during the approximately three seconds it takes to pass through the machine. This compares with the 9 seconds crop dwell time through a conventional cylinder and walkers.
- Compact machine envelope leaves room for a large grain bin. (145 Bu with optional extensions for 190 Bu capacity).

International Harvester's
Three Axial Flow Combine Models

International Harvester's Models 1440, 1460 and 1480 combines represent 15 years of development. One million hours of engineering and development work were expended on this series of machines. The series featured a single axial flow rotor, stationary rotor housing with "forwarding vanes" and separating grates, to replace the cylinder, rear beater and straw walkers of the conventional combine design. The models 1440 and 1460 have a 24 in. diameter, 9 ft long rotor, while the model 1480 has a 30 in. diameter rotor of the same length. The same IH 436 CID engine was used in all three models, but at different power ratings. Respective engine power and grain bin sizes are 1440: 135 HP, 145 Bu bin; model 1460: 170 HP, 180 Bu bin; and Model 1480: 190 HP, 208 Bu bin size.

Cross-sectional view of the IH Axial-Flow combine.

Upper Left: The IH Model 1460 Axial-Flow combine in wheat. The gathering and feeding systems are conventional. A stone retarding lower conveyor drum was available as optional equipment; Above: The sides of the IH Axial-Flow are comparatively uncluttered so that the thresher can be cleaned out or the concaves removed through an access door on the left side of the machine.

Conclusion: What are the Advantages and Drawbacks of the New Generation of Rotary Combines?

There may be times where the gathering front may be the constraint on the capacity of a combine, especially in light crops, and it may prove impossible to load an axial flow separator to its full capacity. Other drawbacks of this new generation of combines:

- Straw breakup is generally much more severe. This may be a drawback where the straw is valued for baling.
- Manufacturers have acknowledged that the first generation axial designs were not yet ready in 1977 for the rice crop in North America or for certain European long-strawed crop conditions.
- Under some tough, damp-straw conditions, straw-wrapping and relative power consumption may be higher with the axial designs.

Features have emerged in the axial designs that have been proven in the market place:

- More compact design.
- Larger grain bin for given machine size.
- Reduced number of working parts, resulting in lower manufacturing costs in mass production. IH for example, cited their 1460 design as having 17 percent higher capacity at only 10 percent higher cost than their conventional 915 model.

- Improved serviceability and accessibility making the machines easier to maintain and clean out. Removal of crop residues at the end of the season become easier. Time taken to clean out an axial design may be only half that of a conventional design.
- The centrifugal separator of the axials was completely self-emptying and essentially maintenance-free. The high "g" force fields and greater separation surfaces provide still more development potential.
- The air flow created by the rotors caused a draft throughout the machine reducing the dust nuisance at the feeder in dry harvesting conditions.
- Under average conditions, a cleaner grain sample can be obtained with a considerable reduction in grain damage (Moss 1977). This should represent a considerable savings in reducing dockage because of unmillable material in loads at the delivery point. In corn harvesting, the lower damage level attainable is particularly significant. US corn harvested with axial combines grades higher, has less fines, and is easier to dry.

The patent record is indicative of much behind the scenes activity—just the tip of the iceberg—indicating that dedicated engineers and farmer-innovators promise as much design change in the future as there has been in the past.

> "The "New Implement" is still unforged that can convert adversity into prosperity. Collectively, however, modern labour-saving appliances do go some distance in this direction."—James Edwards, J. Royal Agricultural Society, England, 1893.

24

The Grain Harvesters of the Future

Where Do We Go From Here?

The next 200 years may be safer to predict than the next five years, yet certain trends can be seen in the fabric of progress woven into this book and from statistical pointers from within the grain harvester industry.

Corporate Consolidation

The four top combine manufacturers in North America in 1977 were full-line multinational corporations that accounted for 90 percent of the combines sold in North America. The balance of the market was scattered among a half dozen or more other producers and importers.

Prior to WW1 there were 160 manufacturers of grain harvesting equipment. In the years ahead there may be a further reduction in the number of models available and in the number of combine manufacturers.

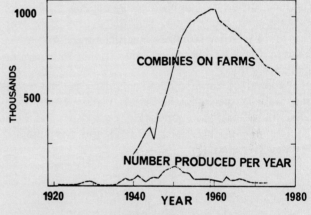

North American combine sales and number of combines on farms since 1920.

Left and Below: The general trend in combine sales has shadowed tractor sales, but on a lower scale (Kulhavy 1977).

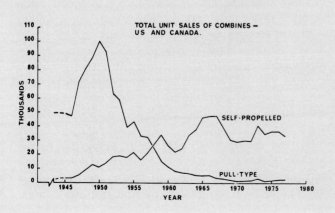

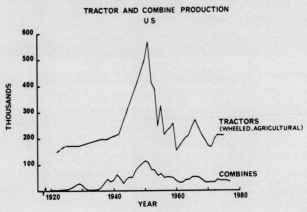

257

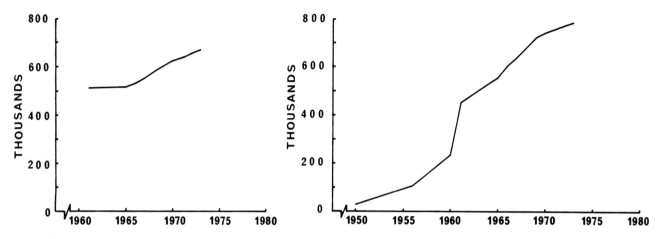

Left: Number of combines (combined harvester-threshers) in use in the USSR 1961-1973; Right: Number of combines in use in Europe, including Eastern Europe, 1950-1973.

Western Europe has tended to follow North America, with one notable exception: the Claas Company was able to capitalize on an early start in the European combine business, establishing a striking lead over all European competition, and for a brief time was the world's largest combine manufacturer. And now, Eastern bloc manufacturers are also a force to be reckoned with.

The total of three million combines in the world has probably peaked. (Fieffer & Fieffer 1969). The US on-farm combine population peaked at just over the one million mark around 1959 and has steadily declined ever since.

Tomorrow's combine will be larger, more reliable, more efficient – and much more expensive. An International Harvester concept of the combine of the future shows a four-wheel drive unit equipped with a 24 row, 30-foot corn head capable of harvesting 160 acres per 11 hour day at 80 percent field efficiency. The clean grain is sent to a separate, but integrated SP robot transport with 475 bushel bin. The unit steers automatically (Baumheckel 1976).

Fewer but Larger

The North American combine market has tended to shadow the tractor market, though on a reduced scale. There were two significant peaks in total US and Canadian combine sales. The first (the all time high) was 105,000 shipments in 1949, mostly pull-types. The second peak came in 1966 at around 53,000 units, about 90 percent self-propelled. The trend in tractors and combines is towards fewer but larger machines. At the present rate of escalating costs of these complex machines, the time may not be too distant when a combine may cost $200,000 on the showroom floor—if there is space inside the show-room for a 300 HP 50 ft cut machine! By comparison, the value of *all* the equipment on the average US farm in 1850 totalled just $115.

Machine Productivity

Today there is but one universal grain harvesting machine in North America—the combine. The modern combine, a field-going factory assembled from over 35,000 components, is a marvel of engineering and production technology. With this one machine and a few minor attachments over 100 different species of seeds from the almost microscopic to broad beans (a 5000 fold range) can be harvested. The combine is expensive, and at 200 average hours of use a year it is "underutilized". But financial logic dictates this insurance against crop loss. If farmers have a choice, they will purchase a more productive (and more expensive) machine so that they can bring their crops in at the best time in the season. Harvest delays are costly.

Yet the combine will only be as productive as the weakest link in the machine, including the operator and his controls. Emphasis in future combine design will continue to be placed on the operator's area,

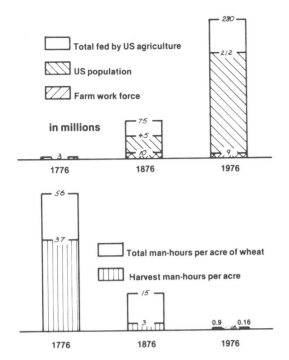

Total fed by US agriculture

US population

Farm work force

in millions

1776 — 3
1876 — 75, 45, 10
1976 — 280, 212, 9

56
37
Total man-hours per acre of wheat

Harvest man-hours per acre

15
3
0.9   0.16

1776   1876   1976

Man and machine productivity in cereal production over the Bicentennial period 1776-1976 (Quick 1976). If the US had to harvest its crops using the hand methods of the 1830's, every American man, woman and child would be needed in the fields every day during the harvest season.

No where does a seed look larger than on the ground after the combine has been through (J.A. Heir, Aas, Norway).

for operator's comfort and convenience remain major concerns. More functions are being monitored electronically because of the operator's isolation in the enclosed cab and the greater complexity of the machine itself. Both functional and process monitoring will become more sophisticated. Crop moisture and grain loss monitors that can signal the percentage loss over the back of the machine and automatically control combine functions according to crop conditions are being tested. (Van Loo 1977; Reed 1977). Improvements such as these enable the operator to reap the machine's full potential.

A driverless combine has been field tested by the Iseki Company in Japan. The experimental machine cut out small rice fields unattended, completing all functions, including automatic bagging of the grain. (Kanetoh 1977).

Dealers have been an integral and inseparable component of the grain harvesting equipment industry. The whole concept of dealerships and of extended credit for equipment purchases, now so widespread in manufacturing industries, was developed by the harvesting machinery business. The need for ready availability of spares at critical times has been recognized and most manufacturers maintain sophisticated parts inventory and mobile service to get parts to the customer quickly. Some companies have mobile parts vans that follow the custom cutters' route Northwards to backup their dealers in the "heartland" States. "Opportunity costs" of a broken-down combine at harvest time can run as high as $110 an hour for custom operators.

Interchangeability of parts for large equipment distributed by manufacturers and their agents was instituted by mower and reaper manufacturers half a century before Henry Ford brought standardization to the automotive industry. Even the production line concept was developed around a grain handling process in Philadelphia. Oliver Evan's patented automatic grain mill became operational in 1784. (Giedion

The 1977 SED grain loss monitor sensed grain losses over walkers and sieves. The piezo-electric crystal detectors and associated electronics were tuned to distinguish the "signature" of certain grains from m.o.g. SED Systems Ltd., Saskatoon, Saskatchewan.

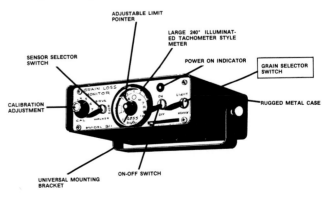

1969).Farmer-innovators have made and will continue to make major contributions to the industry and some of the leaders and innovators of modern industry were themselves farmers—McCormick, Harris, Massey, Baldwin...

Crop Processing Technology

Machine productivity includes not only size, but the ability to recover crops efficiently and keep on going at harvest time with minimum downtime. As more powerful engines are employed (with the swing almost exclusively to diesels) and improvements are made in threshing, separation and cleaning technology, there will be greater emphasis on crop gathering performance. In many crop conditions the gathering head is the capacity-limiting component of the combine.

Crop losses and grain damage will be reduced by adapting and controlling machines more precisely to varying terrain and crop conditions under the higher speeds likely to be encountered. There will be larger grain tanks and faster unloading rates with greater "up time", increasing machine availability (Douglass 1976).

The increasing resistance of insects to chemical control measures may have repercussions, driving

Shortline manufacturers form an important link in promoting farmer inventions, often perfecting designs that major farm equipment manufacturers adopt later. Hardy combines, produced at College City, California were produced as specialized rice combines. They featured a double-knife sickle, draper platform, spike tooth cylinder, and full tracks (courtesy E. Hengen).

The Iseki driverless rice combine has automatic control functions on steering, cutting height, stalk feeding, bagging off, forward motion and crop runout which stops the machine. Iseki Agricultural Machinery Mfg. Co. Ltd., Tokyo, Japan.

Above: More on-the-farm processing of products would help expand the farmer's economic base (Griffin 1973, Deere and Company); Below: Integrated and once-over operations. Binding, discing and fertilizer spreading in one pass, 1920.

engineers back to the drawing boards. Self-cleanout or ease-of-cleaning features to rid combines of insect-harboring crop residues have become an important design requirement in sub-tropical regions such as Australia.

The harvesting and use of the whole plant will receive more attention from designers as ways are found to capture some of the enormous fuel potential in crop residues in an energy-generating agriculture. There will be a tendency for more on-farm processing

of products that will see farmers expanding their financial base by adding value to their crops or even completing certain food processing routines on the farm.

Future Crops

Wheat is the most widely-grown grain crop. The golden grain is being harvested somewhere in the world every month of the year. For many however, rice means survival and rice will probably continue to

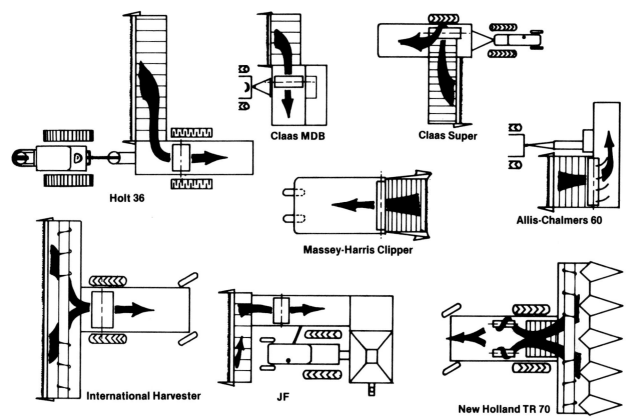

Combine crop processing flow paths, 1900-1977.

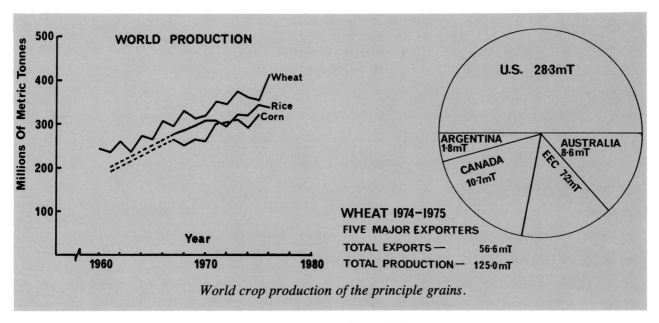

World crop production of the principle grains.

hold its place in a world economy where increasing reliance is placed on foods of vegetable origin. There have been steady increases in production of most of the grain crops in the 1970's. This trend must continue just to maintain the world's present nutritional status. It is almost certain designers and innovators will face challenges to harvest a greater diversity of crops as new sources of food, fiber and energy are sought from the world of seeds. Jojoba, guar, guayule—now just exotic names—will be commonplace in the future.

Contradictions

"Wild acclaim" greeted this combine engineer's prognostication:

"This machine (the combine) is to evade all trouble from weeds and other green material, as well as eliminating the mechanical complication of the reciprocating knife and its driving mechanism, by doing its cutting with an electrically heated wire which will burn through the dry stalks but not the green material, leaving only dry stuff to enter the threshing mechanism. The difficulties now encountered in navigating combines in wet fields are to be eliminated by suspending the machine from a gas bag above instead of wheels beneath.

Above: Every period of history has its contradictions. While an Oklahoma custom cutter's machine carved out 100 acres a day, this Calabrian peasant still laboriously threshed grain with a stick (Rasmussen 1969); Below: Intermediate technology. A bicycle-operated winnowing fan reduces the chore of winnowing and ensures an even breeze for these West Malaysian farmers (Loofs 1976).

Above: Possibly more than half the world's population survives on grain harvested by hand in a subsistence agriculture. This photo was shot in Ghana (courtesy W.F. Buchele).

Above: These Philippino farmers have erected a sturdy bamboo platform to make use of gravity and wind to aid in threshing and winnowing their rice (courtesy of J.A. McMennamy, IRRI, Manila); Below: Harvesting machines produced by local industry in the Philippines provide an effective and vital contribution to the country's economy.

The specter of famine has been with us throughout history, as attested by this copy of a limestone relief at Saqqara showing a group of Bedouin women suffering from the dire effects of severe famine, Late Dynasty V, ca 2350.

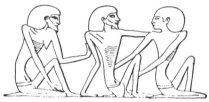

Floating jauntily over fences and streams all transport difficulties will be solved. Propulsion is by a fan in the rear. The design embodies an atomic cylinder in which the enormous and irresistible power of the atom is invoked. Removal of threshed grain is to be by airplane—The customary conservatism of engineers was cast aside and the ideal design greeted with wild acclaim."

—so reported *Agricultural Engineering* in 1928!

Conclusion

Rosy pictures of seemingly inevitable progress must be tempered with the reminder that possibly more than half the people on overcrowded planet Earth are living on grain still harvested by hand in a subsistence level of agriculture. (Thompson 1965).

To expect Western technology to resolve all the farmers' problems in other cultures has been shown by bitter experiences to be folly. Nevertheless, there are believed to be significant lessons to be learned from the patterns of harvest mechanization that have been outlined here.

The adequacy of our food supply is the number one global priority. As long as "seedtime and harvest shall remain" there will be the grain harvesters.

References and Selected Bibliography

ADAMS, E. A. 1976. Anecdotes from 43 years in the farm machinery business. Personal communication.

AGRICOLA, Georgius. 1556. De re metallica. Transl. from 1556 Latin Ed. London, 1912.

ALDRED, Cyril. 1951. New Kingdom art in ancient Egypt during the 18th Dynasty, 1570 to 1320 BC. London.

ALLEN, J. T. 1886. Digest of agricultural implements patented in the United States from 1789 to July 1881. US Patent Office. J. C. Von Arx., New York.

AMERICAN SOCIETY OF AGRICULTURAL ENGINEERS. 1977. Terminology for combines and grain harvesting. ASAE Std. S343. Agricultural Engineers Yearbook. ASAE, St. Joseph, Michigan.

AMERICAN THRESHERMAN. 1927. Queer combines for strange purposes. American Thresherman 30(12):8.

AMERICAN THRESHERMAN. 1925. Threshers first made in Scotland. American Threshman 27(10):9.

ANDREWS, G. H. 1853. Modern husbandry; a practical and scientific treatise on agriculture. Nathaniel Cooke Publ., London.

ARDREY, R. L. 1973. American agricultural implements; a review of invention and development in the agricultural implement industry of the United States. Parts 1 and 2. 1894 ed. reprinted. Arno Press, 330 Madison Ave., New York, New York.

ARNOLD, J. H. and R. R. SPAFFORD. 1919. Farm practice in growing wheat. Agric. Yrbk. 123-152.

ARNOLD, R. E. 1966. Trends in combine development. Jour. Min. Agr. 73(3):118-121.

ARNOTT, P. D. 1972. An introduction to the Roman World. Sphere Books Limited. MacMillan.

ARTHUR, E. and D. WITNEY. 1972. The barn. A vanishing landmark in North America. A & W Visual Library, Toronto, Canada.

AVITSUR, S. 1967. Implements for harvesting and similar purposes used in the traditional agriculture of Eretz Israel. Israel.

BAILEY, W. 1772. The advancement of arts, manufactures and commerce, or a description of the useful machines and models contained in the repository of the Society for the Encouragement of Arts, Manufactures and Commerce. London.

BAINER, R. 1974. Science and technology in western agriculture. Symposium on Agriculture in the Development of the Far West. Davis, California, June.

BAKER, E. J. 1955. Pneumatic combines. Farm Implement News 70(10):26, 71, 72, 73.

BAKER, E. J. 1961. Combine reflections. Implement and Tractor 76(4):35,36.

BALDWIN, Curtis C. 1930. Inside facts. Curtis Harvesters Inc. Ottawa, Kansas.

BAUMANN, Hans. 1960. The world of the Pharaohs. Pantheon Books Inc., New York, New York.

BERESFORD, H. and E. N. HUMPHREY. 1930. Bulk handling grain from the hillside type combine. Idaho Agr. Stn. Bull. 175, 19pp.

BERG, Gösta. 1976. The introduction of the winnowing-machine in Europe in the 18th Century. Tools and Tillage 3(1):25-46.

BETTMANN, Otto L. 1974. The good old days—they were terrible. Random House, New York, New York.

BICHEL, D. C. and G. K. CORNISH. 1974. The sidehill 6600 combine—a new concept for the corn belt. ASAE Paper No. 74-1599. ASAE, St. Joseph, Michigan.

BICHEL, D. C., E. J. HENGEN and R. E. MOTT. 1975. New concept header for combine harvester. ASAE Paper No. 75-1534. ASAE, St. Joseph, Michigan.

BICHEL, D. C., E. J. HENGEN and G. W. ROHWEDER. 1976. Row-crop head for combine harvesters. ASAE Paper No. QSC 76-14.

BLAUSER, I. P. 1927. Combine proves itself in soybeans. Power Farming 36:6-7.

BLAUSER, I. P. 1927. Field tests on the combine in Illinois. Agricultural Engineering 8(4):88-89.

BORDAZ, Jacques. 1965. The threshing sledge. Natural History 74(4):26-29.

BOSWELL, V. R. 1961. What seeds are and do: an introduction. 1961 USDA Yearbook of Agriculture: 1-10. Washington, D.C.

BRAIDWOOD, R. J. 1975. Prehistoric men. Scott, Foresmen & Co., Glenview, Illinois.

BRENNER, W. G. 1969. Geschichte des Mähdreschers (Direktverfahren). Die Geschichte der Landtechnik im 20 Jahrhundert. DLG-Verlag GmbH., Frankfurt, Germany.

BRODELL, A. P. 1942. Machine and hand methods in crop production. F.M.18. USDA Bureau of Agric. Econ. and Agric. Mktg. Serv. Washington, D. C.

BROWN, J. J. 1967. Ideas in exile. A history of Canadian invention. McClelland & Stewart Ltd., Toronto, Canada.

BROWN, W. T. and G. H. VASEY. 1967. Wheat harvester survey. Agr. Eng. Dept., Univ. Melbourne, Victoria, Australia.

BRUMFIELD, Kirby. 1968. This was wheat farming. A pictorial history of the farms and farmers of the northwest who grow the nation's bread. Bonanza Books, New York.

BUCHELE, W. F. and W. F. LALOR. 1963. Design and testing of a threshing cone. TRANSACTIONS of the ASAE 6(1):73-76.

BUESCHER, W. M. 1970. Farm machinery and the soybean. Soybean Digest 30(11):84-86.

BULL, J. W. 1884. Early experiences of life in South Australia and an extended colonial history. 2nd Ed. E.S. Wigg & Son. Adelaide, South Australia.

BUTTERWORTH, B. 1888. The growth of industrial art. Edition reproduced by G. B. Gunlogson. Chicago, 1935.

BYERLY, A. E. 1928. Dr. Patrick Bell — inventor of the reaper. The Bystander, Guelph, Ontario, Canada.

CAESAR, Julius. 1967. The conquest of Gaul. Transl. by S. A. Handford. Penguin Books, England.

CAIRNS, Trevor. 1974. People become civilized. The Cambridge Introduction to History. Lerner Publications Company.

CAMPBELL, J. K. 1975. Development and manufacture of a thresher for developing countries of Southeast Asia. ASAE Paper No. 75-1539. ASAE, St. Joseph, Michigan.

CAPPIE, K. M. 1972. The Massey-Ferguson harvest brigade. Paper No. 72-313. Presented at CSAE Annual Meeting, June.

CARROLL, T. 1948. Basic requirements in the design and development of the self-propelled combine. AGRICULTURAL ENGINEERING 29(3):101-105.

CASE, J. I. 1942. Case wood engravings: Reminders of a past art. Form A-89542-F. J. I. Case Co., Racine, Wisconsin.

CASE, J. I. 1949. Jerome Increase Case, The man and his works. Form A-84249-J. J. I. Case Co., Racine, Wisconsin.

CASE CENTENNIAL. 1942. Case completes century of service to agriculture. Farm Implement News 63(1):22-27, 54, 56.

CASE CENTENNIAL. 1942. Case reaches century mark. Implement & Tractor. 58(1):12-14, 30-34.

CASPERS, L. 1969. Systematik der Dreschorgane. (Review of threshing mechanisms). Grundlagen der Landtechnik 19(1):9-17.

CATERPILLAR TRACTOR CO. 1954. Fifty years on tracks. Caterpillar Tractor Co., Peoria, Illinois.

CHAMPOLLION, F. M. 1878. Egypte Ancienne. Paris.

CHANCELLOR, W. J. 1965. An experiment on force and energy requirements for cutting padi stalks. The Malaysian Agric. Jour. 45(2):200-203.

CHILDE, V. G. 1951. The balanced sickle. In Aspects of Archeology in Britain and Beyond, Ed. by W. F. Grimes. London, England.

CHURCH, L. 1949. Partial history of the development of grain threshing implements and machines. Slightly rev. by Dieffenbach, E. M. USDA-AES. Agr. Eng. Inf. Ser. No. 73.

CHURCHILL, A. 1841. Churchill and Danford's harvesting machine. The Union Agriculturalist and Western Prairie Farmer 1(8):61, 1(6):1.

CLARKE PUBLISHING. 1929. Combine year book—United States

and Canada. Clarke Publishing Co. Madison, Wisconsin.

CLYMER, E. 1969. The second greatest invention. Search for the first farmers. Holt, Rinehart & Winston of Canada Ltd., Toronto, Canada.

COLLINS, E. J. T. 1969. Sickle to combine. A review of harvest techniques from 1800 to the present day. Reading, England.

COLLINS, E. J. T. 1972. The diffusion of the threshing machine in Britain, 1790-1800. Tools and Tillage 2(1):16-33.

COMMON, R.F.J. 1907. The history of the invention of the reaping machine (with special reference to the inventor, John Common). London, England.

CONNER, G. F. 1906. The science of threshing. The Thresherman's Review Company, Michigan.

CRAWFORD, O. G. S. 1935. A primitive threshing-machine. Antiquity 9(35):335-339.

CRAWFORD, A. and J. L. OLD. 1972. The new M. F. knotter. TRANSACTIONS of the ASAE 15(4):644-649, 652.

CÜPPERS, H. 1964. Gallo-römische mähmaschine (Gallo-Roman reaper). Trierer Zeitschrift 27:151-152.

DAVIDSON, J. B. 1931. Agricultural machinery. John Wiley & Sons, New York, New York.

DAVIES, N. D. 1917. The tomb of Nakht at Thebes. Vol. I of series. Metropolitan Museum of Art, New York, New York.

DAVIES, N. M. 1936. Ancient Egyptian paintings. In 3 Vols. Chicago, Illinois.

DEERING HARVESTER COMPANY. 1900. The official retrospective exhibit of the development of harvesting machinery, Paris. Chicago, Illinois.

DENHAM, H. J. 1934. Some trends in mechanized farming. 1. The grain harvest. Scottish Jour. Agr. 8(4).

DENISON, M. 1949. Harvest triumphant. Dodd, Mead & Company, New York, New York.

DIDEROT, D. 1959. A Diderot pictorial encyclopedia of trades and industry. Dover Publications Inc., New York, New York.

DODDS, M. E. 1967. Grain harvesting with the windrower. ASAE Paper No. 67-151. ASAE, St. Joseph, Michigan.

DONATH, E. J. and J. McGARRITY. 1976. Wheat in Australia. Oxford University Press.

DOUGLAS, D. L. 1975. How big can tractors get? Implement & Tractor 90(5):12, 40.

DUNCAN, J. S. 1971. Not a one-way street. Clark, Irwin & Co. Ltd., Toronto, Canada.

DUNLAP, M. L. 1863. Agricultural Machinery.

DURAN, F. D. 1971. Book of the gods and rites and the ancient calendar. Univ. of Oklahoma Press.

EVERETT, C. E. 1946. Self propelled combines to open grainfields. Agricultural Engineering 27(6):261.

FARM IMPLEMENT. 1884. Cyrus Hall McCormick. A sketch of the origin and development of a great invention. The Farm Implement 4(4):1,2.

FARM IMPLEMENT NEWS. 1888. Twine knots of sheaf binders. The Farm Implement News 9(12):20-22.

FARM IMPLEMENT NEWS. 1928. The development of the combine. Farm Implement News 49(49):30, 31.

FARM IMPLEMENT NEWS. 1930. Fleming-Hall baby combine. Farm Implement News 51( ):38.39.

FARM MECHANIZATION. 1951. Mowing and reaping. Farm Mechanization 3(25):1975-177, May.

FELDHAUS, F. M. 1914. Die Technik der vorzeit, der geschichtlichen zeit und der naturvölker. Leipzig & Berlin, Germany.

FIEFFER, P. and R. FIEFFER. 1969. Technical fundamentals. The combine-harvester and its operating conditions. Edition Leipzig, E. Germany.

FINLEY, M. I. 1965. Technical innovation and economic progress in the ancient world. The Economic History Review 18(1):29-45.

FISCHER, G. 1910. Die Entwicklung des landwirtschaftlicken Maschinwesens in Deutschland. Festschrift zum 25 jahrigen Bestehen der Deutschen Landwirtschafts-Gesellschaft. Heft 177. Berlin, Paper 77-1050. Pres. ASAE Annual Mtg. Raleigh, North Carolina.

FISCHER, G. 1928. Landmaschinenkunde. Stuttgart, Germany.

FORBES, R. J. 1958. Studies in ancient technology. Vol. VI. E.J. Brill.

FORSYTH, R. 1805. Principles and practice of agriculture. England.

FOUSS, E. P. 1958. Le "vallus" ou la moissoneuse des Trèvires. Le Pays Gaumais 19:125-136.

FRANKE, O. 1913. Këng Tschi Tu, Ackerbau und Seidengewinnung in China. Ein Kaiserliches Lehr-und-Mahn Buch. Hamburg, Germany.

FRICK COMPANY. 1952. A history of Frick Company. Waynesboro, Pennsylvania.

FRUSHOUR, G. V. and C. C. JOHNSON. 1974. Development of a new concept edible bean combine. ASAE Paper No. 74-1578. ASAE, St. Joseph, Michigan.

FUSSELL, G. E. 1952. The farmer's tools — AD 1500-1900. The history of British farm tools, implements and machinery. Andrew Melrose, London, England.

FUSSELL, G. E. 1966. Farming technique from prehistoric to modern times. Pergamon Press Ltd. Oxford, England.

FUSSELL, G. E. 1969. Science and practice in eighteenth-century British agriculture. Agricultural History 43(1):7-18.

GHOSH, B. N. 1970. The performance of a bicycle operated winnower-grader. Jour. Agricultural Engineering Research 15(3):274-282.

GIEDION, Siegfried. 1969. Mechanization takes command. A contribution to anonymous history. W. W. Norton & Co., Inc. New York, New York.

GILES, G. W. The reorientation of agricultural mechanization for the developing countries. Agricultural Machanization in Asia 6(2).

GRAY, A. 1814. Explanation of the engravings of the most important implements of husbandry used in Scotland.

GRAY, L. C. 1941. History of agriculture in the southern United States to 1860. Vols. 1 and 2.

GRAY, R. B. 1975. The agricultural tractor: 1855-1950. Revised Ed. ASAE, St. Joseph, Michigan.

GREENO, F. L. 1912. Obed Hussey, who, of all inventors, made bread cheap. Rochester, New York.

GRIFFIN, G. A. 1973. Fundamentals of machine operation — combine harvesting. John Deere Service Publications, Moline, Illinois.

GROVE, Noel. 1972. North with the wheat cutters. National Geographic Magazine 142(2):194-217.

GUAMAN POMA de Ayala, Felipe. 1966. La nueva cronica y buen gobierno. L. Bustios Galvez, Lima.

HANDLEY, J. E. 1953. Scottish farming in the eighteenth century. Faber and Faber Ltd., London, England.

HANSON, H. 1966. History of swathing and swath threshing. The Western Producer 43(28):21.

HARCUS, W. 1876. South Australia—its history, resources and productions. London, England.

HARRIS, B. S. 1928. New engineering developments in combines. Agricultural Engineering 9(1):13.

HEITSHU, D. C. 1956. A marvel of engineering—the self-propelled hillside combine. Agricultural Engineering 37(3):182, 183, 187.

HENDRICK, J. 1928. Patrick Bell and the centenary of the reaping machine. Trans. Highland and Agr. Soc. of Scotland. 5th Ser. 60:51-69, Edinburgh, Scotland.

HERBSTHOFER, F. J. 1974. Wo stehen wir im Mähdrescherbau und wie geht es weiter? Grundlagen der Landtechnik 24(3):94-102.

HIGGINS, F. H. 1930. The Moore-Hascall harvester centennial approaches. Michigan History Magazine 14(7):415-437.

HIGGINS, F. H. 1930. A woman dreamed the combine harvester. Pacific Rural Press Aug. 9:132, 133.

HIGGINS, F. H. 1937. The reapers: yesterday and today. Pacific Rural Press 133:233, 277.

HIGGINS, F. H. 1949. The combine parade. Farm Quarterly (4): 42, 47, 94, 96, 98, 100. Summer.

HIGGINS, F. H. 1952. California invented the "Pusher" combine. California Farmer 196(1):28, 29.

HIGGINS, F. H. 1958. John M. Horner and the development of the combined harvester. Agricultural History 32(1):14-24.

HIGGINS, F. H. 1967. The farm machine in the Sacramento Valley. Mimeo Rpt. Colusi County Historical Society. Feb.

HOBSBAWM, E. J. and J. RUDÉ. 1969. Captain Swing. London, England.

HODGES, H. 1970. Technology in the ancient world. Allen Lane. The Penguin Press, London, England.

HOLT, P. E. 1925. The development of the track-type tractor. Agricultural Engineering 6(4):76-79.

HOMMEL, R. P. 1937. China at work. John Day, New York, New York.

HOMO, Juan. 1918. Agricultural implements and machinery in Australia and New Zealand. Special Agent Series, No. 166. US Dept. of Commerce, Bureau of Foreign and Domestic Commerce. Government Printing Office, Washington, D.C.

HOPFEN, H. J. 1928. The harvester-thresher and its adaptation to European conditions. International Review of Agriculture 19:811-814.

HOPFEN, H. J. 1960. Farm implements for arid and tropical regions. FAO Development Paper No. 67. Rome, Italy.

HORINE, M. C. 1924. Farming by machine. In: A popular history of American invention 2:246-309. Charles Scribner's Sons, New York, New York.

HORN, S. H. 1960. Seventh-Day Adventist Bible Dictionary. Review and Herald Publ. Assoc., Washington, D.C.

HUDSON, D. & K. W. LUCKHURST. 1954. The Royal Society of Arts 1754-1954. John Murray, London, England.

HURLBUT, L. W. 1955. More efficient corn harvesting. Agricultural Engineering 36(12):791, 792.

HUTCHINSON, W. T. 1930. Cyrus Hall McCormick, seed-time, 1809-1856. The Century Co., New York.

HUTCHINSON, W. T. 1935. Cyrus Hall McCormick. Harvest, 1856-1884. D. Appleton-Century Co., New York.

IMPLEMENT & Machinery Review. 1928. A new type of harvester-thrasher. The Implement & Machinery Review 53(636):1305.

IMPLEMENT & Machinery Review. 1929. An English owned harvester-thrasher at work. Details of its performance under ordinary working conditions. The Implement & Machinery Review 55:496-502, 592-594.

IMPLEMENT & Machinery Review. 1929. Origin of the combined harvester-thresher. The Implement & Machinery Review. 54(3):1163-1164.

IMPLEMENT & Tractor. 1930. Kansas again is proving ground. Implement & Tractor 45(7):

INGLETT, G. 1974. Wheat: production and utilization. AVI. Publ. Co. Inc., Connecticut.

INGPEN, R. 1972. Pioneer settlement in Australia. Rigby Limited, Adelaide, Australia.

INTERNATIONAL HARVESTER CO. 1971. McCormick reaper centennial source material. International Harvester Co., Chicago, Illinois.

ISSELSTEIN, R. and H. SCHWARTZ. 1965. Die geschichte des Mähdreschers. Feld und Wald. Essen, Germany.

JACKSON, R. W. 1950. History of corn harvesting machinery. Unpubl. M.S. Thesis. Iowa State College, Ames, Iowa.

JACKSON, R. T. 1976. The soybean producer and the world market situation. ASAE Paper No. 76-1551. ASAE, St. Joseph, Michigan.

JENNINGS, D. C. 1968. Days of steam and glory. Northern Plains Press, Aberdeen, South Dakota.

JOHNSON. 1895. Reaping and mowing machines. Johnson's Universal Cyclopaedia 7:17-19.

JOHNSON, P. C. 1976. Farm inventions in the making of America. Wallace-Homestead Book Company, Des Moines, Iowa.

JOHNSON, W. H. and B. J. LAMP. 1966. Principles, equipment and systems for corn harvesting. ACA, Ohio.

JOHNSTONE, P. H. 1937. In praise of husbandry. Agricultural History 11(2):80-95.

JOHNSTONE, P. H. 1938. Turnips and romanticism. Agricultural History 12(3):224-255.

JONES, E. R. 1927. Twine knotter history. Wisconsin Magazine of History 12(2):225-227, Dec.

JOUR. D'AGRICULTURE. 1930. Adaptation of the harvester-thrasher to European conditions. Jour. d'Agriculture Pratique 2:293-297.

KANETOH, Y. 1977. Driverless combine harvester. Proc. Int. Grain & Forage Harvesting Conf. Ames, Iowa. September.

KASPAREK, M. V. 1960. Die "getreideworfmühle"—eine oberpfälzer erfindung? In Die Oberpfalz. Regensberg.

KENDALL, F. J. 1972. H. V. McKay—A pioneer industrialist. The Victorian Historical Magazine 43(3):885-895.

KENNEDY, J. C. G. 1862. Preliminary report on the eighth census, 1860. Gov't. Printing Office, Washington, D.C.

KERBER, D. R. and O. W. JOHNSON. 1977. Advances in grain header developments. ASAE Paper No. 77-1547. ASAE. St. Joseph, Michigan 49085.

KERRIDGE, E. 1973. The farmers of old England. Rowan and Littlefield Inc., Totowa, New Jersey.

KHAN, A. U. 1970. Mechanizing the tropical farm. Agricultural Engineering 51(11):640-642.

KHAN, A. U., Bart DUFF, D. O. KUETHER and J. A. McMENNAMY. 1975. Rice machinery development and mechanization research. IRRI Report No. 21. Manilla, Philippines.

KLINE, C. K., D. A. G. GREEN, R. L. DONAHUE, and B. A. STOUT. 1969. Agricultural mechanization in equatorial Africa. Inst. of Int. Agr., College of Agr. & Nat. Res., Res. Rpt. No. 6, Michigan State Univ., Dec.

KOPP, K. A. 1977. Electronic technology—a changing force in harvest machinery monitor and control systems. Proc. Int. Grain & Forage Harvesting Conf. Ames, Iowa. September.

KULHAVY, J. T. 1977. Food and Energy—the tractor engineer's role. Paper 77-1050. Pres. ASAE Annual Mfg. Raleigh, North Carolina. ASAE. June.

LAMP, B. J., W. H. JOHNSON, K. A. HARKNESS and P. E. SMITH. 1962. Soybean harvesting — approaches to improved harvesting efficiencies. Ohio Agr. Exp. Stn. Res. Bull. 899.

LANNING, E. P. 1967. Peru before the Incas. Prentice-Hall, Englewood Cliffs, New Jersey.

LASTEYRIE, C. 1820. Collection de machines, d'instrumens, etc. Paris, France.

LAW, T. W. M. 1975. How to choose a combine harvester. Power Farming 54(12):46-49.

LEE, N. E. 1960. Harvests and harvesting through the ages. Cambridge Univ. Press.

LEFEBVRE, G. 1924. Le tombeau de Petosiris. III service des antiqueties de L'Egypte.

LEONARD, J. N. 1973. The first farmers. Time-Life Books, New York, New York.

LERCHE, G. 1968. Observations on harvesting with sickles in Iran. Tools and Tillage 1(1):33-49.

LESTER, W. 1811. A history of British implements and machinery applicable to agriculture, with observations on their improvement. Longman et al., London, England.

LEWIS, L. 1941. John S. Wright, prophet of the prairies. Prairie Farmer.

LIU, H.—C. 1963. History of invention of agricultural implements of ancient China. Peiping, China.

LOGAN, C. A. 1931. The development of a corn combine. Agricultural Engineering 12(7):277-278.

LOOFS, H. H. E. 1976. A new winnowing method from Northern West-Malaysia. Tools & Tillage 3(1):20-24, 46.

LOUNSBERRY, D. 1957. Field shellers are not new. Implement and Tractor 72:62-63.

LUBEN, A. 1929. Zur Geschichte und Entwicklung des Mähdreschers. Landmaschinen 9(6):427-430, 445-448.

LUCAS, A. 1948. Ancient Egyptian materials and industries. Edward Arnold & Co.

LÜDTKE, H. 1932. Die arbeit des ersten deutschen Mähdreschers in der Ernte 1932. Die Technik in der Landwirtschaft 14(3):70-72.

MacGREGOR, W. F. 1925. The combine is 100 years old. Farm Mechanics. May.

MacGREGOR, W. F. 1925. The combined harvester-thresher. Agricultural Engineering 6(5):100-103.

MacGREGOR, W. F. 1925. The combined harvester-thresher. Transactions of the ASAE 19:40-47.

MacKENZIE, J. K. 1927. The combined reaper-thresher in western

Canada. New Ser. Pamph. 83, Domin. Can. Dept. Agr.

MacKENZIE, J. K. 1929. The combine in Saskatchewan. Agricultural Engineering 10:57-58.

MacNAUGHT, J. B. 1977. Universal Power Flowtable. Proc. Int. Grain & Forage Harvesting Conf. Ames, Iowa. September.

MAGNUS, Olaus. 1555. Historia de gentibus septentrionalibus—Libr. 22. Rome, Italy.

MANGELSDORF, P. C. 1974. Corn—its origin, evolution and improvement. The Bellknap Press of Harvard University Press, Cambridge. Massachusetts.

MARE, E. de. 1972. London 1851. The year of the great exhibition. The Folio Society, London, England.

MARSH, C. W. 1887. Corn shellers. Farm Implement News 8(7):12-13.

MASSEY-FERGUSON. 1964. No sheaves in the field. Massey-Ferguson Inc., Detroit, Michigan.

MASSEY-FERGUSON. 1972. Heritage 1847-1972. Massey-Ferguson Limited, Toronto, Canada.

MASSEY-HARRIS. 1947. 100 years of progress in farm implements. Massey-Harris Company, Toronto, Canada.

McCALL, M. A. 1926. Some factors to be considered in extending the use of the combine harvester. Transactions of the ASAE 20:83-88.

McCLUNG, N. 1964. Clearing in the west. Thos. Allen & Son.

McKAY, S. S. 1971. An historical review of Australian development of farm machinery. Agricultural Engineering Australia 2(2):7-18, June.

McKIBBEN, E. G. 1929. Harvesting corn with a combine. Agricultural Engineering 10(7):231, 232.

McMANIGAL, J. W. 1974. Farm town. A memoir of the 1930's. Vermont.

McMENNAMY, J. A. and J. S. POLICARPIO. 1977. Development of a portable axial-flow thresher. Proc. Int. Grain & Forage Harvesting Conf. Ames, Iowa. September.

MELLAART, J. 1965. Earliest civilizations of the near east. Thames & Hudson Ltd.

MELLOR, J. H. 1956. From sickle to harvester. Mimeo. address to Agr. Eng. Soc. (Australia). Melbourne. February 1956.

MEYER-LUBKE, W. 1909. Zur Geschichte der Dreschgeräte. Worter und Sachen I., Heidelberg, Germany.

MICHIGAN STATE AGRICULTURAL SOCIETY. 1849. Great harvesting machine invented by Hiram Moore, Esq. Trans. Michigan State Agr. Soc.: 107.

MILLAR, G. H., et al. 1974. Report of the Deere technical delegation to Russia., Sept. 1974. Oral presentation at ASAE Winter Meeting.

MILLER, M. F. 1902. The evolution of reaping machines. Bulletin 103. Government Printing Office, Washington, D.C.

MONTET, P. 1925. Scenes de la vie privee dans les tombeaux Egyptiens de l'Ancien Empire. Strasbourg, France.

MONTET, P. 1958. Everyday life in Egypt in the days of Ramesses the great. Edward Arnold, London, England.

MORPHETT, G. C. 1945. C. B. Fisher — pastoralist, studmaster and sportsman. An epic of pioneering (limited edit.) Adelaide, Australia.

MORSE, F. 1913. Threshing from standing grain. Technical World Magazine, March.

MORTON, J. C. 1856. A cyclopedia of agriculture. In 2 vols. Blackie & Sons, Glasgow, Scotland.

MUMFORD, L. 1963. Technics and civilization. Harcourt, Brace & World Inc. Harbinger Books, New York, New York.

MURRAY, D. A., R. A. DEPAUW, R. L. FRANCIS and K. D. JOHNSON. 1977. Recent development in grain threshing and separating mechanisms. Proc. Int. Grain & Forage Harvesting Conf., Sept. ASAE, St. Joseph, Michigan.

NACHTWEH, A. 1911. Rekonstruktion der ältesten gallischen Mähmaschine. Jour. für Landwirtschaft 59(1):1-8, 367-370.

NEAL, A. E. 1977. The evolution of the flexible floating cutterbar. Proc. Int. Grain & Forage Harvesting Conf. Ames, Iowa. September.

NEEDHAM, J. 1965. Science and civilization in China. Vol. 4. Physics and physical technology. Part II, Mechanical engineering. Cambridge, at the Univ. Press.

NEUFELD, E. P. 1969. A global corporation—a history of the international development of Massey-Ferguson Limited. University of Toronto Press.

NEWMAN, J. E. and J. H. BLACKABY. 1928. Report of trials of the combine-harvester-thresher in Wiltshire, 1928. Inst. Res. Eng., Univ. Oxford Bull. 3.

NICHOLLS & SHEPARD. 1929. As the golden grain grows from the seed. Nicholls & Shepard Company, Battle Creek, Michigan.

NYGERG, C. 1957. Highlights in the development of the combine. Agricultural Engineering 38(7):528, 529, 535.

OLMSTEAD, A. L. 1975. The mechanization of reaping and mowing in American agriculture, 1833-1870. J. Econ. Hist. 35(2):327-352.

PARSONS, J. R. 1893. Who is the inventor of the reaping machine? Farm Implement News: 21-23. July.

PARSONS, J. R., L. MILLER and J. F. STEWARD. 1897. Overlooked pages of reaper history. Chicago, Illinois.

PARTRIDGE, B. and O. BETTMANN. 1946. As we were—family life in America, 1850-1900. Whittlesey House, McGraw-Hill Book Co. Inc.

PARTRIDGE, M. 1969. Early agricultural machinery. Hugh Evelyn Limited, London, England.

PARTRIDGE, M. 1973. Farm tools through the ages. Osprey Publishing Ltd., Reading, England.

PAYNE, R. 1966. The horizon book of ancient Rome. American Heritage Publishing Co. Inc., New York, New York.

PEEL, L. J. 1971. The first hundred years of agricultural development in western Victoria. Proc. Symp. on The Natural History of Western Victoria: 74-84. Horsham, Victoria. Australian Institute of Agricultural Science.

PERKINS, D. H. 1969. Agricultural development in China, 1368-1968. Aldine Publ. Co.

PERKINS, J. A. 1976. Harvest technology and labour supply in Lincolnshire and the East Riding of Yorkshire, 1750-1850. Part one. Tools & Tillage 3(1):47-58, 64.

PETRIE, W. M. F. 1917. Tools and weapons. First Publ. by the British School of Archeology in Egypt, 1917. Reprint 1974 by Aris Phillips Ltd., Wiltshire.

PHILLIPS, W. G. 1956. The agricultural implement industry in Canada. A study of competition. Univ. of Toronto Press.

PICKARD, G. E. and D. F. HOPKINS. 1953. Corn shelling with a combine cylinder. Agricultural Engineering 24(7):461-464.

PICKARD, G. E. and H. P. BATEMAN. 1954. Combining corn. Agricultural Engineering 35(7):500-504.

PICKETT, L. K. 1974. Harvesting edible beans—a challenge for the agricultural equipment industry. ASAE Paper No. 74-1559. ASAE, St. Joseph, Michigan.

PIDGEON, D. 1892. The evolution of agricultural implements. Part 1. Jour. Royal Agricultural Society of England, 3rd Ser. 3(9): 49-70.

PIERCE, J. E. 1964. Life in a Turkish village. Holt, Rhinehart & Winston, New York, New York.

PINCHES, H. E. 1960. Revolution in agriculture. In Power to Produce—1960 Yearbook of Agriculture. USDA.

POOLE, B. W. 1928. Tribute to Scottish inventors of farm equipment. Farm Implement News. 49(11):26, 27.

POSTAN, M. M. 1966. The Cambridge economic history of Europe. Vol. 1. The agrarian life of the middle ages. Cambridge, at the University Press.

POWER FARMING. 1969. Cut-to-order combines. Power Farming 34(10):34, 35.

POWER FARMING. 1975. Price guide to combine harvesters. Power Farming 54(12):50, 51, 53.

PRAIRIE AGRICULTURAL MACHINERY INSTITUTE. 1976. Evaluation Reports. Prairie Agricultural Machinery Institute, Humbolt, Saskatchewan, Canada.

QUICK, G. R. 1971. On the use of cross-flow fans in grain harvesting machinery. Transactions of the ASAE 14(3):411-416, 419.

QUICK, G. R. 1972. Analysis of the combine header and design for the reduction of gathering loss in soybeans. Ph.D. Thesis, Iowa State Univ. University Microfilms, Ann Arbor, Michigan.

QUICK, G. R. 1976. Grain harvesting 1776—a bicentennial review. ASAE Paper No. 76-1555. ASAE, St. Joseph, Michigan.

QUICK, G. R. 1977. Development of rotary and axial thresher/separators. Proc. Int. Grain & Forage Harvesting Conf. Ames, Iowa. ASAE, St. Joseph, Michigan.

QUICK, G. R. 1978. World's first maize header ahead of its time. Power Farming 87(6):33, 37.

QUICK, G. R. and W. F. BUCHELE. 1974. Reducing combine gathering losses in soybeans. Transactions of the ASAE 17(6):1123-1129.

QUICK, G. R. and W. M. MILLS. 1977. Narrow-pitch combine cutterbar design and appraisal. Proc. Int. Grain and Forage Harvesting Conf., Ames, Iowa. ASAE, St. Joseph, Michigan.

QUICK, G. R. and G. F. MONTGOMERY. 1974. Bibliography on combines and grain harvesting. ASAE, St. Joseph, Michigan.

QUODLING, H. C. 1924. Maize reaper thresher. Queensland Agricultural Jour. 22(4):318-321.

RANSOME, J. A. 1843. The implements of agriculture. London, England.

RASMUSSEN, H. 1969. Grain harvest and threshing in Calabria. Tools and Tillage 1(2):93-104.

RASMUSSEN, W. D. 1962. The impact of technological change on American agriculture, 1862-1962. Jour. Econ. Hist. 578-599. Dec.

RAWLINSON, G. 1880. A history of ancient Egypt. Vol. 1. The Nottingham Society, New York, New York.

RAY, W. and M. RAY. 1974. The art of invention. Patent models and their makers. The Pyne Press, Princeton.

RENARD, M. 1959. Technique et agriculture en pays trévire et rémois. Latomus 18:77-109.

REED, W. B. 1977. A review of monitoring devices for combines. Proc. Int. Grain & Forage Harvesting Conf., Ames, Iowa. ASAE, St. Joseph, Michigan.

REED, W. B. and F. W. BIGSBY. 1977. A cleaning system aspirator for combines. Proc. Int. Grain & Forage Harvesting Conf. Ames, Iowa. September.

REED, W. B., GROVUM, M. A. and A. E. KRAUSE. 1969. Combine harvester grain loss monitor. Agricultural Engineering 50(9): 524, 525, 528.

RIDGWAY, I. G. 1977. The resilient tapered thresher. Proc. Int. Grain & Forage Harvesting Conf. Ames, Iowa. September.

RIDLEY, A. E. 1904. A backward glance — the story of John Ridley. J. Clarke & Co., London, England.

RIES, L. W. 1930. Mähdrescher in der Ernte 1930. Technik in der Landwirtschaft 11:259-260.

RIES, L. W. 1930. Mechanisierung der Getreideernte. Technik in der Landwirtschaft 11:207-213.

ROBERSTON, H. 1974. Salt of the earth—the story of the homesteaders in western Canada. Jas. Lorimer & Co., Toronto, Canada.

ROGIN, L. 1931. The introduction of farm machinery in its relation to the productivity of labor in the agriculture of the United States during the nineteenth century. Univ. of California Publications in Economics., Vol. 9, Berkeley, California.

ROWLAND-HILL, E. W. 1975. Twin rotor combine harnesses potential of rotary threshing and separation. ASAE Paper No. 75-1580. ASAE, St. Joseph, Michigan.

SAHAGUN, Bernardino de Florentine Codex. 1950-1970. General history of the things of New Spain. Transl. and edited by C. E. Dibble and A. J. O. Anderson. The School of American Research and the University of Utah.

SAIJPAUL, K. K., DREW, L. O. and D. M. BYG. 1977. New design combine effects on soybean seed quality. Proc. Int. Grain & Forage Harvesting Conf. Ames, Iowa. September.

SCHLAYER, F. 1928. Dreschmaschine "Schlayer-Heliaks". Technik in der Landwirtschaft 9(11):258-259.

SCHLAYER, F. 1931. Neue versuche mit der "Schlayer-Heliaks" Dreschmaschine. Aus den arbeiten des RKTL (Latest developments with the Schlayer-Heliaks thresher. RKTL tests). Technik in der Landwirtschaft 12(5):148-149.

SCHLEBECKER, J. T. 1969. Bibliography of books and pamphlets on the history of agriculture in the U.S., 1607-1967.

SCHLEBECKER, J. T. 1975. Whereby we thrive. A history of American farming, 1607-1972. Iowa State Univ. Press.

SCHOENLEBER, L. G. and L. F. BOUSE. 1963. Progress in mechanization for harvesting castor beans. ASAE Paper No. 63-132. ASAE, St. Joseph, Michigan.

SCOTT, H. G. and D. J. SMYTH. 1970. Demand for farm machinery—Western Europe. Royal Commission on Farm Machinery. Study No. 9. Ottawa, Canada.

SCOTT, W. O. and S. R. ALDRICH. 1970. Modern soybean production. The Farm Quarterly. Cincinnati, Ohio.

SEEBOHM, M. E. 1927. The evolution of the English farm. Harvard Univ. Press, Cambridge, Massachusetts.

SHAVER, J. L. and W. F. TEMPLE. 1972. Development of the AC dual outlet transverse flow fan. ASAE Paper No. 72-636. ASAE, St. Joseph, Michigan.

SINGER, C., E. J. HOLMYARD, A. R. HALL and T. I. WILLIAMS. 1956. A history of technology, Vol. 2. Oxford, at the University Press.

SKROMME, L. H. 1977. Progress report on TWIN ROTOR combine concept of rotary threshing and separation. Proc. Int. Grain & Forage Harvesting Conf. Ames, Iowa. September.

SLIGHT, J. and R. SCOTT-BURN. 1858. The book of farm implements and machines. Blackwood & Sons, Edinburg, Scotland.

SLOANE, E. 1974. An age of barns. Ballantine, New York, New York.

SMITH, H. C. 1945. 21 years of soybean harvesting. Soybean Digest 5(9):11-12.

SMITH, T. C. 1959. The agrarian origins of modern Japan. Stanford University Press.

SMITH, W. D. 1928. Harvesting of rice with combines. Agricultural Engineering 9(3):73-74.

SOMERVILLE, R. 1805. General view of the agriculture of East Lothian. London, England.

SPENCER, A. J. and J. B. PASSMORE. 1930. Agricultural implements and machinery. Science Museum, London, England.

SPLINTER, W. E. 1974. Harvesting, handling and storage. In Wheat: production and utilization, edited by George E. Inglett: 52-71. The AVI Publishing Co. Inc., Connecticut.

STEENSBERG, A. 1943. Ancient harvesting implements. Transl. by W. E. Calvert. Copenhagen, Denmark.

STEENSBERG, A. 1971. Drill-sowing and threshing in southern India compared with sowing practices in other parts of Asia. Tools and Tillage 1(4):241-256.

STEPHENS, H. 1851. The book of the farm. Blackwood, Edinburgh, Scotland.

STIRNIMAN, E. J. 1926. Grain handling methods in relation to combine harvesting. Transactions of the ASAE 20:227-243.

SUTTON, G. L. 1937. The invention of the stripper. Jour. Agr. W. Australia 14(3):194-247.

SWIFT, R. B. 1897. Who invented the reaper? Publ. by the author, Chicago, Illinois.

TERRY, R. 1949. The combined harvester-thresher. Development and industry outlook. Bus. Inf. Serv., Dept. of Commerce, Washington, D.C. July.

TERRY, R. 1949. The combined harvester-thresher. Implement and Tractor 64:45, 46, 58, 59, 84, 85.

THOMAS, J. J. 1886. Farm implements and farm machinery and the principles of their construction and use. Orange Judd Co., New York, New York.

THOMPSON, A. 1976. The origins of a harvesting revolution: the development of the combine harvester in Australia, Canada and U.S.A. Paper pres. Fourth Int. Cong. Agr. Museums. Reading, England.

THOMPSON, H. S. 1852. Report on the exhibition and trial of implements at the Lewes meeting. Jour. Royal Agricultural Society of England 13(24):301-347.

THOMPSON, L. M. 1965. Iowa agriculture, world food needs and educational response. Center for Agr. & Econ. Dev. AAC 156. Iowa State Univ., Ames, Iowa.

THOMPSON, L. M. 1972. The world food situation. Jour. Soil & Water Conservation. 27(1):4-7.

THYER, J. E. 1952. Early Australian inventions. The reaper-thresher and the stripper. Power Farming (Australia). 61(3):11, 110.

TRACTOROEXPORT. 1975. KOLOS: A new highly efficient grain combine harvester. Tractoroexport Today 11(1):3-11.

TRANT, G. I. 1975. The hungry world's stake in Canadian agriculture. Report of the Canadian Agricultural Outlook Conference. Economics Branch, Agriculture Canada, Ottawa, Canada.

TRITTON, W. 1936. The origin of the thrashing machine. Reprint from the Lincolnshire Magazine. Publ. 735.

TUCKER, J. 1944. The self-propelled combine. Agricultural Engineering 25(9):333.

TUNIS, E. 1957. Colonial living. The World Publishing Co., Cleveland, Ohio.

TUNIS, E. 1959. Indians. The World Publishing Co., Cleveland, Ohio.

TURNER, H. S. and R. W. IRWIN. 1974. Ontario's threshing machine industry. A short history of these pioneer companies and their contribution to Ontario agriculture. Publ. 126-30. School of Eng., Univ. of Guelph, Ontario, Canada.

TURTON, J. G. 1911. The reaping of the harvest. The College Quarterly. The official organ of the working-men's college, Melbourne, Australia (8), Oct.

US BUREAU OF AGRICULTURAL ECONOMICS. 1940. Technology on the farm. US Government Printing Office, Washington, D.C. Aug.

US CENSUS OFFICE. 1860. Progress of invention in threshing instruments. Prelim. Rep., 8th Census US:90-100.

VAN DER POEL, J. M. G. 1967. Hondered jaar landbouwmechanisatie in Nederland. Wageningen, The Netherlands.

VAN VOLKENBURG, L. R. 1919. Thresher methods. Transactions of the ASAE 13:117-124.

VAN WAGENEN, J. 1927. The golden age of homespun. Bull. 203. Dept. of Agric. and Markets, State of New York. June.

VICAS, A. G. 1970. Research and development in the farm machinery industry. Royal Commission on Farm Machinery Study No. 7. Ottawa, Canada.

WAKEMAN, L. 1910. Threshing machines in 1828. Canadian Thresherman and Farmer 15(2):77.

WALLEN, N. L. 1930. Combine performance in Germany. Agricultural Engineering 11(6):216.

WALLEN, N. 1930. Die arbeit des Mähdreschers in Deutschland im sommer 1929 (The operation of combines in Germany, summer 1929). Technik in der Landwirtschaft:617.

WALLEN, N. 1931. Ein paar wichtige Mähdrescherzahlen (Some important combine figures). Technik in der Landwirtschaft:220.

WANG CHENG. 1313. Nung Shu. The book of agriculture, China, 1313 AD. Reprinted 1956.

WATSON, J. A. S., M. E. HOBBS. 1937. Great Farmers. Selwyn & Blount, London, England.

WATTS, W. A. 1972. Thoughts on combine design. ASAE Paper No. 72-635. ASAE, St. Joseph, Michigan.

WHEELHOUSE, F. 1966. Digging stick to rotary hoe—men and machines in rural Australia. Cassell Australia Ltd., Melbourne, Australia.

WERTH, E. 1954. Grabstock, Hacke und Pflug. Ludwigsburg.

WHITE, H. E. 1945. The self-propelled combine. Agricultural Engineering 26(5):195.

WHITE, K. D. 1968. Agricultural implements of the Roman world. Cambridge University Press, New York, New York.

WHITE, K. D. 1970. Roman farming. Thames & Hudson.

WIENEKE, F. 1964. Einleitende betrachtungen über-Dreschsysteme, Einflussgrössen und Bewertungsmasstäbe beim Mähdrusch (Introductory observations on threshing systems, parameters and standards of evaluation of combine-harvesters). Grundlagen der Landtechnik 21:5-7.

WIK, R. M. 1953. Steam power on the American farm. Univ. of Pennsylvania Press, Philadelphia, Pennsylvania.

WILEMAN, R. H. 1927. Field tests of the combine in Indiana. Agricultural Engineering 8(5):118.

WILKINSON, J. G. 1883. The manners and customs of the ancient Egyptians. S. E. Cassino & Co., Boston, Massachusetts.

WILSON, M. L. 1943. Thomas Jefferson and agricultural engineering. Agricultural Engineering 24(9):299-303.

WITZEL, H. D. & B. F. VOGELAAR. 1955. Engineering the hillside combine. Agricultural Engineering 36(8):523-525, 528.

WOOD, H. T. 1913. A history of the Royal Society of Arts. John Murray, London, England.

WOODCROFT, B. 1853. Appendix to the specifications of English patents for reaping machines. London, England.

WRIGHT, J. S. 1853. Atkins' reaper and mower. The Prairie Farmer 13(6):209-212.

WRUBLESKI, P. D., E. O. NYBORG. 1977. Field evaluation of grain combines. Proc. Int. Grain & Forage Harvesting Conf. Ames, Iowa. ASAE, St. Joseph, Michigan.

YERKES, A. P. 1929. Windrower—offspring of combines. In Combine Yearbook:24, 25. Clarke Publishing Co., Madison, Wisconsin.

YOUNG, A. 1778. The Farming Kalendar.

ZINTHEO, C. J. 1907. Corn harvesting machinery. Bulletin 173. US Government Printing Office, Washington, D.C.

ABOUT THE SOCIETY

THE American Society of Agricultural Engineers (ASAE) is pleased to offer this historic work for those interested in the preservation of our American heritage.

The ASAE is a nonprofit engineering society which concentrates its efforts in service to agriculture. Currently, ASAE has some 8,000 members located primarily in the United States, and in some seventy other countries of the world. Its engineer-members concentrate their efforts in five major technical areas: (1) power and machinery, (2) soil and water, (3) structures and environment, (4) electric power and processing, and (5) food engineering. ASAE publishes a monthly magazine, *Agricultural Engineering*, which reports on present engineering developments and concepts. It also publishes research papers in *Transactions of the ASAE* and each year develops special publications on various subjects such as livestock waste management, field modification of tractors, dairy housing, compaction of agricultural soils, and so on.

ASAE also develops and publishes the voluntary standards which serve agriculture. These standards are published annually in *Agricultural Engineers Yearbook*. Included among the ASAE standards are those relating to power take-offs (specifications, terminology, speed), slow-moving vehicle emblem, moisture measurement (grain and seeds), and various standards relating to safety, uniform terminology, machine interchangeability, and the like.

ASAE works cooperatively with agricultural engineering departments in the various states, with the USDA and other branches of the federal government, and with engineers in private practice and those employed in industry, particularly the farm machinery industry. Through ASAE, these agricultural engineers work cooperatively toward further improvements in agricultural efficiency, productivity, and in the conservation and wise use of our soil and water resources.

Other Publications in the Historic Series
The Agricultural Tractor: 1855-1950
Rural America a Century Ago

Metric Conversion
The conversion chart below is for the convenience of readers more familiar with metric units (SI) than with customary or English units.

When You Know	Multiply By	To Find
inches (in.)	*25.4**	*millimeters (mm)*
feet (ft)	*0.3048**	*meters (m)*
yards (yd)	*0.9144**	*kilometers (km)*
acres	*0.404 686*	*hectares** (ha)*
ounces (oz) mass	*28.349 5*	*grams (g)*
pounds (lb) mass	*0.453 592*	*kilograms (kg)*
gallon-US (gal)	*3.785 412*	*liter (l or L)*
bushel-US (bu)	*0.035 239 07*	*cubic meters (m³)*
horsepower (hp)	*0.745 699 9*	*kilowatts (kW)*

*exact
**10 000 sq. meters

269